L'Organisation de Coopération et de Développement Économiques (OCDE), qui a été instituée par une Convention signée le 14 décembre 1960, à Paris, a pour objectif de promouvoir des politiques visant :
— à réaliser la plus forte expansion possible de l'économie et de l'emploi et une progression du niveau de vie dans les pays Membres, tout en maintenant la stabilité financière, et contribuer ainsi au développement de l'économie mondiale;
— à contribuer à une saine expansion économique dans les pays Membres, ainsi que non membres, en voie de développement économique;
— à contribuer à l'expansion du commerce mondial sur une base multilatérale et non discriminatoire, conformément aux obligations internationales.

Les Membres de l'OCDE sont : la République Fédérale d'Allemagne, l'Australie, l'Autriche, la Belgique, le Canada, le Danemark, l'Espagne, les États-Unis, la Finlande, la France, la Grèce, l'Irlande, l'Islande, l'Italie, le Japon, le Luxembourg, la Norvège, la Nouvelle-Zélande, les Pays-Bas, le Portugal, le Royaume-Uni, la Suède, la Suisse et la Turquie.

L'Agence de l'OCDE pour l'Énergie Nucléaire (AEN) a été créée le 20 avril 1972, en remplacement de l'Agence Européenne pour l'Énergie Nucléaire de l'OCDE (ENEA) lors de l'adhésion du Japon à titre de Membre de plein exercice.

L'AEN groupe désormais tous les pays Membres européens de l'OCDE ainsi que l'Australie, le Canada, les États-Unis et le Japon. La Commission des Communautés Européennes participe à ses travaux.

L'AEN a pour principaux objectifs de promouvoir, entre les gouvernements qui en sont Membres, la coopération dans le domaine de la sécurité et de la réglementation nucléaires, ainsi que l'évaluation de la contribution de l'énergie nucléaire au progrès économique.

Pour atteindre ces objectifs, l'AEN :
— *encourage l'harmonisation des politiques et pratiques réglementaires dans le domaine nucléaire, en ce qui concerne notamment la sûreté des installations nucléaires, la protection de l'homme contre les radiations ionisantes et la préservation de l'environnement, la gestion des déchets radioactifs, ainsi que la responsabilité civile et les assurances en matière nucléaire ;*
— *examine régulièrement les aspects économiques et techniques de la croissance de l'énergie nucléaire et du cycle du combustible nucléaire, et évalue la demande et les capacités disponibles pour les différentes phases du cycle du combustible nucléaire, ainsi que le rôle que l'énergie nucléaire jouera dans l'avenir pour satisfaire la demande énergétique totale ;*
— *développe les échanges d'informations scientifiques et techniques concernant l'énergie nucléaire, notamment par l'intermédiaire de services communs ;*
— *met sur pied des programmes internationaux de recherche et développement, ainsi que des activités organisées et gérées en commun par les pays de l'OCDE.*

Pour ces activités, ainsi que pour d'autres travaux connexes, l'AEN collabore étroitement avec l'Agence Internationale de l'Énergie Atomique de Vienne, avec laquelle elle a conclu un Accord de coopération, ainsi qu'avec d'autres organisations internationales opérant dans le domaine nucléaire.

FOREWORD

The concept of geologic disposal of radioactive waste relies on the capability of many geological formations to provide long-term containment of the waste. This capability is demonstrated by numerous geological and geochemical observations. However, these observations refer to the geological past and to undisturbed conditions.

Concern has been expressed about the future, particularly since the disposal operations could significantly modify the original conditions. In addition to the underground excavations and the thermal input of the waste there is the problem of boreholes and shafts that constitute a potential by-pass of the geological barriers. It is therefore essential to develop techniques and procedures for effective plugging of all penetrations connecting the disposal zone with the surface or with water bearing layers. Borehole plugging is routinely performed in the oil industry by means of cement plugs ; however in relation to radioactive waste disposal it will be necessary to produce plugs which effectively restore the original characteristics of the isolating formations. In addition these plugs must be chemically stable in the existing geochemical environment in order to remain effective for very long periods of time. Finally disposal holes in a variety of repository concepts require plugging. This particular problem might turn out to present peculiar aspects, since the plugs of disposal holes can be exposed to high temperatures and radiation doses.

All countries with geologic disposal programmes will have to face the problem of borehole and shaft plugging. As part of its programme on geologic disposal of radioactive waste the NEA, in co-operation with the United States Department of Energy, organised a Workshop, with the following objectives :

- to review data and experience obtained from ongoing investigations and experimental activities ;

- to exchange views among experts on various approaches to the design of long-lived plugs ;

- to assist those now planning R & D activities ;

- to promote contacts and co-operation among experts working in the field of borehole and shaft plugging.

These proceedings represent a record of the papers and discussions at the meeting.

Proceedings of the Workshop on

BOREHOLE
AND SHAFT PLUGGING

Columbus, 7th-9th May 1980

Compte rendu d'une réunion de travail sur le

COLMATAGE DES
FORAGES ET DES PUITS

Columbus, 7-9 mai 1980

jointly organised by the
OECD NUCLEAR ENERGY AGENCY
and the
UNITED STATES DEPARTMENT OF ENERGY

organisée conjointement par
l'AGENCE DE L'OCDE POUR L'ÉNERGIE NUCLÉAIRE
et le
DÉPARTEMENT DE L'ÉNERGIE DES ÉTATS-UNIS

ORGANISATION FOR ECONOMIC CO-OPERATION AND DEVELOPMENT
ORGANISATION DE COOPÉRATION ET DE DÉVELOPPEMENT ÉCONOMIQUES

The Organisation for Economic Co-operation and Development (OECD) was set up under a Convention signed in Paris on 14th December, 1960, which provides that the OECD shall promote policies designed:

— to achieve the highest sustainable economic growth and employment and a rising standard of living in Member countries, while maintaining financial stability, and thus to contribute to the development of the world economy;
— to contribute to sound economic expansion in Member as well as non-member countries in the process of economic development;
— to contribute to the expansion of world trade on a multilateral, non-discriminatory basis in accordance with international obligations.

The Members of OECD are Australia, Austria, Belgium, Canada, Denmark, Finland, France, the Federal Republic of Germany, Greece, Iceland, Ireland, Italy, Japan, Luxembourg, the Netherlands, New Zealand, Norway, Portugal, Spain, Sweden, Switzerland, Turkey, the United Kingdom and the United States.

The OECD Nuclear Energy Agency (NEA) was established on 20th April 1972, replacing OECD's European Nuclear Energy Agency (ENEA) on the adhesion of Japan as a full Member.

NEA now groups all the European Member countries of OECD and Australia, Canada, Japan, and the United States. The Commission of the European Communities takes part in the work of the Agency.

The primary objectives of NEA are to promote co-operation between its Member governments on the safety and regulatory aspects of nuclear development, and on assessing the future role of nuclear energy as a contributor to economic progress.

This is achieved by:

— *encouraging harmonisation of governments' regulatory policies and practices in the nuclear field, with particular reference to the safety of nuclear installations, protection of man against ionising radiation and preservation of the environment, radioactive waste management, and nuclear third party liability and insurance;*
— *keeping under review the technical and economic characteristics of nuclear power growth and of the nuclear fuel cycle, and assessing demand and supply for the different phases of the nuclear fuel cycle and the potential future contribution of nuclear power to overall energy demand;*
— *developing exchanges of scientific and technical information on nuclear energy, particularly through participation in common services;*
— *setting up international research and development programmes and undertakings jointly organised and operated by OECD countries.*

In these and related tasks, NEA works in close collaboration with the International Atomic Energy Agency in Vienna, with which it has concluded a Co-operation Agreement, as well as with other international organisations in the nuclear field.

D
622·24
WOR

AVANT-PROPOS

Le concept de l'évacuation des déchets radioactifs dans des formations géologiques repose sur la possibilité offerte par un grand nombre de ces formations de confiner les déchets pendant de très longues périodes. Cette possibilité a été démontrée par de nombreuses observations géologiques et géochimiques. Il convient toutefois de noter que ces observations ont été réalisées dans des milieux géologiques qui n'avaient subi aucune perturbation.

Des préoccupations existent en ce qui concerne l'avenir, en particulier dans la mesure où les opérations d'évacuation pourraient sensiblement modifier la situation initiale. Les excavations souterraines et la chaleur dégagée par les déchets posent des problèmes mais, en outre, les forages et les puits risquent de servir de voies de passage au travers des barrières géologiques naturelles. Il est donc essentiel de mettre au point des techniques et des méthodes permettant d'obturer de façon efficace toutes les pénétrations faisant communiquer la zone de dépôt avec la surface ou avec des couches aquifères. Dans les puits de pétrole, les forages sont généralement obturés à l'aide de bouchons de ciment ; toutefois, s'agissant de l'évacuation de déchets radioactifs, il faudra réaliser des bouchons reconstituant effectivement les couches géologiques isolantes, avec leurs caractéristiques initiales. De plus, ces bouchons devront demeurer chimiquement stables dans le milieu géochimique environnant afin de conserver leur efficacité pendant de très longues périodes. Enfin, dans certains types de dépôts, il faudra également obturer les cavités où sont déposés les déchets, opération qui risque de poser des problèmes particuliers, étant donné que les bouchons utilisés à cette fin pourront être exposés à des températures élevées et à de fortes doses de rayonnement.

Tous les pays disposant de programmes d'évacuation dans des formations géologiques devront résoudre le problème du colmatage des forages et des puits. Dans le cadre de son programme sur l'évacuation de déchets radioactifs dans des formations géologiques, l'AEN, en collaboration avec le Département de l'Energie des Etats-Unis, a organisé une réunion de travail dont les objectifs furent les suivants :

- étudier les données et l'expérience acquises dans le cadre des recherches et des activités expérimentales en cours ;

- permettre des échanges de vues entre spécialistes au sujet de différents types de bouchons de colmatage de grande durabilité ;

- apporter une aide à ceux qui prévoient actuellement d'entreprendre des activités de R-D ;

- favoriser les contacts et la coopération entre spécialistes travaillant dans le domaine du colmatage des forages et des puits.

Ce compte rendu comprend les communications présentées ainsi que les discussions dont elles ont fait l'objet.

TABLE OF CONTENTS

TABLE DES MATIERES

SESSION 3 - SEANCE 3

Chairman - Président : Mr. R.D. ELLISON (United States)

SESSION 4 – SEANCE 4

Chairman – Président : Mr. N.A. CHAPMAN (United Kingdom)

SESSION 5 – SEANCE 5

Chairman – Président : Mr. T.O. HUNTER (United States)

SESSION 1

Chairman - Président

J. HAMSTRA

(The Netherlands)

SEANCE 1

REPOSITORY SEALING -- INTERACTIONS WITH REPOSITORY DEVELOPMENT*

S. J. Basham, Jr.
Manager, Engineering Development Department
Office of Nuclear Waste Isolation
Battelle Memorial Institute
Columbus, Ohio 43201

ABSTRACT

The development of geologic disposal has as its major goal a repository resting
on a sound design basis. To have a sound basis requires developments in several
technologies, of which repository sealing is one, in a timely fashion to meet the
intermediate goals of the repository development. The development plans for
repository sealing must be both responsive to and contribute to larger goals.
The larger goals range from site characterization and qualification of a variety
of sites to be completed by December, 1985, to the earliest repository operation
in July, 1997. Repository sealing activities in the areas of plug and sealing
designs for (1) boreholes, (2) shafts, and (3) repository horizon tunnels and
emplacement chambers must be concluded in a time commensurate with the dictates
of the overall schedule.

RESUME

Obtenir un dépôt géologique fondé sur une étude sérieuse est le but principal
d'un tel développement. Une étude de ce genre exige l'application de plusieures
technologies, dont l'une est l'obturation du dépôt. Ceci se fait au moment
opportun, afin d'atteindre les buts intermédiaires du développement du dépôt en
question. Les plans de développement d'obturation de dépôt doivent correspondre
et contribuer à des fins de plus grande envergure, lesquelles consistent d'une
part à caractériser et à déterminer la convenance d'une variété d'emplacements,
qui seront terminés vers décembre 1985, et d'autre part à atteindre la mise en
service du dépôt en juillet 1997 au plus tôt. Les activités d'obturation de
dépôt, en ce qui concerne le bouchage et l'obturation de (1) forages, (2) puits,
et (3) tunnels repères de dépôts et chambres d'emplacement, doivent se terminer
en une périods qui corresponde aux exigences de l'ensemble du plan d'exécution.

Work supported by U.S. Department of Energy Assistant Secretary for Nuclear
Energy, Office of Nuclear Waste Management, under Contract No. DE-ACO6-
76RLO1830-ONWI.

The basic goals for radioactive waste management have been given in the President's policy statement /‾1‾7 in his message to Congress in February 1980. There are seven key items in the message. However, the repository development schedule is set by the second item. Quoting in part from this item, we note "Second, for disposal of high level radioactive waste, I am adopting an interim planning strategy focused on the use of mined geologic repositories capable of accepting both waste from reprocessing and unreprocessed commercial spent fuel... When four to five sites have been evaluated and found potentially suitable, one or more will be selected for further development as a licensed full-scale repository." The two key points are <u>mined geologic repositories</u> and <u>four to five sites.</u> All development of tasks and schedules is set by this selected approach. It is beyond the scope of this paper to address the state and local interactions, need for rapid and broad disclosure of all information to the public and the technical community and other key items.

The recognition of the environmental impacts of technological decisions and the worldwide distrust of major institutions, including government, coupled with the increasing world population and rapid dissemination of information have all contributed to the need for deliberate, carefully documented approaches to new developments that have a perceived major impact on society as a whole. Radioactive waste management is one of these developments and must follow the legal and societal steps in the evolution towards a solution. It is sufficient here to recognize that all the steps are not technical ones.

With these complex factors in mind, the functions and schedule for developing repositories may be examined. Eight key functions have been defined and are presented in Table I.

Table I
Functions in the Repository Development Process

- Public Interaction

- Site Characterization and Qualification

- Environmental Impact Statement Preparation

- Site Acquisition

- Technology and Testing

- Licensing

- Design Engineering and Construction

- Operation

In carrying out the functions these are many intermediate goals that are signalled by certain documents. The acronyns used for these documents and several other terms used in the repository reference schedule are defined in Table II.

Having defined the functions and certain of the less obvious intermediate goals the summary logic network for repository development may be presented. Figure 1 shows the activities leading to geologic repository operation in the late 1990's /‾2‾7.

By following the critical path it is evident that the technology and testing (where repository sealing is located) must treat the stratigraphy of the following potential sites:

- Los Medanos and other bedded salt

- Basalt (Hanford)

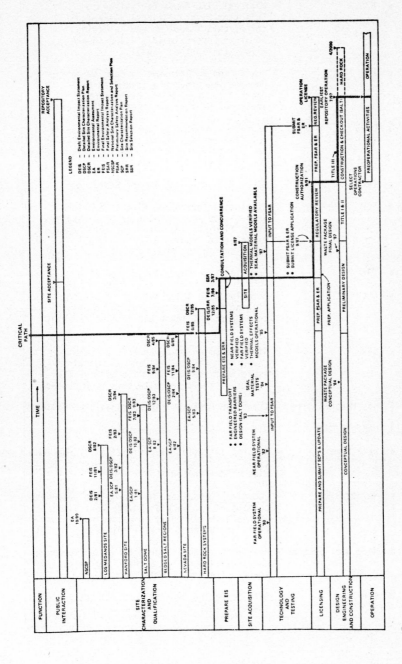

Figure 1: Summary Logic Network Activities Leading to Geologic Repository Operation (Reference Schedule) (Figure III-2 from Reference 2)

Table II
Definitions

- **NSCSP**   - National Site Characterization and Selection Plan: national screening for potential regions and selection of areas (approximately 1,000 square miles) for further study

- **EA**   - Environmental Assessment: concise public document to provide evidence to decide whether to prepare an environmental statement or find no significant impact

- **DEIS**   - Draft Environmental Impact Statement: a detailed written statement as required by Section 102(2)(c) of the National Environmental Policy Act (NEPA)

- **FEIS**   - Final Environmental Impact Statement

- **SCP**   - Site Characterization Plan: report on area (1,000 square miles or less) exploration findings and plans for further site exploration studies at recommended locations in the area

- **DSCP**   - Detailed Site Characterization Plan

- **DSCR**   - Detailed Site Characterization Report: finding from site exploration necessary to support a license application for that site

- **SRR**   - Site Recommendation Report: document covering environmental and technical aspects of the four to five banked sites, a part of the final site selection process

- **Los Medanos Site** - Site in bedded salt in southeast New Mexico

- **Hanford Site** - Site in basalt in Washington

- **Salt Dome** - Natural intrusive diapiritic salt formation typified by those in the Gulf Interior Region of the United States

- **Nevada Site** - Location in southern Nevada on the southwest portion of the Federal Nevada Test Site; emphasis is currently on welded volcanic tuffs as the host rock at this site

- **Hard Rock Systems** - Host rock including granitic intrusive rocks and argillaceous rocks; both are in early stages of study at this time

- **SSR**   - Site Selection Report: updated version of SRR after Federal and State consensus

- Dome salt (Gulf Interior Region)

- Volcanic tuffs (Nevada)

- Hard rock both granitic and argillaceous (areas to be determined.

Repository sealing /¯3¯/ is required for penetrations into the host rock to ensure that they are sealed to prevent significant amounts of ground water from entering the waste emplacement region of the repository and to prevent radio-nuclides from reaching the biosphere in amounts exceeding acceptable levels. Another definition of the basic requirements is given by the U. S. Nuclear Regulatory Commission in a draft of the Technical Criteria which states, "The sealed shafts and boreholes provide a barrier to ground water and radionuclide migration which is equivalent to or better than the barrier provided by the undisturbed section of rock through which they pass."

These end objectives are met by requirements for repository sealing given in Table III.

Table III
Repository Sealing Requirements

- Adequate sealing of all openings at the repository site; boreholes, shafts, tunnels and emplacement chambers

- Care in use and location of boreholes at the repository site

- Location of existing boreholes at the repository site

- Care in borehole, shaft and tunnel construction

With the end objectives and basic requirements set, the program is structured to meet these objectives by arriving at plug and seal designs and specifications. All other activities support this end item. This hierarchy is shown in Figure 2. Therefore the end items of the Repository Sealing Program are designs of repository seals for a specific stratum and host rock at a specific location. Note that generic studies may be performed but there are no generic designs. Each borehole plug and shaft seal must be tailored for a specific site and consist of the appropriate mix of individual material and geometries called for by the specific hydrology, geochemical and other environmental factors of a particular site.

These design milestones are listed in Table IV.

In summary, the repository sealing program must be structured to have preconceptual designs for each host rock medium available at the time of sub-mission of the DSCR, conceptual designs approximately one year later but no later than the time of SSR submission, and final designs for the selected site at the time of submission of the PSAR, ER and License Application.

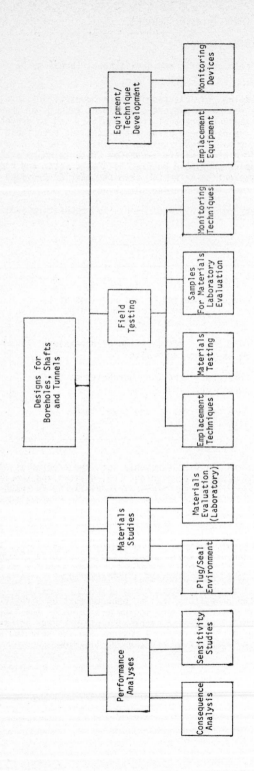

Figure 2: Repository Sealing Program Activities

Table IV
Repository Sealing Milestones

- Complete Preconceptual Designs of Repository Seals for Bedded Salt (Los Medanos) - 8/82

- Complete Preconceptual Designs for Dome Salt (Gulf Interior Region) - 9/83

- Complete Preconceptual Designs for Basalt (Hanford Site) - ~3/84*

- Complete Preconceptual Designs for Bedded Salt (site to be selected) - 4/85

- Complete Preconceptual Designs for Tuffs (Nevada Site) - ~6/85

- Complete Preconceptual Designs for Hard Rock (to be selected) - 9/84

The above meet or precede the corresponding upper level milestone, the DSCR submission, for each host rock (ranging from 8/82 to 12/85).

- Complete Conceptual Designs of Repository Seals for Bedded Salt (Los Medanos) - 9/83

- Complete Conceptual Designs for Dome Salt - 9/84

- Complete Conceptual Designs for Bedded Salt (site to be selected) - 11/85

- Complete Conceptual Designs for Basalt - ~4/85*

- Complete Conceptual Designs for Tuffs - ~6/86*

- Complete Conceptual Designs for Hard Rock - 10/85

The above precede the upper level milestone, SSR issuance on 3/87.

- Complete Final Design for Selected Site - 9/87

The above meets the upper level milestone, submit PSAR, ER, and License Application on 9/87.

* Exact dates not yet available.

REFERENCES

1. The President's Message to Congress, 12 February 1980.

2. Statement of Position of United States Department of Energy - Proposed Rulemaking on the Storage and Disposal of Nuclear Waste, DOE/NE-0007, 15 April 1980.

3. Office of Nuclear Waste Isolation, Program Plan and Current Efforts in Repository Sealing for the NWTS Program, ONWI-54, Battelle Memorial Institute, Columbus, Ohio, October 1979.

NWTS REPOSITORY SEALING PROGRAM -- AN OVERVIEW*

F. L. Burns
Project Manager, Engineering Development Department
Office of Nuclear Waste Isolation
Battelle Memorial Institute
Columbus, Ohio 43201

ABSTRACT

The purpose of the NWTS Repository sealing program is to produce detailed drawings and specifications for an entire sealing system including seals, materials, equipment, and construction techniques. In order to accomplish that purpose for each host rock type under consideration (bedded salt, dome salt, basalt, granite, shale) the program has been divided into three fundamental activities: Engineering, Laboratory Studies (material) and Field Testing. All three activities will be directed toward all kinds of penetration, including boreholes, shafts and tunnels.

RESUME

Le but du programme d'obturation de dépôts de terminaux de déchets nucléaires consiste à produire desd sessins et spécifications de détails pour un système complet d'obturation, y compris les bouchons, matériaux, appareillage, et les techniques de construction. Pour arriver à ce but dans le cas de chaque type de roche réceptrice en question (sel stratifié, sel en dôme, basalte, granit, argile schisteuse), on a divisé le programme en trois activités fondamentales: les tâches d'ingénieurs, les études de laboratoire (matériau) et l'expérimentation sur le terrain. On orientera les trois activités vers toutes sortes de pénétration, y compris les forages, les puits et les tunnels.

Work supported by U.S. Department of Energy Assistant Secretary for Nuclear Energy, Office of Nuclear Waste Management, under Contract No. DE-AC06-76RL01830-ONWI.

OFFICE OF NUCLEAR WASTE ISOLATION

REPOSITORY SEALING PROGRAM

AN OVERVIEW
MAY 7, 1980

FLOYD BURNS, PROJECT MANAGER

MAIN TOPICS

1. ONWI ASSIGNMENT

2. PROGRAM ORGANIZATION

3. CONTRACTOR ORGANIZATION

4. SCHEDULE

This presentation on the ONWI repository sealing program explains ONWI's mission, how the program is organized to carry out our mission, how the sub-contractor's work is organized, and the schedule for completion of our work.

ONWI ASSIGNMENT

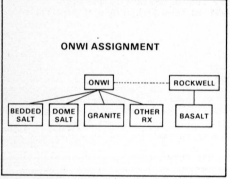

Except for basalt, at the Hanford facility, all host rock types under investigation for a repository are the responsibility of ONWI.

PROGRAM ORGANIZATION

1. GEOLOGIC NATURE OF REPOSITORY

2. TYPE OF HOLE

3. SEALING ACTIVITIES

The three fundamental elements the program is organized around are explained in the next series of slides.

PROGRAM PURPOSE

1. SEALING SYSTEM

2. SPECIFICATIONS

3. DRAWINGS

The program's purpose is to develop drawings and specifications for a complete sealing system in each designated host rock formation. This will include material, equipment, and any special construction techniques.

PROGRAM ORGANIZATION
HOST ROCKS

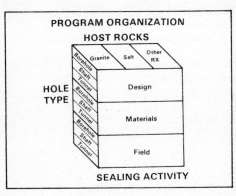

SEALING ACTIVITY

Engineering design will achieve the purpose of the program. Materials studies will identify proper materials for use. Field testing will measure how effectively the materials can be used.

Studies began in bedded salt; ONWI will include dome salt and granite in FY 81. Other rocks, such as argillite and tuff, are not yet scheduled.

Each of three hole types will require a separate seal design. They are: boreholes (narrow diameter, vertical); shafts (large diameter, vertical), and tunnels (large diameter, horizontal).

CONTRACTOR ORGANIZATION

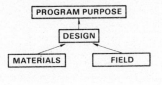

During the design stage of repository development, engineering will have full responsibility for achieving the program's overall purpose. Support will be supplied by laboratory and field test contractors.

CONTRACTOR ORGANIZATION

DESIGN—D'APPOLONIA CONSULTING ENGINEERS

MATERIALS—PENN. STATE UNIVERSITY
        WATERWAYS EXPERIMENT STATION

Those organizations now under contract to ONWI are identified in their respective functional units.

CONTRACTOR ORGANIZATION

FIELD—SANDIA LABORATORIES—BEDDED SALT

WOODWARD CLYDE—BEDDED SALT

LAW ENGINEERING—DOME SALT

?   —GRANITE

Those organizations now under contract to ONWI are identified in their respective functional units.

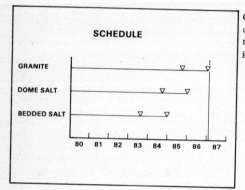

SCHEDULE

Conceptual designs, then preliminary designs will be completed before 1/87 for each of the three host rocks. Our greatest need at this time is for work in granite.

Discussion

T.O. HUNTER, United States

Would you comment on the status of licensing by the NRC for borehole and shaft seals ?

F.L. BURNS, United States

At the present time the NRC requires any man-made penetrations to be sealed so permeability of the formations will be unchanged from the natural state. We believe this requirement will be changed in order to relate to specific limits on radionuclide releases to the biosphere.

PENETRATION SEALING

DESIGN PROGRAM UPDATE

MAY 1980

R. D. ELLISON,[1]   *D. K. SHUKLA,*[2]   *D. STEPHENSON*[2]

[1] D'APPOLONIA CONSULTING ENGINEERS, INC.   [2] D'APPOLONIA CONSULTING ENGINEERS, INC.
PITTSBURGH, PENNSYLVANIA   ALBUQUERQUE, NEW MEXICO

GOALS OF PENETRATION SEALING PROGRAM

PRIMARY

- DEVELOP DESIGNS WHICH PROVIDE SAFE SEALS AT REASONABLE COST

- OBTAIN LICENSING APPROVAL ON SCHEDULE TO SATISFY OVERALL REPOSITORY LICENSING

SECONDARY

- DEVELOP DESIGN REQUIREMENTS
  - —DETERMINE DESIGN CONDITIONS (CHARACTERIZATION)
  - —DEVELOP SUITABLE MATERIALS (MATERIALS)
  - —DEMONSTRATE INSTALLATION AND FUNCTION (VERIFICATION)

TECHNICAL ACTIVITIES

- ENGINEERING DESIGN

- SEAL HOST CHARACTERIZATION

- MATERIALS STUDIES

- SYSTEMS/CONSEQUENCE ANALYSIS

- FIELD TESTING

- PROCESS AND EQUIPMENT DEVELOPMENT

- INSTRUMENTATION

- LICENSING

- TECHNICAL INTERACTION

- QUALITY ASSURANCE

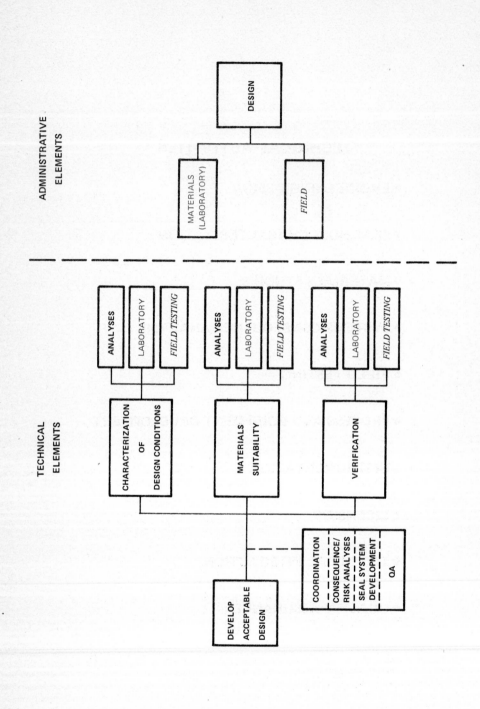

RECENT HISTORY

ITEM	YEAR			
	FY77	FY78	FY79	FY80
MELTED SALT, HYDROTHERMAL TRANSPORT, COMPACTED CLAY, PRELIMINARY INSTRUMENTATION STUDIES 1973–1976				
CEMENT GROUT STUDIES				
FIRST ONWI SEALING WORKSHOP (ONWI-3)				
LITERATURE STATUS REPORT (ONWI-15)				
DESIGN CRITERIA 1979 (ONWI-55)				
WIPP RELATED CEMENT STUDIES				
WIPP RELATED FIELD TESTING (BELL CANYON TEST)				
BWIP MATERIALS AND EQUIPMENT STUDIES				
EARTH SCIENCE TECHNOLOGY PLAN FOR PENETRATION SEALS				
DEVELOPMENT OF NWTS REPOSITORY SEALING PROGRAM PLAN				
COMPARISON OF BOREHOLE SHAFT AND TUNNEL REQUIREMENTS				
DETERMINATION OF MATERIALS RESEARCH NEEDS				
DEVELOPMENT OF FIELD TESTING (VERIFICATION) PROGRAM FOR DOME AND BEDDED SALT				

1979 STATUS REPORT (ONWI - 15)

INDUSTRIAL CEMENTING TECHNOLOGY

— OIL AND GAS

— DEEP WELL DISPOSAL

— GEOTECHNICAL

SHAFT AND CAVERN SEALING

— NUCLEAR EXPLOSIONS

— CHEMICAL WASTE DISPOSAL

— MINING

PAST BOREHOLE PLUGGING EFFORTS

INSTRUMENTATION

IDENTIFICATION OF RESEARCH NEEDS

DESIGN APPROACH —1979 (ONWI-55)

- DEFINITIONS
    - APPLICATIONS
    - FLUID AND RADIONUCLIDE FLOW ZONES
    - SEALING LOCATIONS

- DESIGN GOAL — ACCEPTABLE CRITERIA

- DESIGN LIFE PHASES

- CHARACTERIZATION FACTORS

- POTENTIAL FOR MULTIPLE GEOMETRIES

- BENEFITS FOR MULTIPLE MATERIALS

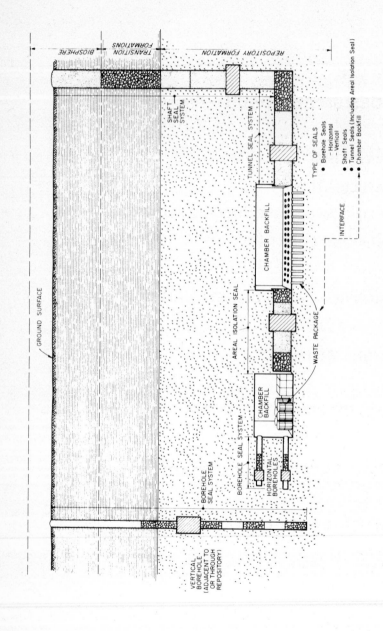

DEFINITION OF SEALING APPLICATIONS

TYPE OF SEALS

• Borehole Seals
  - Horizontal
  - Vertical
• Shaft Seals
• Tunnel Seals (Including Areal Isolation Seal)
• Chamber Backfill

GROUND SURFACE

BIOSPHERE

TRANSITION FORMATIONS

REPOSITORY FORMATION

SHAFT SEAL SYSTEM

TUNNEL SEAL SYSTEM

CHAMBER BACKFILL

AREAL ISOLATION SEAL

CHAMBER BACKFILL

INTERFACE

WASTE PACKAGE

BOREHOLE SEAL SYSTEM

HORIZONTAL BOREHOLES

VERTICAL BOREHOLE (ADJACENT TO OR THROUGH REPOSITORY)

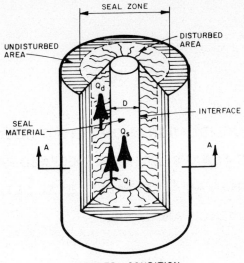

SEALED CONDITION

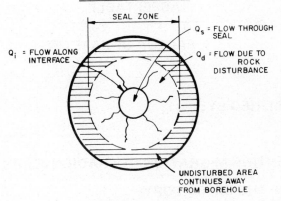

SECTION A-A

Q_t = TOTAL FLOW THROUGH SEAL ZONE = $Q_s + Q_i + Q_d$

R_t = RADIONUCLIDE MIGRATION RATE THROUGH SEAL ZONE
$= R_s + R_i + R_d$

SEAL ZONE COMPONENTS

DESIGN GOAL

THE RADIONUCLIDE MIGRATION RATE THROUGH THE
SEAL ZONE IS ALWAYS LESS BY A SPECFIIED FACTOR
OF SAFETY THAN AN ACCEPTABLE LEVEL DETERMINED
BY A CONSEQUENCE ANALYSIS, i.e.

$$R_t \; < \; \frac{R_{ACCEPTABLE}}{FACTOR\ OF\ SAFETY}$$

ACCOMPLISHED BY:

- PREVENTING MIGRATION OF RADIONUCLIDES
  OUT OF THE REPOSITORY

- PREVENTING MOVEMENT OF WATER INTO THE
  REPOSITORY

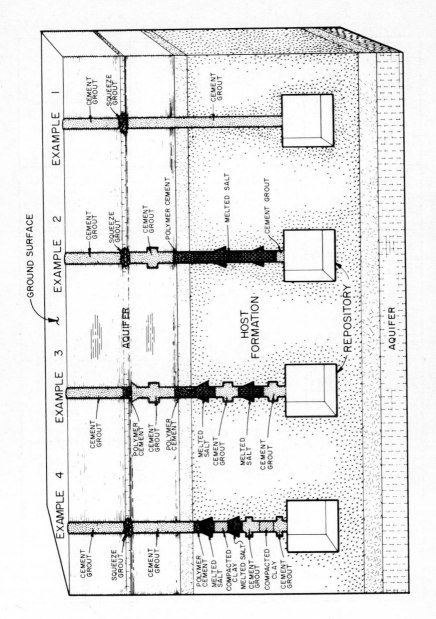

MULTIPLE SHAPES AND MULTIPLE MATERIALS

DESIGN VERIFICATION (VALIDATION) PROGRAM
(PLAN BEING DEVELOPED FOR DOME AND BEDDED SALT AND GRANITE)

METHOD \ ITEM	CHARACTERIZE BOREHOLE (ETC.) CONDITIONS					EVALUATE DISTURBED ZONE				INTERFACE CONDITIONS				EVALUATION OF PLACEMENT METHODS	EVALUATION OF CANDIDATE MATERIALS	PERFORMANCE MEASUREMENTS			
	STRATIGRAPHY	HYDROLOGY	SIZE	SHAPE	GENERAL	EXTENT	DRILLING METHOD	PLACEMENT METHOD	TIME & ENVIRONMENT	GENERAL	DRILLING METHOD	PLACEMENT METHOD	TIME & ENVIRONMENT			INSTALLATION	HYDROLOGIC	GEOCHEMICAL	THERMOMECHANICAL
ANALYTICAL																			
LABORATORY																			
FIELD																			

METHODS TO BE DEVELOPED AND PRIORITIES

MATERIALS RESEARCH PROGRAM IMPLEMENTATION

ACTIVITY	FISCAL YEAR											
	1980		1981				1982					
	QTR	3RD	4TH	1ST	2ND	3RD	4TH	1ST	2ND	3RD	4TH	
DEVELOPMENT OF PROGRAM PLAN TO SUIT DESIGN NEEDS (ONWI & D'APPOLONIA)												
DEVELOPMENT OF TESTING/ANALYSIS PROGRAMS (MATERIALS LABORATORIES)												
REVIEW OF TESTING/ANALYSIS PROGRAMS (ONWI & D'APPOLONIA)												
EXECUTE TESTING/ANALYSIS PROGRAMS (MATERIALS LABORATORIES)												
REVIEW SESSIONS (ALL PARTICIPANTS)												
REVISE PROGRAM PLAN TO SUIT CONTINUING DEVELOPMENT OF DESIGN NEEDS (ONWI & D'APPOLONIA)												
REVISE TESTING/ANALYSIS PROGRAMS												

Discussion

<u>K. MATHER</u>, United States

The question of the existence of the disturbed zone around a diamond drilled core hole needs to be seriously considered. I have examined 6 in (150 mm) diameter diamond drilled cores overcored on a drill-hole about 37 mm in diameter in which strain gages had been initially installed. These were cores of concrete containing 1 ½ in (62 mm) maximum sized crushed aggregate, from a dam affected by alkali-silice reaction, in which compressive stresses up to 1000 psi had been measured. Thus there were chances for cracking in the 150 mm core surrounding the 63 mm core hole from more than one cause ; one of the causes that should have been operating was the formation of a disturbed zone produced by and around the initial 37 mm core hole. However there was no evidence of cracking at the outer surface of the 56-mm-thick walls of the outer annulus. This suggests to me that even in an inhomogeneous anisotropic material such as concrete, with its weakness at the underside of large coarse aggregate particles, the disturbed zone must be thinner than 1.5 diameters of the inner core hole. The elastic modulus of the concrete concerned was approximately 6×10^6 psi. Given proper drilling technique there should be a disturbed zone much smaller than the maximum possible in these cores.

<u>R.D. ELLISON</u>, United States

This is valuable information. However, before drawing firm conclusions, it will be necessary to look at each case on a site specific basis. Rock type, in situ stresses, environmental conditions and time can cause different conditions in different cases. The burden to determine extent of disturbance will always remain with the designer for each case.

<u>M. GYENGE</u>, Canada

It seems to me that the problems of borehole and shaft sealing are investigated under the same heading. Because of the differencies in dimensions, emplacement technologies, etc. the Canadian program handles them separately.

It seems to me that ONWI-15 clearly made a similar differentiation.

<u>R.D. ELLISON</u>, United States

The talk given by Mr. Shukla will address this issue.

<u>J. HAMSTRA</u>, Netherlands

Do you have any dialogue with the licensing authorities with regard to their definitions about the acceptability of the design requirements you specified and the field verification testing you defined ?

<u>R.D. ELLISON</u>, United States

ONWI has had discussions with NRC on these issues. In the future, it may be necessary to increase direct contact between technical investigators on each side. It is a very important point and must be satisfactorily resolved.

J. HAMSTRA, Netherlands

Your design requirements are based on a radionuclide release rate that is stated to be an acceptable release rate. Have you identified what is acceptable relative to future radionuclide release rates ?

R.D. ELLISON, United States

Quantitative allowable release rates will likely be institutional decisions by the Environmental Protection Agency or NRC. Our suggested factor of safety can be used to allow for future changes.

REPOSITORY SEALING: EVALUATION OF
MATERIALS RESEARCH REQUIREMENTS

D. Meyer[1], R. H. Goodwin[1], J. C. Wright[2]

[1]D'Appolonia Consulting Engineers, Inc.   [2]D'Appolonia Consulting Engineers, Inc.
Albuquerque, New Mexico   Pittsburgh, Pennsylvania

ABSTRACT: A program for continued materials research in support of the NWTS program for repository sealing is based upon 1) the establishment of engineering properties of seals which are useful in determining the applicability and in situ performance of a penetration seal or material, and 2) a range of candidate sealing materials potentially capable of providing one of the basic seal functions: prevention of fluid flow or sorption of contaminants. Research should be accomplished according to priorities. Priorities are assigned to engineering properties research for each candidate material group according to the need for future research and the importance of the property in determining the suitability of a material for penetration sealing, and upon an appraisal of the likelihood of successfully sealing with the material.

INTRODUCTION

Materials research is an important aspect of the NWTS program for repository sealing. The overall objective of the repository sealing program is to design acceptable seals for penetrations (boreholes, shafts and tunnels) associated with nuclear waste repository sites, as well as to design backfillings for repository chambers. As such the materials research program includes the following elements:

o Definition of pertinent engineering properties of candidate seal materials which pertain to seal design and in situ performance, and which will form the basis for the comparison and final selection of materials for use in given penetration

o Identification of candidate seal materials capable of fulfilling performance requirements

o Development of a study program to quantitatively evaluate the performance of materials.

This report represents a summary of progress toward satisfying these requirements. The program described is dynamic, allowing for review and redirection as necessitated by the NWTS program.

IDENTIFICATION OF PROPERTIES PERTINENT TO SEAL DESIGN

The engineering properties of a seal are expected to determine the applicability and performance of a seal or seal material in a specific host environment. In general, engineering properties relate to the function of a seal in the host environment, its compatibility with the host environment and its ability to persist in the host through time.

ENGINEERING PROPERTIES OF SEALS AND SEAL MATERIALS

Engineering properties of interest to the penetration sealing effort can be categorized into five general groups, to assist in communication within the technical community and to provide a basic format for analyzing and designing penetration seals. The categories and the specific engineering properties which belong in each category are:

o Chemical properties
   - Sorptivity
   - Longevity
   - Reaction/reactivity with the host environment

o Hydraulic properties
   - Internal permeability of the seal
   - Permeability of the seal-host interface and any permeability effects in the host material surrounding the seal

o Mechanical properties
   - Ductility/elasticity
   - Bond strength
   - Internal strength

o Thermal properties
   - Differential thermal expansion
   - Differential thermal conductivity
   - Differential thermal diffusivity

o Emplacement methods
   - Feasibility
   - Technological requirements
   - Quality control of emplaced seal

Clearly, consideration of emplacement is not an engineering property in the sense of the other four groups; however, emplacement is expected to be critical to the evaluation and application of materials in penetration sealing. Thus, it is necessary to study this aspect of sealing in the context of materials research since there is an interrelationship between emplacement methods and seal properties. Perhaps a significant proportion of this research area would be performed in coordination with field testing programs.

Functions of a Seal and Related Engineering Properties

To better understand the relative significance of various engineering properties, it is necessary to relate them to the basic functions that a seal component must perform. All penetration seals are to be designed to (1) limit within acceptable levels, the migration of radionuclides or other contaminants out of the repository through the penetration and into the biosphere, or (2) prevent or limit the influx of groundwater into the repository through associated penetrations. Penetration seals will perform as designed because seal components provide one of the following basic seal functions:

o Prevention or retardation of fluid (including gas) flow: the ability of a seal component to prevent or retard the transmission of associated fluids. An effective barrier will prevent or retard fluid transmission through the seal-host interface and through permeable, disturbed zones in the host rock adjacent to the seal, as well.

o Sorption of contaminants: the ability of a material to remove or retain dissolved radionuclides or other contaminants from associated fluids by means of adsorption, absorption, precipitation or participating in ion exchange reactions.

Reduction of fluid flow by a seal operates in a two fold manner: by limiting the movement of contaminated fluids through the penetration to the biosphere and by limiting groundwater movement through a penetration into the repository. A sorptive seal component will provide retardation of dissolved radionuclide migration by filtering radioactive contaminants by surficial or chemical means.

The most critical engineering properties of a seal relative to the prevention of fluid flow are the hydraulic properties: internal permeability of the seal, and seal-host interface and surrounding zone permeability. The sorption function of a seal is related to the seal's engineering property, sorptivity. Evaluation of these seal properties must consider actual host conditions, including, for example, fluid viscosity and composition, and hydraulic gradient. Intrinsic physical and chemical properties of the sample or seal will also determine the functions of seals and the engineering properties used to measure them.

Other Significant Engineering Properties

Other significant engineering properties, which are not functions of the seal, but are as important as permeability and sorptivity for adequate performance of a seal include:

o Intrinsic longevity

o Reactions and reactivity (geochemical compatibility) with host environment

o Properties related to thermomechanical durability

o Properties related to thermal and mechanical compatibility between the seal and host, and among adjacent seals

These engineering properties are not seal functions, but are important in tandem with the ability of the seal to perform one of the seal functions.

Longevity is the property of a material which allows it to persist and maintain its basic intended function through time under the range of temperature, pressure and geochemical conditions anticipated in the host environment. Because the effective lifetime of a seal must be greater than the practical time limitations of experimental methods, longevity studies will necessitate a theoretical assessment of seal longevity. This assessment may include an examination of the phase chemistry and thermodynamics of phases present in seals, and a study of the kinetics of possible reactions. This analysis is anticipated to allow a scientific determination of seal stability in a time frame consistent with the licensing requirements.

Seal-host environment reactivity (or geochemical compatibility) incorporates consideration of reactions among seals, host rocks, and fluids present in a penetration as well as consideration for the effects of earth currents, electropotentials and biological activity. Reaction products will be analyzed and characterized. Such reactivity is of prime significance in establishing the behavior of the bond or interface between a seal and host rock, and may also impact on the overall stability of the seal. Special consideration should be allotted to seal-host reactivity during emplacement when the presence of excessive pressures and temperatures, or hydrothermal processes may enhance reactivity.

Thermomechanical durability under the host environment conditions, including response to earth movements and any other stresses (e.g. fluid or lithostatic pressures), and the consequences thereof to seal integrity are largely dealt with under mechanical properties, including the elasticity and ductility of the seal, its internal strength and the strength of the bond between the seal and the host rock. In addition, thermal compatibility is an important consideration. Differential response, most notably expansion or contraction of seals and host due to heating or cooling of the emplacement environment, must be evaluated for any seal-host rock combinations.

As is the case with engineering properties related to seal functions, engineering properties of seals mentioned here can only be assessed with respect to the environment of emplacement. In addition, the intrinsic properties of seal materials will affect or determine the engineering properties. Both host environment and intrinsic material properties of seals must be considered in materials research planning.

Host Environment Conditions

Engineering properties of a penetration seal will be greatly affected by the conditions existing in the penetration to be sealed, and further affected by changes in these conditions expected during the period of required sealing. Host environmental characterization therefore, assumes critical importance to material research and selection, and must at least include determination of:

     o The range of fluid (including gas) compositions as well as the quantities which contact penetration seals through time.

     o The range of expected temperature and pressure regimes in which a penetration seal must perform.

     o The stratigraphy, mineralogy and lithology encountered in specific penetrations at potential nuclear waste repository sites and the material and engineering properties of all host rocks.

     o Condition and shape of the penetration to be sealed including host rock disruptions caused by construction of the penetration.

It is imperative that materials research programs continually interface with site characterization studies and, if necessary, alter their procedures, experimental condition or goals as increasingly detailed and specific host environment data become available. In addition, potential alteration of the host environment due to the repository or natural events should be estimated during site selection and characterization so that a range of probable host conditions, both geological and geochemical, can be incorporated into materials testing.

In the absence of thorough site characterizations, including characterization of specific penetrations, generic geologic and hydrologic studies of potential host rock types will be useful in the development of seal material requirements and seal designs.

Intrinsic Material Properties

Engineering properties of a seal may be strongly affected by certain intrinsic properties of the material with which it is constructed. Intrinsic properties include:

   o Elemental composition

   o Mineralogy of phase chemistry

   o Microtexture (grain size, intergranular
      relationships)

   o Porosity, permeability (intrinsic)

   o Microstructure (microcracks, microjoints)

   o Heat capacity

   o Coefficient of thermal expansion

   o Thermodynamic properties of all phases

This list is not exhaustive. It must be emphasized that seal engineering properties are affected by or derived from intrinsic properties. Thus, the goal of all intrinsic property evaluations must be to enhance our knowledge of engineering properties. One of the initial steps in developing a detailed research program for an engineering property will be to develop thoroughly any relationships between the engineering property of interest and intrinsic material properties. The latter must be determined as part of the research effort to quantify the engineering property.

IDENTIFICATION OF CANDIDATE SEALING MATERIALS

The selection of materials for use in seal designs are to be based upon the critical evaluation of all available materials and their potentially desirable properties, particularly those related to seal functions. The maximum number of candidate materials to be studied in detail must be restricted by consideration of reasonable resource and time limitations of the repository sealing program. It is obvious that a broad, but select range of candidate materials must continue to be considered for the following reasons:

   o Studies to date have not indicated a clear superiority of
      any single material type for use in penetration seals
      particularly under all of the repository conditions and
      environments that will be encountered.

   o Concepts of penetration seal designs have been developed
      and reported in ONWI-55. These conceptual designs indicate
      that it is advantageous to use multiple material seals in
      some penetrations to incorporate the desirable functions
      of different materials and to provide barriers compatible
      with a variety of environmental conditions.

o Licensing activities are anticipated to require that final
  seal designs are the most reasonable and effective in
  light of a thorough examination of alternatives.

The range of materials and material groups which have been identified as potential
seal materials with desirable properties for sealing penetrations includes:

o cements (Portland + additives, hydrothermal, polymer)
o clays
o crushed host rocks or minerals
o salt
o sorptive materials (e.g. zeolites, charcoal)
o chemical grouts
o synthetically produced 'rock-like' materials
o metals
o ceramics

The above materials have been selected based on the current thoughts
within the scientific community involved in repository sealing (Taylor,
1979; Burns, 1979), the identification of sealing functions, (e.g.,
ONWI-55), the possibility of achieving sealing functions with these
candidate materials, and a comprehensive survey of existing knowledge of
materials research for penetration sealing, as summarized in ONWI-15.

Different modes of emplacement may cause dissimilar seal properties and behavior
from identical starting materials. Thus, emplacement options which may be
pertinent include:

o Transport of a (probably aqueous) slurry of material
o Hydrothermal transport of material
o Transport of dry granular material
o Transport of molten material

After transport, the material may be set with or without presssure, under compression, or with compaction. Conceivably, emplaced material could be altered
chemically or physically after setting, perhaps by heat or induced chemical
reactions. Not all emplacement methodologies appear to be appropriate to all
materials. Nonetheless, the variety of possible emplacement options adds another
dimension to materials research considerations.

DEVELOPMENT OF A TESTING PROGRAM

Identification of Needed Research

The status of engineering properties research is evaluated on Table I. For each
generic material group, an assessment of the knowledge of each engineering
property needed in order to design successful penetration seals is compared to
the existing level of knowledge as determined from the literature. Three distinc-
tive categories to describe the current status of knowledge are shown on Table I.

Priority Rankings of Needed Research

The materials research program proposed here is based on the designation of
research needs and goals, and a priority ranking thereof. This program should
serve two general purposes:

o To aid in the communication within the technical community
  and to coordinate their efforts toward common goals

o To provide needed materials research which can address a
  range of options and alternatives in penetration sealing
  within the reasonable resource limits of the program.

Principles of Priority Ranking

Priority research designations are key to meeting materials research goals.
Priorities are assigned to candidate materials groups and the various engineering
properties on the basis of the following considerations:

o The amount and quality of existing data and knowledge,
  achieved as part of the past or ongoing penetration
  sealing effort, and from the scientific and technical
  literature, relative to needed knowledge to design penetration seals. In this regard, the expenditure of effort
  required to meet needs is estimated.

o The pertinence of an engineering property on the determination of a candidate material's suitability for penetration sealing, and upon decisions to accept or reject a
  material for a given design application.

o The evaluation from past experience or research, along
  with a subjective appraisal, regarding the probability of
  achieving acceptable seal designs using a particular
  material, within realistic time constraints and the
  resources of the research effort.

The first consideration, the assessment of existing knowledge, addresses whether
further research is needed to meet the needs of penetration sealing. Table I is
the result of this consideration, and shows where research appears to be most
needed. The second consideration, given the results shown on Table I, is intended
to assess which engineering properties, for each material, are most critical to
acceptability of the material in a seal (for example, its ability to perform a
function), or which engineering properties form the most effective basis for the
comparison of material performance. The third consideration is most directly
based upon the expected need for technological development and intensive research
before a seal incorporating a given material can be licensed and installed at a
repository site. Weighing heavily in such a consideration is emplacement feasibility, including matters related to in place melting, chemical or physical
alteration of seals, quality control and the likelihood of adverse effects on
host formations.

In providing an effective and efficient means for material comparison and selection, the priority ranking of research should also provide a systematic screen
for materials and research needs. Materials can be eliminated from consideration
based on quantitative assessments and comparisons of seal engineering properties
judged to be the most important or critical to seal design and performance. In
many cases, research of secondary importance may be avoided.

Definition of Priority Rankings

Research priorities for nine material groups are discussed on Tables II - X and
summarized in Table XI. Priorities are ranked on a 1 to 4 scale. The significance
of the numerical rankings follow:

Priority 1 - Research of immediate importance in determining the overall
             feasibility of utilizing a candidate material and to
             provide information necessary for continued activities
             associated with the overall penetration seal design effort.

Priority 2 - Areas where the overall research effort may be extended
             over a longer period of time but where an initial "phase
             one" effort is required and should be investigated immediate-
             ly in order to develop answers to very basic information needs.
             Usually the phase one effort can be accomplished from
             literature surveys, and extractions from other types of
             activities or analyses without extensive laboratory or
             field testing.

Priority 3 - Areas where it is anticipated that research will be required
             to develop final design and licensing documents but where
             the results of that research are not expected to impact
             decisions with regard to the feasibility of using materials
             or developing basic seal design and configurations which
             incorporate that material. These research efforts can be
             delayed until priority 1 and 2 issues for that material are

substantially resolved or until the priority of a particular
issue is modified due to increases in overall program
knowledge.

Priority 4 - At this time it is anticipated that no substantial research
effort will be required for any item given a priority 4
ranking. The priority 4 ranking indicates that either
the particular property is not of importance for the
material being considered or that available information
about that material is adequate to answer all questions
about that property.

Engineering property research ranked Priority 4 is not shown in Tables II
through XI.

The priority rankings proposed herein are not immutable. Changes will be necessary as research is completed, and materials progress through the program.

DISCUSSION

Tables II through XI form a logical basis for the development of a materials
research program that will be of greatest benefit to the overall penetration seal
design effort. These tables can also provide the basis for the development of
detailed analytical laboratory testing or field testing procedures and methods.
It is considered most crucial to the materials research program that steps be
taken to implement regular review procedure which include all groups involved in
penetration sealing activities. These reviews should assess research accomplish-
ments and, if necessary, revise and/or redirect certain research efforts and
reassign research priorities. A first order work breakdown, showing frequent
reviews, periodic redirection of research and reassignment of priorities is
proposed on Figure 1. Advocation of frequent reviews is based upon recognition
of the importance that each new piece of research information has on determining
the feasibility and applications of materials and in revealing additional unknowns
that may have to be resolved. Coordination among the involved parties will help
assure that important issues will not be overlooked while at the same time the
available resources will continually be directed toward development of the
information necessary to provide reasonable designs for penetration seals that
can be licensed and properly installed.

LIST OF REFERENCES

Burns, F., compiler, 1979, "Needed Research and Development in Repository Seal-
ings: Recommendations of the Earth Science Technical Plan (ESTP) Sealing
Subgroup," Battelle Memorial Institute, Office of Nuclear Waste Isolation,
Columbus, Ohio, 90 pp.

D'Appolonia, C. E., 1979a, "Repository Sealing Design Approach - 1979," ONWI-55,
Battelle Memorial Institute, Office of Nuclear Waste Isolation, Columbus, Ohio.

D'Appolonia, C. E., 1979b, "The Status of Borehole Plugging and Shaft Sealing for
Geologic Isolation of Radioactive Waste," ONWI-15, Battelle Memorial Institute,
Office of Nuclear Waste Isolation, Columbus, Ohio.

Taylor, C. L., G. J. Antonnen, J. E. O'Rourke, and M. R. Niccum, 1979, "Borehole
Plugging of Man-Made Accesses to a Basalt Repository: A Preliminary Study,"
Draft: RHO-BWI-C-49, to be published as RHO-BWI-6, Woodward-Clyde Consultants,
San Francisco, California, 157 pp. and appendices.

TABLE I

STATUS OF EXISTING KNOWLEDGE REGARDING SEAL ENGINEERING PROPERTIES

ENGINEERING PROPERTIES	CANDIDATE SEALING MATERIALS → CEMENT			CLAY	CRUSHED HOST ROCK		SALT		CHEMICAL GROUTS	SORPTIVE MATERIALS	SYNTHETIC ROCK OR MINERALS	METALS	CERAMICS
	PORTLAND (+ADDITIVES)	HYDRO-THERMAL	POLYMER	MATERIAL & EMPLACEMENT TO BE DETERMINED	COMPACTED	SLURRY	MOLTEN	COMPACTED	SOLUTION/SLURRY	MATERIAL SELECTION TO BE COMPLETED	MATERIAL SELECTION TO BE COMPLETED	MATERIAL SELECTION/EMPLACEMENT OPTION TO BE DETERMINED	MATERIAL SELECTION/EMPLACEMENT OPTION TO BE DETERMINED
CHEMICAL — Sorptivity	S	S	S	U	U	U	S	S	S	U	N	S	N
Longevity	N	U	N	U	U	U	S	S	N	N	N	N	N
Reactions/Reactivity (emplaced in host)	U	N	N	U	U	U	S	S	N	N	N	N	N
HYDRAULIC — Permeability (internal)	S	S	S	U	N	N	S	S	S	S	N	S	N
Permeability/Porosity of bond and surrounding zone	N	N	S	N	N	N	N	S	N	S	N	N	N
MECHANICAL — Ductility/Elasticity	U	U	U	S	U	U	U	S	U	S	N	U	N
Bond Strength	U	U	S	S	U	U	S	S	U	S	N	S	N
Internal Strength	S	S	S	S	U	U	S	S	U	S	N	S	N
THERMAL — Differential Thermal Expansion	U	U	U	U	U	U	S	S	U	N	N	S	N
Differential Thermal Conductivity and Diffusivity	U	U	U	S	U	U	S	S	U	N	N	S	N
EMPLACEMENT — Methodology Technology Quality Control	U	U	U	S	N	N	S	S	U	N	N	S	N

Legend

Level of existing knowledge relative to knowledge necessary for design of seals

S. Knowledge from current or past research approaches that necessary for the design of penetration seals.

U. Current knowledge is unsatisfactory for the design of penetration seals. For the most part, existing data provides only a generic understanding of material behavior

N. Negligible knowledge to meet needs for the design of penetration seals.

TABLE II
Proposed Priorities for Research Programs
for Candidate Sealing Materials--

CEMENTS
Portland and additives, hydrothermal, polymer

ENGINEERING PROPERTIES	PRIORITY RANKING	DISCUSSION
Longevity (determined by thermodynamic properties and reaction kinetics)	2.	Longevity of Portland cements will be important to seal design. A long-term research program will be needed, beginning with a preliminary priority study to include literature studies, field investigation and experimental designs for determining cement longevity. Longevity may be one of the stronger potential advantages of hydrothermal cements over Portland cement. Further consideration of hydrothermal cements hinges on assurance of their longevity, beginning also with an immediate preliminary investigation. Polymer cements may not be long-lived. It is unlikely that any design incorporating these materials will depend on polymer cements for long term function. Longevity studies will be conditional on high priority research results, comparison with similar materials (Portland cements, chemical grouts), especially regarding seal function, and seal designs including polymer cements (that is, if the longevity property is critical to the design of the seal component). Thus, longevity should be assigned the conditional (3) priority for polymer cements.
Reactions and reactivity (under host environment conditions, determined experimentally; includes behavior/properties of reaction products; description of all host-seal reactions)	2.	Important to seal design, and may limit use of some cements in certain environments (for example, acidic, aqueous environments). Interaction during emplacement of hydrothermal cements may be critical to their applicability in penetration sealing. Study must interface with emplacement study.
Permeability and porosity of bond and surrounding zone (effect on disturbed zone adjacent to seal)	1.	Low bond/interface permeability is critical to the continued consideration of cements for repository seals. Major laboratory effort (simulation experiments) and field testing is required. It will be necessary to coordinate this research with emplacement studies.
Ductility and elasticity	3.	Important to mechanical compatibility of seal and host. Research effort is justified only for samples which are judged satisfactory during high-priority testing.
Bond strength	3.	May not be important to seal design, except as a possible effect on bond zone permeability (especially with time). Further study will interface or follow permeability and reactivity research.
Differential thermal expansion	2.	Potentially critical for cement-host compatibility and seal integrity in changeable environmental conditions. Early study phase must generically identify compatible materials.

TABLE II
(Continued)

CEMENTS

Portland and additives, hydrothermal, polymer

ENGINEERING PROPERTIES	PRIORITY RANKING	DISCUSSION
Differential thermal conductivity and diffusivity	3.	Possibly important for the use of cements in some host materials. Research need depends on higher priority studies.
Methodology and technology of emplacement and quality control	2.	Very important aspect of seal design, feasibility, constructability. The direction for this research will depend on results of surrounding zone and seal-host interface permeability studies. Long-term research should be initiated with study of improved methodology, technology requirements and quality control methods including generic and historic perspectives. Initial phase may be addressed in the field-testing program. Emplacement will be critical to the application of hydrothermal cements. Initial research phase must interface with longevity study and address practical means to achieve long-lived cements and the consequences, particularly on the host, of the hydrothermal emplacement methods.

TABLE III

Proposed Priorities for Research Programs
for Candidate Sealing Materials--

CLAYS

ENGINEERING PROPERTIES	PRIORITY RANKING	DISCUSSION
Sorptivity (including variability with time, temperature, physico-chemical condition)	2.	Importance of clays as sorptive media in seal designs will depend on results of this study. Initial study should accumulate existing data, select promising species/mixes, plan research on artificially compacted clay bodies.
Longevity (determined by thermodynamic properties and reaction kinetics)	1.	Property is ascribed to certain clays. Under conditions expected in penetrations (design conditions) longevity has not been proven. Research must determine clay species or mixes, which are long-lived under expected conditions, for further study.
Reactions and reactivity (under host environment conditions, determined experimentally; includes behavior/ properties of reaction products; description of all host-seal reactions)	1.	Chemical stability of clays under design conditions not assured. Includes type and rate of reactions, nature of reaction products (particularly regarding the alteration of seal functions) under host conditions (fluids, geologic media, waste). Includes dehydration studies and properties of anhydrous phases.
Permeability (internal)	2.	Importance of clays as impermeable media in seal designs will depend on the results of these studies; that is, designs will incorporate clay plugs as impermeable barriers only upon completion of at least the initial phases of this study, with affirmative

TABLE III
(Continued)
Proposed Priorities for Research Programs
for Candidate Sealing Materials--

CLAYS

ENGINEERING PROPERTIES	PRIORITY RANKING	DISCUSSION
		results. Special regard for the permeability of clay seals compacted under feasible pressures will require interfacing with emplacement studies.
Permeability and porosity of bond and surrounding zone (effect on disturbed zone adjacent to seal	3.	Important for clay seals proven to have low permeability after compaction and proven feasible to emplace. May be coordinated into the impermeability study (thereby changing priority)
Differential thermal expansion	2.	May be critical to application of clays and seal-host compatibility in seals subject to variable conditions. Study must interface with reactivity study because chemical/mineralogical shrinkage or expansion could be critical. Initial studies should identify compatible materials, and plan needed research.
Methodology and technology of emplacement and quality control	2.	Initial feasibility study of clay compaction completed. Constructability and equipment problems remain unresolved. Effect of compaction on adjacent host formation or seal segments undetermined. Long-term effort is required, beginning with preliminary study of technical needs, experimental program design, etc. Slurry or loosely-packed clay seals present less critical emplacement problems. Only upon affirmation of primary plug functions, active research should begin.

TABLE IV
Proposed Priorities for Research Programs
for Candidate Sealing Materials--

CRUSHED HOST ROCK/MINERALS
(exclusive of salt and clay)

ENGINEERING PROPERTIES	PRIORITY RANKING	DISCUSSION
Sorptivity (including variability with time, temperature, physicochemical condition)	1.	Ability of crushed (and possibly compacted) host rock to serve the sorptive function is undetermined for relevant lithologies (basalt, granite). Study should be coordinated with reactivity studies, since reaction products (e.g. clays, zeolites) may provide additional sorption.
Longevity (determined by thermodynamic properties and reaction kinetics)	2.	A presumed advantage of using host rock in the seals, which has not been proven under physical/chemical conditions in host environment. Critical property only if sorptive or impermeable function can be established for a given material. Long-term study should begin with preliminary studies of mineralogic thermodynamic studies and existing experimental results.

TABLE IV
(Continued)
Proposed Priorities for Research Programs
for Candidate Sealing Materials--

CRUSHED HOST ROCK/MINERALS
(exclusive of salt and clay)

ENGINEERING PROPERTIES	PRIORITY RANKING	DISCUSSION
Reactions and reactivity (under host environment conditions, determined experimentally; includes behavior/ properties of reaction products; description of all host-seal reactions)	1.	With respect to intact host rock, crushing increases the potential reaction surface area as well as increasing the number of flow paths for fluids and reagents. Therefore, the apparent steady state between host rock and associated fluids, possibly due to insulating effects of reaction layers or altered surfaces, may not occur in the same material after crushing. Low pressure-temperature alteration products may affect seal functions and longevity. Thus, results of this research should be input into function and longevity studies.
Permeability (internal)	1.	Low permeability is possible, but unestablished for crushed host rock seals. Research must be completed before designs are formulated which incorporate these materials for the impermeable function. Function may be affected by grain size, mineralogy and the ability to compact crushed materials. Thus, some coordination with emplacement study is required. Reaction products, if any, may affect the seal function. Requires coordination with the reactivity study.
Permeability and porosity of bond and surrounding zone (effect on disturbed zone adjacent to seal)	3.	Important only if internal permeability studies affirm crushed (and compacted) host rock can function as an impermeable material.
Ductility and elasticity	3.	Importance depends on completion and results of high-priority studies.
Bond strength	3.	Importance depends on completion and results of high-priority studies.
Internal strengh of seal	3.	Importance depends on completion and results of high-priority studies.
Differential thermal expansion	3.	Importance depends on completion and results of high-priority studies.
Differential thermal conductivity and diffusivity	3.	Importance depends on completion and results of high-priority studies.
Methodology and technology of emplacement and quality control	3.	Study will benefit from higher priority emplacement studies for clay and salt research. Additional studies, if needed, will depend on special requirements of crushed rock materials, including pretreatment of materials to achieve seal functions, or large-scale mineral separation methods. Research needs will be dictated by high priority seal function research.

TABLE V
Proposed Priorities for Research Programs
for Candidate Sealing Materials--

SALT

ENGINEERING PROPERTIES	PRIORITY RANKING	DISCUSSION
Permeability and porosity of bond and surrounding zone (effect on disturbed zone adjacent to seal)	1.	Molten salt may cause thermal shock in host salt. Compaction may initiate host rock fractures. Extent of host formation disruption and subsequent healing are critical to seal soundness and applicability of salt as seal material. The study must interface with the emplacement methodology and quality control studies.
Bond strength	3.	May be affected by emplacement method and could affect bond zone permeability. Study may be incorporated into bond permeability study or follow it.
Methodology and technology of emplacement and quality control	1.	Important research area for determining the applicability of salt to the repository sealing effort. Unsolved technological problems, including equipment life and corrosion, should be addressed. Must achieve seals which are as similar as possible to host, without extensive disruption of host. Limited applicability of salt seals narrows the scope of emplacement research to rock salt hosts.

TABLE VI
Proposed Priorities for Research Programs
for Candidate Sealing Materials--

CHEMICAL GROUTS

ENGINEERING PROPERTIES	PRIORITY RANKING	DISCUSSION
Longevity (determined by thermodynamic properties and reaction kinetics)	2.	Preliminary study required to identify the potential for grout longevity. Comparison with cement grouts and polymer cements will be important for continued consideration of chemical grouts.
Reactions and reactivity (under host environment conditions, determined experimentally; includes behavior/properties of reaction products; description of all host-seal reactions)	2.	Important preliminary studies should begin to assess grout-host reactivity, particularly with respect to existing experimental and field evidence of reactivity with geologic materials.
Permeability and porosity of bond and surrounding zone (effect on disturbed zone adjacent to seal)	2.	Major application of grouts may be in the treatment of downhole disturbed zones. Preliminary research (laboratory or field) should assess the potential for chemical grout application in disturbed zone or interface.
Ductility and elasticity	3.	Importance depends on results of high-priority research.

CHEMICAL GROUTS

ENGINEERING PROPERTIES	PRIORITY RANKING	DISCUSSION
Bond strength	3.	Importance depends on results of high-priority research.
Internal strengh of seal	3.	Importance depends on results of high-priority research.
Differential thermal expansion	2.	Preliminary research should identify compatible host and grout materials generically. Further research depends on results of high-priority research.
Differential thermal conductivity and diffusivity	3.	Importance depends on results of high-priority research.
Methodology and technology of emplacement and quality control	3.	Well-known technology. Additional study may be needed to address technical aspects of downhole emplacement or downhole grouting of disturbed zones in the annulus of a penetration.

TABLE VII
Proposed Priorities for Research Programs
for Candidate Sealing Materials--

SORPTIVE MATERIALS

ENGINEERING PROPERTIES	PRIORITY RANKING	DISCUSSION
Sorptivity (including variability with time, temperature, physico-chemical condition)	1.	All generic sorptive materials should be thoroughly examined and compared on the basis of their ability to perform sorptive functions, particularly through a range of physical-chemical conditions for an extended period of time. This program should result in initial sorptive material selection for all further research.
Longevity (determined by thermodynamic properties and reaction kinetics)	3.	Important criteria for materials which are able to serve sorptive function. Sorptivity study must be completed before initiation of longevity studies.
Reactions and reactivity (under host environment conditions, determined experimentally; includes behavior/properties of reaction products; description of all host-seal reactions)	3.	Important criteria for materials which are able to serve sorptive function. Effects of reactions between sorbers and host environment (as well as reaction products) on sorptive behavior of materials must be addressed primarily.
Permeability and porosity of bond and surrounding zone (effect on disturbed zone adjacent to seal)	3.	Potentially important if seal is more sorptive than host formation. Thus, limiting flow in the interface will force flow through sorptive material.

TABLE VII
(Continued)
Proposed Priorities for Research Programs
for Candidate Sealing Materials--

SORPTIVE MATERIALS

ENGINEERING PROPERTIES	PRIORITY RANKING	DISCUSSION
Differential thermal expansion	3.	May be critical to seal and host compatibility and seal function. Conditional on higher priority research.
Differential thermal conductivity and diffusivity	3.	May be critical to seal and host compatibility and seal function. Conditional on higher priority research.
Methodology and technology of emplacement and quality control	3.	Important research area, but will be initiated only after high-priority sorptivity studies and initial material selection is complete.

TABLE VIII
Proposed Priorities for Research Programs
for Candidate Sealing Materials--

SYNTHETICALLY-PRODUCED ROCK-LIKE MATERIALS

ENGINEERING PROPERTIES	PRIORITY RANKING	DISCUSSION
Longevity (determined by thermodynamic properties and reaction kinetics)	3.	A possible advantage of synthetic 'rock' for use in seals although formation of insulating reaction surfaces may contribute more to seal soundness than thermodynamic equilibrium. Therefore, importance of research depends on results of high-priority studies, especially reactivity.
Reactions and reactivity (under host environment conditions, determined experimentally; includes behavior/ properties of reaction products; description of all host-seal reactions)	2.	Particularly important with regard to reactions during high-temperature or hydrothermal emplacement. Efforts to reduce interaction of seal and host during emplacement must interface with emplacement studies.
Permeability (internal)	2.	Presumed property of some synthetic rock seals. Preliminary studies should be directed toward measurement of the simplest laboratory-prepared synthetic rocks.
Permeability and porosity of bond and surrounding zone (effect on disturbed zone adjacent to seal)	3.	Important study for synthetic rocks which serve impermeability function, particularly if seal-host reactions affect permeability of bond and adjacent zones. Study must interface closely with reactions study.
Ductility and elasticity	3.	Importance is conditional on results of high-priority studies.
Bond strength	3.	Importance is conditional on results of high-priority studies.
Internal strength of seal	3.	Importance is conditional on results of high-priority studies.

TABLE VIII
(Continued)
Proposed Priorities for Research Programs
for Candidate Sealing Materials--

SYNTHETICALLY-PRODUCED ROCK-LIKE MATERIALS

ENGINEERING PROPERTIES	PRIORITY RANKING	DISCUSSION
Differential thermal expansion	3.	Importance depends on the ability to match seal to host rock, physically and chemically. Conditional on high priority studies.
Differential thermal conductivity and diffusivity	3.	Importance conditional on results of high-priority studies.
Methodology and technology of emplacement and quality control	2.	Priority effort on generic feasibility and constructability of seals using synthetic rock-like materials. Studies should address candidate materials, physical state of material during transport, transport and emplacement technology/methods (slurry, crushed, molten, hydrothermal transport, hot or cold pressing), with consideration for the minimization of damage to the host formation. Preliminary phase of research should include consideration of the feasibility of earth melting and hydrothermal transport methods developed for other material groups. Further research depends on the results of the preliminary phase of all priority research efforts.

TABLE IX
Proposed Priorities for Research Programs
for Candidate Sealing Materials--

METALS

ENGINEERING PROPERTIES	PRIORITY RANKING	DISCUSSION
Longevity (determined by thermodynamic properties and reaction kinetics)	3.	Important property which should be researched only upon completion of higher priority studies.
Reactions and reactivity (under host environment conditions, determined experimentally; includes behavior/ properties of reaction products; description of all host-seal reactions)	2.	Study of the effect of emplacement of molten metals in (probably wet) geologic formations, including alteration of host and interface or metal contamination. Initial preliminary studies must closely coordinate with emplacement technology study and bond and surrounding zone permeability study.
Permeability and porosity of bond and surrounding zone (effect on disturbed zone adjacent to seal)	2.	Important criteria for use of metals. Effect of metal emplacement on host rock permeability will be important for continued consideration of metal seals. Close coordination with reactivity and emplacement studiess should address minimization of host rock disruption and permeability due to chemical and physical alteration. Initial study phase must research any existing relevant information and design the experimental program for continued research.

TABLE IX
(Continued)
Proposed Priorities for Research Programs
for Candidate Sealing Materials--

METALS

ENGINEERING PROPERTIES	PRIORITY RANKING	DISCUSSION
Bond strength	3.	Importance depends upon results of bond zone permeability study and may interface with it.
Methodology and technology of emplacement and quality control	2.	Important factor in use of metal seals. Initial feasibility study is required, including technological requirements for emplacement. Research should be closely coordinated with study of bond/surrounding zone studies to develop criteria of maximum seal effectiveness with acceptable damage to host rock.

TABLE X
Proposed Priorities for Research Programs
for Candidate Sealing Materials--

CERAMICS

ENGINEERING PROPERTIES	PRIORITY RANKING	DISCUSSION
Sorptivity (including variability with time, temperature, physicochemical condition)	3.	An important property of permeable ceramics for use in seals. Research is conditional on results of emplacement study.
Longevity (determined by thermodynamic properties and reaction kinetics)	3.	Possible property of ceramic seals. Research is conditional on results of emplacement study.
Reactions and reactivity (under host environment conditions, determined experimentally; includes behavior/ properties of reaction products; description of all host-seal reactions)	3.	Many ceramics tend to be inert under especially harsh conditions, and can be formulated to optimize this property. Research is conditional on results of emplacement study.
Permeability (internal)	3.	Important characteristic of many ceramics and could be used to advantage in repository seals. Considerable existing data should be investigated conditional on results of emplacement study.
Permeability and porosity of bond and surrounding zone (effect on disturbed zone adjacent to seal)	3.	If ceramic paste can be squeezed into disturbed or porous host rock and if it can be fired adequately without disrupting host, host-seal interface or causing alteration detrimental to seal or host, ceramic may have advantageous properties. Research is conditional on results of emplacement study.
Ductility and elasticity	3.	May be important property of ceramic seals. Considerable existing data. Research is conditional on results of emplacement study.

TABLE X
(Continued)
Proposed Priorities for Research Programs
for Candidate Sealing Materials--

CERAMICS

ENGINEERING PROPERTIES	PRIORITY RANKING	DISCUSSION
Bond strength	3.	May be important property if permeability is affected. Research is conditional on results of emplacement study.
Internal strength of seal	3.	Possibly important seal property. Research is conditional on results of emplacement study.
Differential thermal expansion	3.	May be important property of ceramic seals. Considerable existing data. Research is conditional on results of emplacement study.
Differential thermal conductivity and diffusivity	3.	May be important property of ceramic seals. Considerable existing data. Research is conditional on results of emplacement study.
Methodology and technology of emplacement and quality control	2.	Most critical area of study at present. No serious study of use of ceramics in seals exists. Initial feasibility study required. Need to address: downhole firing and its effects on host; alternatives to downhole firing; feasibility of host rock grouting with ceramics; formulations and procedures for achieving impermeable or sorptive seals. Results of this study determine continued consideration and reassignment of priorities for ceramics.

TABLE XI

Summary of priority materials research

ENGINEERING PROPERTIES	CEMENT — PORTLAND (+ADDITIVES)	CEMENT — HYDRO-THERMAL	POLYMER	CLAY (MATERIAL & EMPLACEMENT TO BE DETERMINED)	CRUSHED HOST ROCK — COMPACTED	CRUSHED HOST ROCK — SLURRY	SALT — MOLTEN	SALT — COMPACTED	CHEMICAL GROUTS — SOLUTION/SLURRY	SORPTIVE MATERIALS (MATERIAL SELECTION TO BE COMPLETED)	SYNTHETIC ROCK OR MINERALS (MATERIAL SELECTION TO BE COMPLETED)	METALS (MATERIAL SELECTION/EMPLACEMENT OPTION TO BE DETERMINED)	CERAMICS (MATERIAL SELECTION/EMPLACEMENT OPTION TO BE DETERMINED)
CHEMICAL													
Sorptivity	●	●		●	●	●				●			○
Longevity	●	●	○	●	●	●			●	○	○	○	○
Reactions/Reactivity (emplaced in host)	●	●	●	●	●	●			●	○	●	●	○
HYDRAULIC													
Permeability (internal)			●	●	●	●			●		●	●	○
Permeability/Porosity of bond and surrounding zone	●		●	○	○	○	●	●	●	○	○		○
MECHANICAL													
Ductility/Elasticity	○		○				○	○	○		○	○	○
Bond Strength	○		○						○		○		○
Internal Strength													○
THERMAL													
Differential Thermal Expansion	●		●	○	○	○			●	○	○		○
Differential Thermal Conductivity and Diffusivity	○		○	○	○	○			○	○	○		○
EMPLACEMENT													
Methodology Technology Quality Control	●	●	●	●	○	○	●	●	○	○	●	●	●

PRIORITY RANKINGS

Priority 1 (●) — Research of immediate importance

Priority 2 (◐) — Long-term study required; completed in phases; early research phase of immediate importance

Priority 3 (○) — Research of importance currently anticipated; but not of immediate importance and depending upon results of higher priority research.

Discussion

M.J. SMITH, United States

      Who established the research priorities in the program outlined ?

D. MEYER, United States

      D'Appolonia staff only, including geologists, geochemists and engineers.

M.J. SMITH, United States

      To what extent have you evaluated what material evaluation studies have been completed and are available in the literature ?

D. MEYER, United States

      We attempted to cover as much of the currently available literature as possible, especially that related specifically to penetration sealing. Unpublished or generally unavailable data exists, we are sure, but it is not always possible for us to consider such information.

A PRELIMINARY EVALUATION OF VARIOUS SEALING CONFIGURATIONS

C. R. Chabannes*, D. E. Stephenson*, R. D. Ellison**
*D'Appolonia Consulting Engineers, Inc. **D'Appolonia Consulting Engineers, Inc.
Albuquerque, New Mexico   Pittsburgh, Pennsylvania

ABSTRACT: The overall design approach being established by ONWI for repository seals consists of sections of various materials placed in various seal geometries suitable for performing specific functions in the penetrations. One investigation is directed toward evaluating the sensitivity of the seal component behavior with respect to several seal configurations and material types. Numerical methods are being used to evaluate the influence of the seal system on fluid flow and nuclide migration through the sealed zone. This study was a preliminary investigation of the fluid flow through and around the seal. These preliminary results indicate that the hydraulic conductivity of the interface and disturbed zone appear to control the fluid flow through and around a sealed penetration; and hence, determination of hydraulic conductivities in these zones from laboratory and field investigation should be given a high priority.

INTRODUCTION

Radionuclides released from the waste in a repository and migrating by fluid transport may not reach the biosphere in any signficant quantities if (1) some are sorbed by the media, and/or (2) if fluid movements are sufficiently slow to allow their decay.

One potential pathway to the biosphere that will be created by man is through penetrations for exploration or access to the repository. As part of the current sealing design studies being performed by D'Appolonia, under a contract with the Office of Nuclear Waste Isolation (ONWI), a repository sealing design approach was developed in ONWI-55 [1]. ONWI-55 presented a conceptual design approach for repository seals. The qualitative approach presented in this report forms the foundation or starting point for future studies. Continued investigations will lead to successful design, licensing and demonstration of seals for specific repositories.

Figures 1 and 2 illustrate the three potential paths for flow and radionuclide migration from a penetration into the repository:

- Through the seal material,

- Along the interface (or bonding zone) between the seal material and the host formation, and

- Through any zone of disturbance in the host forma-tion around the penetration opening.

The combination of these zones has been termed the "seal zone".

Several basic shapes and configurations for seal components have been identified and are illustrated in Figure 3. Some of these shapes will be suitable with only certain types of materials or penetrations, while others may be appropriate to all conditions. Potential uses or functions of each geometry are indicated in the figure.

Detailed analytical evaluations are a means of evaluating the ability of the seal configurations to satisfy the proposed functions, which can be reduced to two basic ones, adsorption and flow barriers. To evaluate the design goals presented in ONWI 55, the analytical consideration involves the estimation of the rate of migration of radionuclides through the seal zone.

As previously discussed, the evaluation of radionuclide migration through the seal zone can be evaluated in two components:

- A preliminary evaluation of fluid flow, and

- A subsequent evaluation of radionuclide retardation
  including sorption

The results presented in this paper deal only with the preliminary evaluation of fluid flow through the seal zone to verify calculational procedures. An evaluation of the influence of various seal materials and geometries on radionuclide migration through the seal zone is a more complex problem which is currently being evaluated.

PURPOSE AND SCOPE

Parametric analyses are in progress to quantitatively assess the functionality of the basic seal geometries given in Figure 1 and using different materials. These type of analyses are useful for evaluating the entire sealing system design to be used in repositories. The initial element of these parametric studies was an evaluation of fluid flow intended to investigate the effectiveness of various seal geometries to function as flow barriers.

To evaluate the numerical techniques being used, basic studies were performed using ideal fluid flow through the seal zone. These studies investigated two basic geometries, the prismatic (or cylindrical) geometry and the cutoff collar or bulkhead.

The sensitivity to hydraulic conductivity and extent of disturbed zone were evaluated for various seal lengths. The models for which results were obtained are presented schematically in Figure 4. The four zones of interest include the seal, the interface or bonding zone, the disturbed host formation, and the immediate undisturbed host formation. From the point of view of fluid flow, the disturbed zone hydraulic conductivity used in the study is such that it includes both the hydraulic conductivity of the disturbed zone and the equivalent effect of any cracks that may exist between the seal and the host rock due to improper bonding.

PRELIMINARY RESULTS

The preliminary results obtained using standard finite element techniques are discussed in the following paragraphs. These results are consistent with preliminary data reported elsewhere [2, 3]. They provide insight and guidance for planning future studies which will be a part of the input to sealing system design. Figures 5 and 6 illustrate the influence of the disturbed zone hydraulic conductivity and area as a function of seal length and presence of a bulkhead. In each of these figures the total flow rate through the seal zone is taken through the central cross section I-I (see Figure 4) that is perpendicular to the borehole axis.

Influence of Dusturbed Zone Hydraulic Conductivity and Cross Sectional Area

The hydraulic conductivity and the cross sectional area of the disturbed zone
have a definite impact on the flow rate through the seal zone as would be expected.
Once the hydraulic conductivity of the disturbed zone is two orders of magnitude
or more greater than that of the host formation and seal, a preferential flow
path around the seal is established in the disturbed zone as shown in Figure 5.

Influence of Bulkhead or Cutoff Collar

Figure 6 presents the limited results available showing the influence of a
bulkhead or cutoff collar. In this figure the percent reduction in total flow
through the seal zone is presented for a given percent reduction in the cross
sectional area of the assumed initial disturbed zone. The results are presented
for two different bulkhead lengths. The longer bulkhead gives a greater reduction
in flow since the hydraulic gradient is decreased in the disturbed zone. The
flow through the seal zone cannot be totally eliminated by removing the disturbed
zone since the seal has some finite hydraulic conductivity. It appears that
reducing the assumed initial disturbed zone cross sectional area beyond 70 to 80
percent only reduces the flow through the seal zone by 10 percent or less. It
should be kept in mind that the disturbed zone hydraulic conductivity used in this
study includes both the hydraulic conductivity of the disturbed zone and the
equivalent effect of any cracks that may exist between the seal and the host rock
due to inadequate bonding.

Influence of Seal Length

For a sufficiently high contrast between the disturbed zone and host formation/
seal hydraulic conductivities the flow through the seal zone for all practical
purposes becomes flow through the disturbed zone and can thus be treated as an
annular pipe flow problem with the flow rate being inversely proportional to the
seal length. When the contrast between the disturbed zone and host formation/seal
hydraulic conductivities is less than or equal to two orders of magnitude, an
increase in seal length provides a larger reduction in flow rate through the seal
zone than would be predicted from the simple pipe flow analogy as shown in Figure
5. This is due to a larger part of the total flow going around the seal occurring
in the undisturbed host formation near the seal. This effect becomes more
pronounced as the length of the seal is increased.

CONCLUSIONS AND RECOMMENDATIONS

The preliminary results obtained to date provide a basis for further sensitivity studies, and provide guidance in planning more in depth studies by indicating the importance of the parameters considered. As expected, the hydraulic conductivity and cross sectional area of any disturbed zone at the seal/rock interface or the surrounding rock have a significant impact on the flow through the seal zone. It appears that preferential flow paths become important as the conductivity of the disturbed zone or interface is two orders of magnitude or more greater than the seal and host formation. The flow through the seal zone can be reduced by increasing the seal length or including a bulkhead or cutoff collar that is emplaced in such a manner that the extent and/or the hydraulic conductivity of the interface and/or disturbed zone is significantly reduced.

A determination of the presence and extent of the interface and disturbed zone and their hydraulic conductivities from laboratory and field investigations and more detailed analysis of their effects on flow and nuclide migration should be given a high priority as well as means of reducing the extent and/or hydraulic conductivities in these zones.

REFERENCES

1. Office of Nuclear Waste Isolation, 1980, Anticipated Release, "Repository Sealing Design -- 1979," ONWI-55, prepared by D'Appolonia Consulting Engineers, Inc., for Battelle Memorial Institute, Program Management Division, Columbus, Ohio, (in publication).

2. Peterson, E. W. and C. L. Christensen, 1980, "Analysis of Bell Canyon Test Results," SAND80-7044C, Sandia National Laboratories, Albuquerque, New Mexico, 87185, 37 pp.

3. Hodges, F. N., J. E. O'Rourke, G. J. Anttonen, 1980, "Sealing a Nuclear Waste Repository in Columbia River Basalt: Preliminary Results," presented at the Workshop on Borehole and Shaft Plugging, Columbus, Ohio, May 7, 1980.

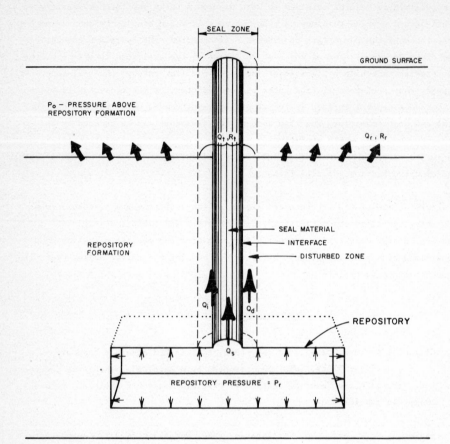

Q_t = FLOW THROUGH SEAL ZONE = $Q_s + Q_i + Q_d$
R_t = RADIONUCLIDE MIGRATION THROUGH THE SEAL ZONE,
    MEASURED AT THE TOP OF THE REPOSITORY FORMATION
Q_r, R_r = FLOW AND RADIONUCLIDE MIGRATION FROM
    THE ENTIRE REPOSITORY

NOTE: SEE FIGURE **2** FOR DEFINITION OF
    FLOW TERMS Q_s, Q_i AND Q_d

FIGURE 1   CONCEPTUAL ILLUSTRATION OF COMPONENTS OF POTENTIAL RADIONUCLIDE
        TRANSPORT FROM A REPOSITORY [1].

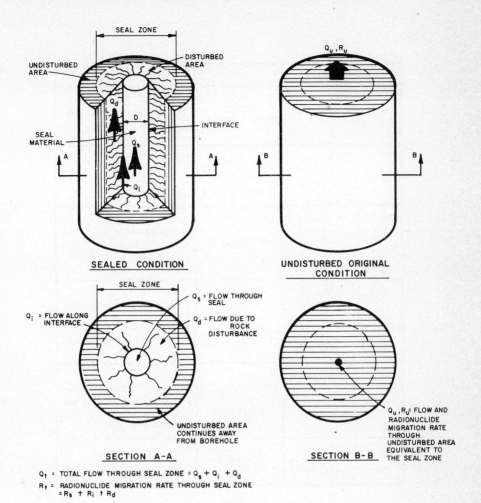

FIGURE 2   COMPONENTS OF POTENTIAL RADIONUCLIDE TRANSPORT THROUGH THE SEAL
ZONE [1].

SEAL GEOMETRY	TYPE	EXAMPLE FUNCTION
	PRISMATIC (OFTEN CYLINDRICAL)	• FLOW BARRIER • ADSORBING COLUMN • INTERRUPT DISTURBED ZONE • FILLER • STRUCTURAL SUPPORT • DRAINAGE
	ANNULAR	• MONITORING POTENTIAL • USED IN TESTING • POSSIBLE CONDITION OF EXISTING LININGS • INTERFACE BONDS
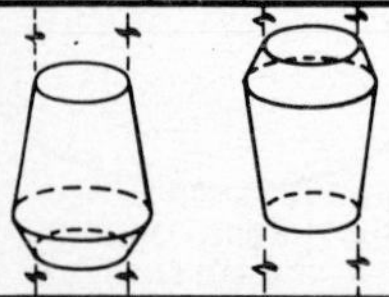	WEDGE KEY	• FLOW BARRIERS • STRUCTURAL SUPPORT • INTERRUPT DISTURBED ZONE
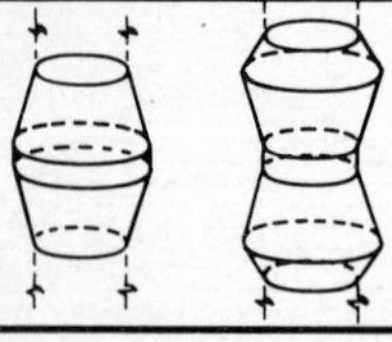	DUAL WEDGE KEY	• IMPEDE FLOW FROM EITHER DIRECTION • STRUCTURAL SUPPORT • INTERRUPT DISTURBED ZONE
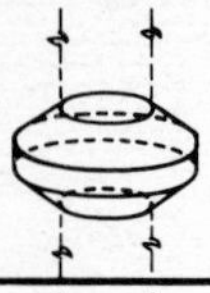	CUTOFF COLLAR (FOR BOREHOLES) BULKHEAD (FOR SHAFTS)	• INTERCEPT DISTURBED ZONE • INCREASE POTENTIAL FLOW PATH • STRUCTURAL SUPPORT
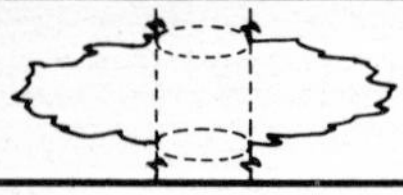	PRESSURE GROUT	• ISOLATION OF AQUIFERS (OR OTHER FLUID/GAS PRODUCING ZONES)

FIGURE 3   POTENTIAL SEAL GEOMETRIES AND FUNCTIONS [1].

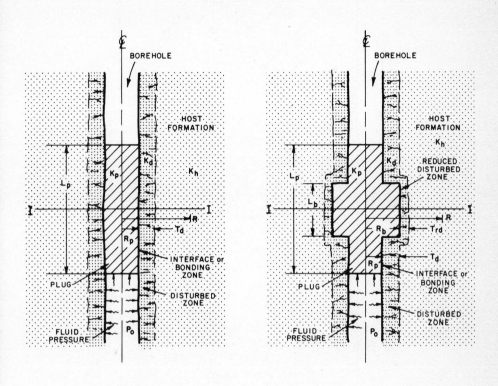

(a) WITHOUT BULKHEAD   (b) WITH BULKHEAD

FIGURE 4   SCHEMATIC OF BOREHOLE SEALS ANALYZED WITH AND WITHOUT
BULKHEAD.

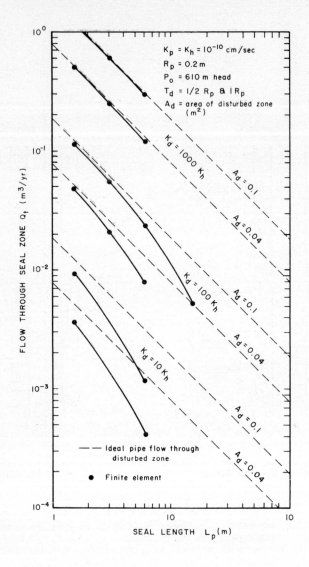

FIGURE 5   SENSITIVITY OF FLOW THROUGH THE SEAL ZONE TO SEAL LENGTH, AND
DISTURBED ZONE HYDRAULIC CONDUCTIVITY AND CROSS SECTIONAL AREA.

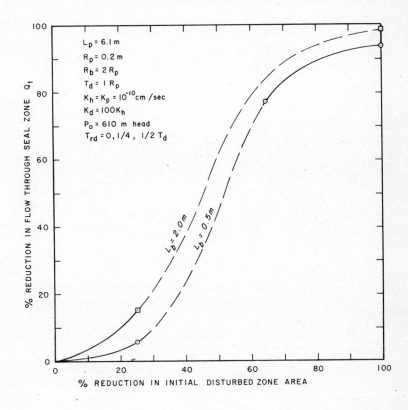

FIGURE 6   INFLUENCE OF BULKHEAD (OR CUTOFF COLLAR) ON FLOW THROUGH THE SEAL ZONE.

Discussion

N.A. CHAPMAN, United Kingdom

Did you use a porous medium model for assessing flow around the plug ?

C.R. CHABANNES, United States

Yes, a porous medium model was used during the study.

N.A. CHAPMAN, United Kingdom

Does the technology exist for reaming out cavities in boreholes with specific geometries for bulkhead plugs, and what is your justification for assuming that this process produces a smaller zone of disturbance than does borehole drilling itself ?

C.R. CHABANNES, United States

The technology for cutting out sections of casing in boreholes to depths greater than 100 meters exists, and it is assumed that this can be extended to cutting the rock. The condition and extent of any disturbed zone may be a result of several factors such as the initial drilling process, time dependent environmental degradation, etc. The extent of this disturbed zone could possibly be reduced by carefully reaming just prior to seal placement. In this way the influence of other factors besides the drilling process could be minimized and thus one might be able to reduce the extent of the zone of disturbance.

T.J. CARMICHAEL, Canada

Have you been able to verify your model test results by field testing ?

C.R. CHABANNES, United States

We have not yet conducted field tests to verify the model.

R. PUSCH, Sweden

The possible existence of "disturbed zones" is such an essential issue that I suggest we use some time in the general discussion for "brain storming" on the subject.

C.R. CHABANNES, United States

This is true. It is commonly accepted that "disturbed zones" exist around penetrations, especially if they are left open for some time allowing degradation to occur. The important question is : what is the size and permeability of the "disturbed zone" ?

GROUTING OF NUCLEAR WASTE VAULT SHAFTS

M. Gyenge
Mining Research Laboratories
Canada Centre for Mineral and Energy Technology
Fnergy, Mines and Resources Canada
Ottawa

ABSTRACT

A nuclear waste vault must be designed and built to ensure adequate isolation of the nuclear wastes from human contact. Consequently, after a vault has been fully loaded it must be adequately sealed off to prevent radionuclide migration which may be provided by circulating groundwater.

Of particular concern in vault sealing are the physical and chemical properties of the sealing materials its long-term durability and stability and the techniques used for its emplacement. Present grouting technology and grout material are reviewed in terms of the particular needs of shaft grouting. Areas requiring research and development are indicated.

RESUME

Un coffre d'emmagasinage des déchets nucléaires doit être concu et construit afin d'assurer l'isolement adéquat de ces déchets de tout contact humain. Dès lors, lorsque ce coffre aura été rempli à capacité, il devra être scellé pour empêcher la migration des radionucléides par les eaux souterraines.

En ce qui a trait au scellage des coffres, on s'intéresse tout particulièrement aux propriétés physiques et chimiques des matériaux de scellage, de leur durabilité et de leur stabilité à long terme et des techniques employées pour la mise en place. La technologie et les matériaux de scellage actuels sont étudiés en fonction des besoins distincts du ciment de scellage des conduits. Les domaines nécessitant la recherche et le développement sont indiqués.

INTRODUCTION

The overall objective of the radioactive waste management program is to develop the technologies for the disposal of high-level radioactive waste in a manner which ensures no significant hazard to the public or environment at anytime. The responsibility for developing methods of disposing of high-level radioactive wastes from Canadian nuclear power stations rests with Atomic Energy of Canada Limited (AECL).

The research and development activities included in the Fuel Cycle Waste Management Program are to be conducted with the following objectives in mind: (a) to verify that disposal of irradiated fuel wastes in deep, stable geological formations will achieve the overall objective, and (b) to develop the necessary capability for safely disposing of these wastes [1].

Several departments and agencies of the federal and Ontario governments are participating in the program, along with sections of private industry and the university community. Energy, Mines and Resources Canada is participating in the geoscience aspects of the overall program. The Mining Research Laboratories (MRL) of the Canada Centre for Mineral and Energy Technology (CANMET), in cooperation with Ontario Hydro, is involved in the vault sealing investigations.

VAULT SEALING REQUIREMENTS

Problem definition

Several concepts of waste disposal in geological formation are being investigated. One of the most promising is that of a vault mined at depth.

At the present conceptual stage the vault will be mined within a suitable crystalline igneous formation of the Canadian Precambrian Shield. The isometric sketch of the proposed vault is shown in Fig. 1 [2].

The waste canisters are to be placed inside the holes drilled in the floor. Parallel rows of disposal rooms are to be excavated at a depth of about 1000 m. The disposal level is to be connected to surface by several vertical shafts all excavation being by conventional methods. The shafts will be lined, whereas it is planned to construct the disposal rooms without lining.

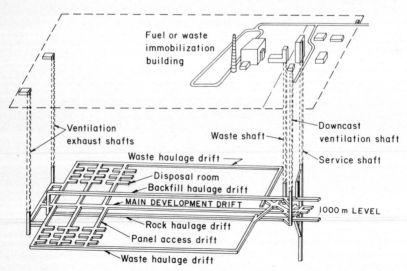

Fig. 1. Conceptual vault design (after Scott and Charlwood [2]).

The nuclear waste vault will have two functions, i.e.: (a) to provide adequate isolation of the nuclear wastes from human environment; and (b) to prevent access to the disposed nuclear wastes.

The vault must therefore be designed and constructed in such a way that these functions are effectively fulfilled. One of several important design and construction aspects is proper vault sealing.

The aim of the research and development activities of the vault sealing task is to identify suitable sealing materials and techniques.

The various aspects of nuclear waste vault sealing are demonstrated by the schematic section in Fig. 2.

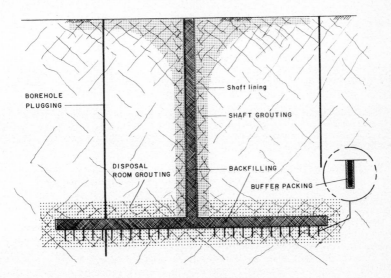

Fig. 2. Vault sealing aspects.

The most likely mechanism of radionuclide release to the biosphere is by dissolution and transport by groundwater, the latter flowing through both man-made and natural pathways of the rock mass barrier between the disposal level and surface. The basic purpose of vault sealing is to seal-off the important pathways by applying appropriate emplacement techniques and using suitable sealing material.

Assessment of the importance of particular pathways is presently in progress, using hydrogeologic modelling techniques. For man-made pathways the likely objective is to fill in all the voids created by mining operations or as a consequence of such activities. In some areas the artificially emplaced material may have to resist radionuclide movement as well as or better than the original undisturbed rock mass.

As for natural pathways, such as joint systems, shear zones and faults, aim is to improve the natural confining capacity of the rock mass surrounding the vault. Natural pathways must be properly filled with appropriately engineered sealing material.

General R/D approach

The various aspects to be considered in emplacing the vault sealing material, are as follows:

(a) Grouting of vault
    (i)   shafts

   (ii) disposal rooms and headings
(b) Sealing of boreholes
(c) Buffer packing
(d) Backfilling of vault
   (i) disposal rooms and headings
   (ii) shafts

   In pursuit of the overall objective, research and development activities are being planned and conducted for each of the above aspects as follows:

(i) establishing the state-of-the-art of existing technology;
(ii) assessing its suitability in meeting requirements; then
(iii) either verifying the suitability of the adopted technology or developing new technology.

   Special requirements of vault sealing involve:

(a) properties of the sealing materials
(b) emplacement techniques
(c) long-term durability.

   The sealing material must have physical and chemical properties compatible with the rock mass. Furthermore the emplaced materials must satisfy the special design requirements under the potential vault loading and environmental conditions.

   The emplacement techniques must take into account the physical and geological environment of the emplacement operation. The techniques must ensure with a high degree of confidence that the quality of the sealing material will fulfill design specifications.

   The entire system of vault sealing must be fully effective over a time span which is well in excess of any reasonable test or monitoring period. Perhaps the long-term performance of the sealing material is the most significant item; but it also represents the major unknown factor in the sealing operation.

   Each of the many problems is relatively complex, containing numerous interrelated variables. The solution can be achieved only through tedious and fairly long-range investigations, which will require sizeable funding and manpower resources. Due to the fact that safe disposal of nuclear wastes is of great interest to the entire international community, effective cooperation among the OECD nations does exist. The results of research and development activities conducted by other nations are readily available. Research priorities within the Canadian program are therefore, being established with existing international cooperation in mind.

   The major emphasis of investigation related to vault sealing presently is placed on vault grouting, particularly shaft grouting.

SHAFT GROUTING

Grouting technology in general

   The prime purpose of grouting in this particular application is to provide an effective barrier to the movement of groundwater transported radio-nuclides. This objective is achieved by filling the voids of the host rock in the immediate vicinity of the shaft.

   A comprehensive review of the current state-of-the-art of rock grouting, both in mining and in other types of work, has been completed [3]. This revealed that the effectiveness of grouting is enhanced by applying the experience gained from previous jobs to the particular task at hand. A problem is that the geological conditions, fissuration, groundwater, and permeabilities are rarely comparable, and consequently, each case must be assessed on its own merits and solutions must be worked out individually. Grouting is therefore on of those technical arts calling for an alert, innovative approach in moving from job to job, and sometimes even from one grout hole to the next.

Fundamental work principles can be laid down, but the essential details of each grout injection, such as its duration, variation in grout consistency and in applied grouting pressure throughout the injection of each hole, require individual judgement. The complexity of the physical and chemical process and particularly of predicting grout behaviour in the ground, both during injection and during its final settling stage, usually makes grouting only a roughly designed procedure. The in situ adaptation, by repeating the cycle -- observation, redesign, regrouting -- of the established procedure, is an essential element of success. Consequently, successful grouting relies heavily on good engineering judgement and experience, and not merely on basic correlations or equations.

The design and execution of a large important grouting operation involve several essential steps. Some of these are related to data acquisition such as field and laboratory investigations, the others to grouting procedures such as grouting methods, equipment and grout selection, grouting cost, etc. The decisions made at any step influence the others, directly or indirectly.

The interactions between steps, or, grouting parameters are shown in Fig. 3. Although, it is necessary to consider all parameters some are more important than others and are shown by large circles. The numbers show one possible sequence of steps in acquiring the information necessary to arrive at the best grouting procedure. The arrows indicate direct interactions between the parameters.

An assessment of available grouting technology suggest that many of the techniques required to provide adequate grout barriers around the shafts and disposal rooms already exist and that these could be adapted to meet specific requirements. However, field tests will be necessary to verify that the procedures and emplacement techniques are effective in achieving the specific objectives.

Some aspects still require further research. As an example instrumentation is required to provide better control over placement of grout barriers. Grout mixes must also be developed to meet specific vault sealing requirements and long-term durability.

Design factors

Design factors represent essential input to the planning and execution stages of the grouting operation. In shaft grouting the required information includes specifications of the shaft and liner, and details of the sinking and liner emplacement methods.

The waste vault will require several service and ventilation shafts. So far in the design stage, shaft specifications are not yet available except for circular shape. This also applies to shaft liners. The specification of the liner material will be based on the results of material testing carried-out by other groups participating in the overall program.

The present conceptual design calls for conventional drill and blast shaft sinking methods. To minimize blast damage of wall rock, a pre-splitting technique is planned. In view of the high drilling costs involved however, and also in light of recent equipment development the full size shaft boring method is seriously being considered. The advantage to be achieved by essentially eliminating wall rock damage could easily off set the still higher cost of shaft boring. As liner emplacement depends on both the liner material selected and the sinking method, it has not yet been specified.

Grouted rock mass properties

Specifications for the grouted rock mass are concerned with two interrelated factors - purpose of the grouting and desired performance of the emplaced grouts.

Grouting a shaft for nuclear waste vault, has the dual purpose of aiding construction and providing a barrier to radionuclide movement.

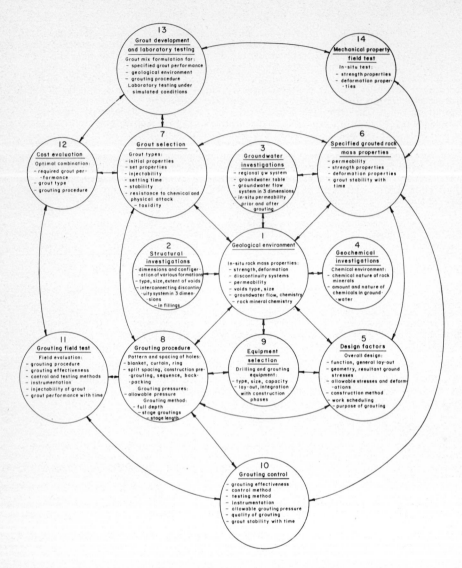

Fig. 3.   Interactions of grouting parameters.

With respect to the first purpose, the grouting requirements are basically the same as those that apply in normal mining practice. Water-bearing zones and formations intercepted by the shaft are to be grouted ahead of sinking operations to prevent water flowing into the working area. Moreover, extensively fractured zones are grouted to improve their strength and resistance to deformation.

One important point must be emphasized: grouts once injected cannot be withdrawn and replaced. Therefore, grouts used during construction must also satisfy the requirements of vault sealing in every respect. In other words, both the appropriate grout material and grouting procedure must be on hand before sinking is started.

The second purpose of shaft grouting can be demonstrated by considering the most likely path for movement of groundwater. Figure 4 shows the schematic cross section of a shaft, following completion of the entire sealing operation. The first phase of groundwater movement - the groundwater inflow phase - is shown at the left. Away from the shaft, groundwater movement is governed by the regional flow pattern which will be considerably disturbed by the sinking operation. Actual sinking and subsequent readjustment of ground stresses will result in a weaker and more permeable annulus of rock mass around the shaft, either by opening the natural fractures or by developing new ones.

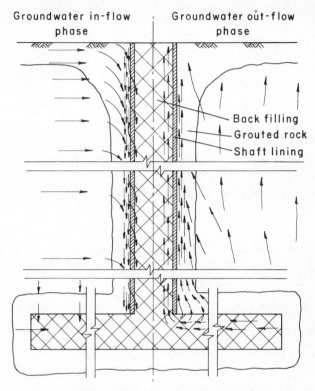

Fig. 4.   Groundwater flow at a sealed shaft.

The purpose of grouting is to fill and seal off these fractures. Ideally, when all fractures have been perfectly sealed, groundwater flow will merely be deflected around the shaft. However, if grouting is ineffective the annulus becomes a major path for groundwater inflow. This may or may not be a problem.

However, the effect of inadequate grouting is clearly serious during the second or groundwater outflow phase, shown at the right in Fig. 4. The fractured and poorly grouted annulus could be the path for radionuclide escape into the human environment. The upward movement of the contaminated groundwater will be concentrated adjacent to the shaft lining.

The ideal achievement of grouting is to create a uniform impermeable grouted skin around the shaft. This requirement could possibly be used to lay out the overall grouting plan, as shown in Fig. 5. The sketch on the left assumes the favourable condition, in which permeability of the host rock decreases continuously with depth along the shaft axis. The shape of the grouted annulus should follow the variation of the in situ permeability values. Similar logic could apply, as shown on the right, in the occurrence of a zone with high permeability values. The possible patterns of grout hole arrangement are also indicated for both cases.

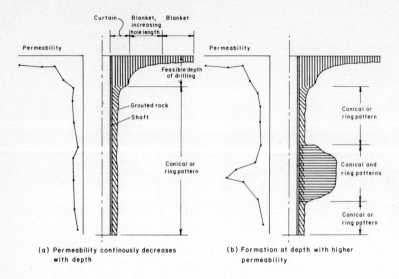

Fig. 5. Examples of shaft grouting.

The grout material must possess physical and chemical properties compatible with the environment of the in situ rock. Furthermore, the emplaced grouts must adequately satisfy the designed performance requirements under the potential vault loading and environmental conditions.

At the present stage of development, the desired performance of the emplaced grouts can only be indicated, not specified. Anticipated requirements are:

(i)   that they be a dense, impermeable solid
(ii)   have low wetting, drying and thermal shrinkage
(iii)   have sufficient deformability to allow for rock movement without developing cracks
(iv)   that the coefficient of thermal expansion has an order of magnitude similar to that of the surrounding rock material
(v)   be inert to sulphate, H^+, chloride ions, and other solutes in the groundwater of the vault site
(vi)   have a low leach rate in the presence of groundwater
(vii)   have good bonding strength
(viii) have ability to retard radionuclides by sorption or other means of immobilization
(ix)   have exceptionally long-term durability.

It is impossible yet to assign meaningful values to most of these desirable properties. First of all, some are more important than others. Furthermore, all are more or less interrelated. The performance specification of the emplaced grouts must therefore, embody the optimal combination of the above properties.

Another difficulty is caused by a lack of tangible data. For example, the in situ permeability of the grouted rock mass can, on the basis of current grouting practice, be specified as 10^{-5} to 10^{-6} cm/s. However, this value seems to be much too high for vault grouting purposes. The attainable lowest permeability value can only be determined by specifically designed grouting field tests.

Physical and chemical environments of grouting

Grouting parameters such as grouting procedures, injection pressures, as well as type and consistency of the grout will depend on the in situ physical and chemical environment of the rock mass located within the fractured annulus around the shaft. Techniques for investigating the grouting environment in a fissured rock mass are available and can be adapted for specific purposes. However, each technique must be verified as to suitability for the special requirements in question.

The in situ grouting environmental data will have to be obtained by three types of investigation - structural, groundwater and geochemical.

The objective of the structural investigation is to obtain all the data required for creating a three-dimensional picture of the prevailing discontinuities within the zone to be grouted. This information must include the type, distribution, approximate size and direction of the discontinuities, as well as the interconnections between them. The infilling materials must also be described, indicating whether they will be permeable or whether they could readily be washed away by groundwater movement. Special care must be taken to identify faults. An analysis of the investigation results must identify the most critical discontinuity system, i.e. one which must be grouted.

The investigation will be based on the three-dimensional rock exposure provided by the shaft sinking operation. It will be supplemented by a borehole survey using the grout holes. If necessary diamond core drilling will also be used for further clarification of the prevailing structures.

The groundwater investigation must be performed to establish in situ permeability of the jointed and fractured rock mass. Permeability of the grouting zone at the level of the continuously progressing shaft bottom will be determined by water pressure tests both prior and after injection. The frequency of water pressure tests will be governed by variations occurring within the geological environment, such as changes in the types of discontinuity, frequency and size, changes in rock type or in groundwater discharge, drastic changes in grout take, etc.

The accuracy of the water pressure tests must be tailored to the requirements of grouting to provide a qualitative measure of the grouting required for any specific injection location. The water pressure test device to be used must be automated for speedy and simple interpretation.

Both the structural and groundwater investigations are conducted to understand the nature of the underground fracture systems which are to be grouted. Therefore, the results of the two investigations must be correlated.

The injected grout will be affected by the chemical environment of the grouted rock mass. The probable immediate effects include an influence on setting time and setting properties. Alternatively, or in addition, there are the long-term effects such as possible complete alteration or even deterioration of the grout material. To ensure that the zone of grouting is injected by the proper grout the geochemical environment, among others, must be known. Using the wrong type of grout will have irreparable consequences, because it is impossible to replace it later with the proper type.

Any major change in the rock minerals constituency is indicated by a change of rock type. This change will be detected by the structural investigation. However, alterations of groundwater chemistry cannot be noticed as easily. A simple device and method will be needed to continuously monitor groundwater chemistry. The frequency of these tests will have to be tailored to prevailing conditions. Tests will be required if the rock type changes or if changes occur in mineralization along the fractures; otherwise routine checks are required at appropriate intervals.

Physical and chemical properties of grouts

Two fundamental requirements for good grout are:

(a) it should confer on the formation the desired properties, and
(b) it should be capable of being injected into the voids of the geological formation.

The overall performance of a grout is thus determined by its initial fluid state, its set state, and by the transition state between the two.

All conceivable physical forms of grout ranging from gases to molten solids have been described in the literature. The vast number of potential grouts represent an extremely wide range of chemical and physical properties. Proper grout selection is based on the following major factors:

(i)   physical and chemical properties of the rock mass to be grouted
(ii)   specified performance of the emplaced grouts
(iii) physical and chemical properties of the potential grouts.

The first two factors have been dealt with previously. However, a meaningful research program, aimed at selecting or developing grouts suitable for vault grouting, must reflect the interdependent nature of all three governing factors. Consequently, the grout selection and development program, presently being formulated, must include both laboratory and field testing.

Two basic types of grouting materials must be investigated i.e. cement-based and non-cement-based grouts. Under the heading of cement-based grouts, testing will include several selected grout mixes of those composite materials in which the binder is one or more of the portland cement types with or without fly ash, pozzolan or slag components, possibly modified to include expansive or high early strength properties.

As for non-cement-based grouts the suitability of several chemical grouts will have to be investigated. Grout types with potential use in vault grouting are to be selected first through a systematic evaluation of the vast number of available chemical grouts. The selected grout types must then be tested both in the laboratory and in the field. A final grout selection can only be made after the test results and subsequent analyses are available.

Shaft grouting procedures

The high quality of grouting required for the nuclear waste vault, can only be achieved if the grouting procedure is truly appropriate for the given geological environment and for the fluid state and transition state properties of the selected grouts.

With respect to grouting procedures, decisions must be made as to:

(a) grout hole pattern and spacings
(b) grout pressure to be applied
(c) grouting methods to be used.

Assuming that the shafts will be sunk by conventional methods, suggested hole patterns are shown in Fig. 6.

Close to the surface of a rock mass, joints are usually open. To seal around the shaft collar the rock mass must be grouted to a relatively shallow depth, but over a relatively large area. For this blanket or area grouting is used and grout holes are drilled either in a square or zig-zag pattern.

The fissured rock around the shaft is grouted by curtain grouting. For the upper section to a depth of 30-50 m, long parallel holes can be drilled from surface either in single line or multiple line patterns.

For the lower section the curtain can be formed through a conical hole pattern, or alternatively by using ring grouting either radial or spin patterns.

In difficult ground conditions, such as in highly fractured or water-bearing zones, shaft sinking must be preceded by pre-grouting. For this the pattern of holes is generally spin-conical, with some additional holes placed randomly.

The irregular space between the shaft lining and the surrounding rock is filled by backpacking. Its main purpose is to provide an effective seal between the grouted annulus and the shaft lining; other important reasons are: to transfer ground stresses and to eliminate major differences in deformability. Backpacking is done through standpipes secured through the lining.

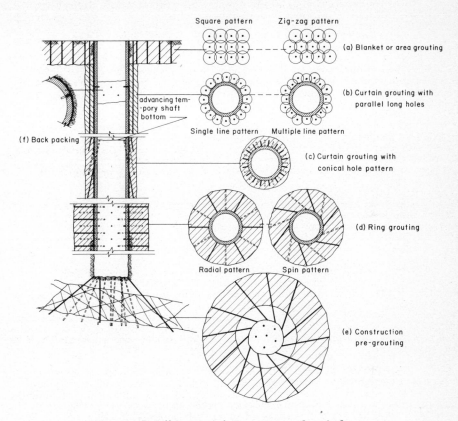

Fig. 6. Possible grout hole patterns for shafts.

The best method of establishing the required spacing is to apply the split spacing technique combined with water pressure tests. This method is based on sequential drilling and grouting of boreholes. Starting with holes in the primary sequence, the distance between holes is gradually decreased by splitting it equally when boring holes of the secondary, tertiary, and so on sequences, until the desired degree of rock impermeability is reached.

The effectivess of fissure grouting depends to a large extent on the applied injection pressure. In general, the higher the pressure the more effective the grouting, particularly where there are fine fissures. Injection pressure, however, must be related to depth of the grout in the hole, to the extent and dip of the discontinuities to be grouted, and to the rate of grout acceptance. The actual allowable or safe pressure must be decided in the field.

Four possible grouting methods can be considered:

(a) downstage without packer
(b) downstage with packer

(c) upstage grouting
(d) circuit grouting downstage.

Each method has its merits and limitations. On the basis of experience through field tests, it will be possible to decide and specify the grouting methods that will best satisfy the requirements of a waste vault.

Equipment and instruments

The equipment, tools and instruments for grouting may be divided into two groups - machinery and grout hole devices.

Drilling for shaft grouting can be done by using suitably sized drilling equipment of required capacity, without the need to modify it.

This is also true for grouting equipment. However, due to the differences in injection procedures, the type of grout must be decided before selecting equipment. The type of grout will influence not only the type of grouting machinery to be used but the entire grouting operation.

Because of the size of the operation, cement-based grout would have to be mixed at a central station located at ground surface. The mixed grout would be pumped from there to the points of injection. Because neither the ingredients nor the mixed chemical grout can be transported through long grout lines, the mixing and pumping machineries for chemical grouts must be located at, or near to the injection faces.

Parts of the grout hole devices, such as valves, pressure gauges, headers, standpipes, inflatable and mechanical packers, etc. are commercially available. The various ready made parts will have to be tested during the grouting field tests, as to their suitability and effectiveness, with the objective of selecting the best types for use in vault grouting.

The grouting experts agree that the quality requirements of vault grouting can be obtained by continuous control of the grouting operation. The essential details of any injection, such as the duration of the injection, the variation in grout consistency and in the applied injection pressure, require individual judgement for each stage of each hole. Therefore it is imperative that the injection process is directed by a foreman with several years of practical experience in fissure grouting, backed up by a site engineer and organization of no less experience.

Furthermore, in view of the high quality requirements of vault grouting, it is also essential that an automatic control system be used. The development of a system which will record the pressure and grout take and which will automatically regulate the grout header pressure and pump speed can be based on a proven system [4]. Another system, separate of or in combination with the first one, is needed to measure and record the grout consistency, as well as to regulate the grout consistency is in line with the requirements of grouting. These systems should be as simple as possible. However, they should also include the most advanced developments in the fields of remote control techniques and of electronic measurements.

Intergration of shaft grouting and construction

The quality of fissure grouting is a function of the injection pressure and experts agree that moderate excess pressure is beneficial. This however, is, not to say that applying high pressure can offset improper viscosity or improper relationship between particle size and fissure width.

The possibility of uplift, disturbance of rock strata through movement along existing fissures or by hydrofracture cracks, and leakage of grout increases with grouting pressure. The allowable or safe grouting pressure must therefore be decided by field tests.

In shaft grouting, a sufficiently high injection pressure can only be applied with proper integration between sinking and grouting operations.

For example, the curtain grouting phase with parallel long holes shown in Fig. 6, must be completed prior to excavation of the shaft core. Also, conically arranged grout holes must be drilled and grouted before advancing the shaft bottom. In both cases back pressure against the applied injection pressure is provided by the intact shaft core.

For a ring pattern, grouting can be performed either prior to or after liner emplacement. In the first instance, the applied injection pressure is governed mainly by length of the grout holes. In the second case, however, the emplaced liner will furnish the necessary resistance against the injection pressure. This also applies during backpacking.

Whichever grouting procedure is used, shaft design and work scheduling of sinking and liner emplacement must accommodate the specific requirements of the grouting procedure. Grouting must be an integral part of the entire shaft construction and design considerations can only be completed after grouting procedures are established.

Grouting effectiveness

Purpose of testing grouting effectiveness is to ensure that specific objectives of the operation have been achieved. The main reason for grouting shafts for a nuclear waste vault is to decrease permeability of the rock mass around the shaft to a specified value. Consequently, the effectiveness of a completed grouting job or of its various phases will be assessed by testing and comparing the in situ permeability of the rock mass prior to and following injection.

To test the effect of grouting around the shaft collar the permeability measurements can be based on pumping tests combined with piezometer tests both techniques being well developed. A limited number of holes should also be drilled and the core examined for grout penetration.

At depth, assessment of grouting effectiveness is based on water pressure tests and on limited core drilling. The water pressure tests are not done separately but are part of the split spacing technique. Sequential grouting with accompanying decrease in hole spacing is continued until water pressure tests yield the specified permeability value.

Grout performance monitoring

The objective of grout performance monitoring is to verify that the emplaced grouts satisfy the specified requirements as to performance under prevailing environmental conditions. Monitoring strategy, monitoring systems and instrumentation must all be designed with due consideration for the changes taking place in the environment and the time span involved.

During the first stage lasting several decades, the disposal room panels will be excavated and loaded with waste canisters during which the shafts must be kept dry. The overall performance of the shaft grouting program can be assessed by measuring the quantity of groundwater inflow. Possible grouting deficiencies can be corrected by regrouting the affected sections.

Durability of the emplaced grouts and changes in their physical and chemical properties can also be measured with relative ease at that stage. Chemical analysis of groundwater inflow is possible. Furthermore, specimens of the grouted rock can be obtained for physical and chemical analyses of the emplaced grout by diamond drilling from the shaft.

All those remotely controlled instruments which are to be used for monitoring performance of the emplaced grouts during the subsequent stages will be installed, tested and calibrated during this period.

After loading the vault with canisters, disposal rooms and shafts must be backfilled. The time required for the vault to fill with groundwater will depend on permeability of the host rock and on the delaying effect of the entire vault sealing system. This phase, previously referred to as the groundwater inflow phase, will last for years or even decades.

When the vault has been filled with water and the enclosing rock formations are saturated with groundwater, pressures on both sides of the grouted annulus will be equal. Movement of groundwater will not be governed by the regional groundwater gradient. This condition however, is only temporary, especially around the bottom section of the shaft. Due at least in part to the heating effect of the emplaced canisters the groundwater may next start to move away from the vault, presumably upwards within and around the shaft. This environmental stage has previously been referred to as the groundwater outflow phase.

The exact nature and mechanism of this groundwater movement is so far not known; it can only be assessed and predicted after intensive research conducted by others within the overall program.

Instrumentation for monitoring performance of the emplaced grouts is so far non-existent. Furthermore, the maximum life span of any type of presently available monitoring device is only about five years when buried in soil or rock.

Before developing any monitoring instrument several difficult decisions must be made such as, above all, the length of the required monitoring time.

Development of instrumentation with a life span of a few decades is seemingly within reach of present day technology. To monitor over longer time spans would seem to be unrealistic.

REFERENCES

1. Boulton, J. (Editor): "Management of radioactive fuel wastes: The Canadian disposal program", Atomic Energy of Canada Limited, AECL-6314, 1978.

2. Scott, J.S., Charlwood, R.G.: "Canadian geoscience research and design concepts for disposal of high-level waste in igneous rocks", International Symposium on the Underground Disposal of Radioactive Wastes, Otaniemi, Finland, 1979, Pap. IAEA-SM-243/168.

3. Gyenge, M.: "Nuclear waste repository grouting", Division Report MRP/MRL 79-41, Unpublished working paper, CANMET, Energy, Mines and Resources, Canada, 1979.

4. Benko, K.F.: "Instrumentation in rock grouting for Portage Mountain Dam", Water Power, 18, 407-415, 1966.

GENERAL DISCUSSION

S.J. LAMBERT, United States

I am a little bit disturbed at the emphasis on sorption as a very desirable property for any kind of open space seal. It has been shown that virtually every grout has some sorption capacity and it has also been shown that in virtually no case sorption can be described by the simple Kd approach, in which sorption is treated as a constant. For example if one relies entirely upon the backfill for all the sorption capacity at the repository, including the surrounding rock, I can imagine a very high differential sorption between rock and backfill if the sorption capacity of the latter is very much higher. In this case one would duplicate natural conditions that have led to the formation of uranium deposits. On the other hand if the material with very high sorption capacity is also characterized by very low permeability, how much contact will actually occur between the moving radionuclides and the sorptive material in the backfill ?

J. HAMSTRA, Netherlands

I think that most talks during this Session made reference to permeable host rocks and not to salt. So there is a flow, and a transport mechanism through the host rock ; therefore radionuclide retardation is possible. If the repository is located in a permeable host rock, there is no need to have seals and plugs with zero permeability. Maybe you are looking at it from the viewpoint of an impermeable host rock such as salt ; in that case you would try to produce impermeable plugs for the boreholes within the repository.

S.J. LAMBERT, United States

I was not thinking of any particular rock type. Virtually all the discussion so far has been generic. I have not heard of this consideration of the consequences of either too much sorption or too little contact with the sorptive material, in cases where the plug has the highest sorption capacity.

J. HAMSTRA, Netherlands

If the plug is the only pathway for the radionuclides to move towards the biosphere then it should be useful to choose a material with high sorption capacity. If the pathway is flow through the host rock, then the situation is different.

R.D. ELLISON, United States

I have two points. The first one is that all comments made so far are applicable and must be included in the analysis. The second point is that a totally impermeable host rock probably does not exist. Even salt does have a finite and measurable permeability.

T.O. HUNTER, United States

I think we must admit that there is a finite permeability to everything. But I want to ask a question to the last speaker. He quoted a permeability of 10^{-5} cm s^{-1}. Could we ask him to clarify first what the use of that permeability was and then why he thinks that that is not adequate for waste isolation purposes.

M. GYENGE, Canada

We have no requirement for the permeability ; but there is the feeling that a lower permeability for example 10^{-7} cm s^{-1}, would be preferable.

J. HAMSTRA, Netherlands

It is my opinion that we would not be here if the permeability of the host rock was such that the pathway were through the host rock. We need to talk about sealing of boreholes only if we see that boreholes and the shaft are the main pathways for the radionuclides to come back to the biosphere.

T.O. HUNTER, United States

I agree. If the reference state is 10^{-7} cm s^{-1}, which I believe is 10^{-4} darcy, then we would conclude that host rocks are impermeable compared with plugs.

J. HAMSTRA, Netherlands

I wish to ask two questions to Mr. Ellison. The aim of borehole plugging is either to achieve complete isolation, or to limit the release of radionuclides within acceptable limits. My questions are : have you defined an acceptable limit of radionuclide release ? Or have you already had a dialogue with the licensing people so that you know what the basis of your design work should be ?

R.D. ELLISON, United States

The regulatory authorities will have to define what is an acceptable release and what is an acceptable risk for a waste repository. That will be an institutional decision and in the United States it will be taken by the Environmental Protection Agency and the Nuclear Regulatory Commission. The numerical values have not been set yet. NRC documents are being revised and are in the process of becoming public.

J. HAMSTRA, Netherlands

The same question applies to field verification. You define it but do licensing authorities accept this as sufficient evidence of the adequacy of the plugs ? Again there is need for dialogue.

R.D. ELLISON, United States

That has to occur ; it will be preferable that it occurs in the planning stage.

K. MATHER, United States

It must be noted that in addition to the institutional limit that may be set the reactions of the public will have to be considered, for example in Mississippi, they do not want radioactive waste repositories and I think that this is going to be increasingly the case within the United States. Someday somebody will have to face up to the relation between energy and waste products. But this has not yet been addressed by the public in a rational way.

J. HAMSTRA, Netherlands

This is an international problem because every country has to face it. There should be a licensing authority that sets the acceptability level ; but, even more important nowadays, public opinion has a say in the matter and the people really decide on what is acceptable or not.

N.A. CHAPMAN, United Kingdom

The root of the problem is deciding what we are trying to achieve. If you try to assess the performance of a seal or a barrier you have got to tie it to some measurable quantity. And if you are not going to set dose limits or release rates, how can you tie the performance of the seal to some measurable value ? How could we define what is acceptable ?

J. HAMSTRA, Netherlands

One practical answer may be to aim for a release as low as reasonably achievable. This implies proving that you have done your utmost ; in other words that the plugs are the best that can be produced with present technology.

N.A. CHAPMAN, United Kingdom

But you cannot just take the most effective approach to penetration sealing and not relate it to the whole release scenario.

J. HAMSTRA, Netherlands

But if you do not have acceptability criteria, the only technical approach is to do your best. As an alternative you may say : well, sorry, I cannot do the job at all as long as you do not give me the criteria.

R.D. ELLISON, United States

You could establish your own criteria if no one else would.

J. HAMSTRA, Netherlands

You could ; but then you would have to defend those criteria.

R.D. ELLISON, United States

The criteria could be based on the utilization of the environment and on acceptable release rates from a site.

T.O. HUNTER, United States

I just want to comment. The United States Department of Energy has just published a document in support of rule making. In that document there is some discussion about acceptable limits. Briefly the result, in terms of exposure to populations or maximum exposure to individuals, is that the limit should be in the order of the variations of natural background, which is tens of mrems per year. From that it would be possible to work your way back and calculate the required performance of borehole plugs.

J. HAMSTRA, Netherlands

But can you defend such a standard ?

T.O. HUNTER, United States

The question of acceptability is one that is hard to answer. But I think that such a number has some appeal since people throughout the United States are exposed to these variations in natural background radiation and nobody seems to worry enough to move to low background areas.

F. GERA, NEA

It is interesting to know that there is a discussion going on about dose limitations, as recommended by ICRP, and how applicable they are to the long-term aspects of geologic disposal. ICRP states that they are entirely applicable and that geologic disposal is in no way different from other nuclear activities. Of course, when you try to apply the ICRP limits to geologic disposal it turns out to be difficult because you do not know about the future environment, or the future population distribution and pathways. Thus an approach like trying to compare the impacts of repositories with the variations of natural background would appear to have a lot of merit. Assuming that agreement has been reached on the release limits for geological repositories the required performance of borehole plugs and shaft seals will follow. The requirements on the plugs and the seals will depend on the release limits but also on the geology of the specific site. Certainly there is no need to have a super plug if the geological formations are relatively permeable. The objective of plugging should be to ensure that the penetrations are not the weak spots in the geological barriers.

J. HAMSTRA, Netherlands

I disagree with you if you feel that the dose limits are applicable to accidental releases from a repository. I have always understood that ICRP dose limits only apply to normal operating conditions while in our analyses we assume accidental releases. An additional point is that dose limits of today are applicable to the man of today ; we cannot know how radiation resistant future man will be and I doubt that ICRP dose limits can be applied to future releases.

R.D. ELLISON, United States

In conjunction with the design of the shaft Rockwell is interested in the possible extent of the disturbed zone ; they have carried out some specific studies on this problem.

M.J. SMITH, United States

We have tried to obtain a feeling for the extent of the disturbed zone by asking the opinion of many rock mechanic experts in the United States. The consensus seems to be that the disturbed zone could extend 1 to 1 ½ diameters. If the permeability of the intact rock were 10^{-9} cm s^{-1}, such as in rock salt, the permeability of the disturbed zone should be as high as 10^{-5} cm s^{-1} before sealing became necessary. Somebody else has stated that if the disturbed zone had a permeability more than two orders of magnitude greater than the permeability of the plug or of the undisturbed rock significant preferential pathways would be established.

With respect to the question about what are acceptable release rates from a repository we are using the recommended radionuclide concentration limits in water. We recognize that the calculated concentrations depend on the release scenario. We tried to simplify the release scenario and be as conservative as possible in making those assessments. A very simple modeling job we did was to assume that the repository was located not in a tight salt rock zone where we would actually locate it but rather in a flow zone characterized by significantly higher permeability.

R.D. ELLISON, United States

I wish to discuss criteria for borehole plugging. NRC is thinking about the problem. It should be possible to make plugs with the same permeability as the host rock, but sealing the disturbed zone might be much more difficult. One sure thing is that there is no point in making plugs much better than the host rock.

R. PUSCH, Sweden

If we consider a strong host rock that behaves as an elastic medium it is difficult to generate a significant disturbed zone around boreholes unless the depth is very great. It is a different matter around shafts or tunnels that are excavated by means of blasting. Explosives will produce the kind of damage likely to result in increased permeability. Therefore it is necessary to distinguish between boreholes and other types of excavations. In Sweden when we choose the blasting technique we know that, by proceeding very carefully, with very small distances between the boreholes where small explosive charges are placed, we can produce excavations surrounded by very limited disturbed zones.

SESSION 2

Chairman - Président

C.L. CHRISTENSEN
(United States)

SEANCE 2

SHAFT AND TUNNEL SEALING CONSIDERATIONS

P. C. Kelsall and D. K. Shukla

D'Appolonia Consulting Engineers, Inc.
Albuquerque, New Mexico

ABSTRACT: Much of the emphasis of previous repository sealing research has been
placed on plugging small diameter boreholes. It is increasingly evident that
equal emphasis should now be given to shafts and tunnels which constitute more
significant pathways between a repository and the biosphere. The paper discusses
differences in requirements for sealing shafts and tunnels as compared with
boreholes and the implications for seal design. Consideration is given to a
design approach for shaft and tunnel seals based on a multiple component design
concept, taking into account the requirements for retrievability of the waste.
A work plan is developed for the future studies required to advance shaft and
tunnel sealing technology to a level comparable with the existing technology for
borehole sealing.

INTRODUCTION

An essential technological requirement for waste repositories in geologic media is the systematic development of design and construction methodology for sealing all penetrations (boreholes, shafts, tunnels) which connect the repository or its immediate environment with the biosphere. The emphasis on most of the past repository sealing research has been on plugging boreholes, but it is increasingly evident that sealing of shafts and tunnels is equally critical to adequate isolation of nuclear waste in underground repositories. Shafts and tunnels constitute definite and large pathways between the repository and biosphere, whereas careful site selection and characterization may eliminate boreholes directly connecting the repository with the biosphere.

A study has been initiated to assess the applicability of existing sealing studies (which have been directed mostly to vertical boreholes) to the requirements for sealing shafts and tunnels and to identify areas in which additional studies are required to obtain design criteria for shaft and tunnel seals (ONWI, 1980). The first stage of this study has been to make a comparison between shafts and tunnels and boreholes and to relate the differences to sealing requirements. Preliminary consideration has been given to conceptual designs but detailed design work has not yet been included in the study. The aims of this paper are to summarize the differences between shafts, tunnels and boreholes and to indicate areas in which the scope of sealing research, testing and design efforts must be expanded to encompass the requirements for sealing shafts and tunnels.

GENERAL COMPARISON OF SHAFTS, TUNNELS AND BOREHOLES

Since the design approach and fundamental research for sealing boreholes are relatively well advanced it is useful to consider the requirements for sealing shafts and tunnels initially in terms of the differences between the various types of openings that might have an impact on seal design and construction.

Shafts differ fundamentally from boreholes in terms of size. Current repository conceptual designs include access shafts with diameters typically in the range of 10 to 30 ft. Boreholes connecting with, or adjacent to, a repository and requiring sealing are unlikely to have diameters greater than 30 inches and typical diameters may be in the range 4 to 12 inches. This size difference has a number of impacts on sealing. As the diameter of the opening increases, for example, the requirements for structural strength within the seal increase, the surface area of the interface between the seal and the host rock increases, the possibility of damage to the host rock during excavation increases, and the overall volume of the seal increases. As the diameter increases, the importance of achieving an adequate seal increases and possibly, due to the factors listed above, the difficulty of achieving an adequate seal may increase also. Conversely, as the diameter of the opening increases it becomes possible to gain direct access to the seal zone for host rock characterization and preparation, and for seal emplacement, and this should facilitate achieving an adequate seal in larger openings.

A second fundamental difference between shafts and boreholes is related to their purpose in a repository. Current repository conceptual designs (for example, Stearns-Roger, 1978) do not include boreholes drilled from the surface for construction purposes. It is also current design philosophy to drill geotechnical test holes on the sites of future shafts and it is thus possible that there may be no boreholes requiring sealing in a repository. Exceptions may be geotechnical test holes drilled from within the repository or preexisting mineral exploration holes encountered unexpectedly. In any case, boreholes will not be required during or after construction and the possibility exists to seal boreholes during the construction phase. On the other hand, shafts must be kept open during construction and waste emplacement and, to some degree yet to be determined, after waste emplacement in order to satisfy requirements for waste retrievability. This point will be given further consideration in a later section.

The obvious distinction between shafts and tunnels relates to inclination which impacts placement techniques and stress distributions. Seal placement may be more difficult in a horizontal tunnel than in a vertical shaft due to the tendency for the seal material to separate from the roof. Seals in horizontal tunnels will be subjected to more complex rock-loadings than will seals in a vertical shaft.

The stresses from rock around tunnel seals will be unsymmetrical, whereas they probably will be symmetrical around a vertical shaft.

A further distinction between shafts and tunnels concerns location since it may be anticipated that tunnels will be located within the repository, entirely within the repository host formation. There is no necessity, therefore, to consider sealing tunnels in an aquifer zone, and the host formation may be assumed to be relatively impermeable, other than in a possible disturbed zone. Seals located within a repository will be more difficult to monitor after decommissioning, allowing little possibiltiy of remedial action should a problem develop. Designs for tunnel seals should take into account this difficulty in accessibility by introducing sufficient redundancy or conservatism that a problem cannot develop.

The distinctions between shafts, tunnels and boreholes are further considered in Table 1 together with the implications that these distinctions have for seal design. These implications related to shaft and tunnel sealing are summarized as follows:

o Shafts will connect directly with the repository, whereas boreholes may not.

o Shaft and tunnel seals will have greater volume, which may prevent the use of certain materials for technical or economic reasons.

o Shaft and tunnel seals will require greater structural strength.

o In shafts and tunnels, the interface between the seal and the host rock will have greater surface area.

o Shaft and tunnel excavation may produce a more extensive zone of disturbed rock beyond the periphery of the opening which will require sealing.

o Direct access to the seal zone in shafts and tunnels will facilitate host rock preparation and seal emplacement.

o Horizontal tunnel seals may be subjected to complex asymmetric loadings; seal emplacement may be more complex in horizontal tunnels.

DESIGN APPROACH FOR SEALING SHAFTS AND TUNNELS

General Requirements for Sealing Repository Openings

The overall objectives for sealing any type of repository opening, shafts, boreholes or tunnels, as to be defined finally in EPA or NRC regulations, will be quantitatively the same. The seals will be required to prevent an unacceptable escape of radionuclides from the repository and to prevent groundwater penetration into the repository. Both functions will be required for a specified time period which is yet to be determined but which will be significantly longer than the design life of typical engineering structures.

Prevention of groundwater penetration will be required during construction as well as during waste emplacement and beyond, although this may be achieved during construction by the shaft lining or by temporary seals rather than by the permanent seal, as considered in this report. Prevention of groundwater ingress is critical in the case of shafts penetrating aquifers, particularly in soluble rock types where leaks behind the lining endanger the integrity of the shaft and can lead to collapse. For this reason, prevention of groundwater ingress in considered to be at least as important a long-term function of the permanent seal as prevention of radionuclide escape.

ONWI (1979b) has previously considered four alternate ways in which design goals may be defined. The following goal was selected as being unambiguous and not impossible to satisfy:

The radionuclide migration rate through the seal zone is
always less by a specified factor of safety than an acceptable level determined by a consequence analysis.

In this context, the factor of safety is applied in such a way as to develop a
design migration rate that is less than the maximum acceptable migration rate.
Other factors of safety may be applied within the design to account for uncertainties in design parameters. An advantage of the criterion is that the overall
factor of safety can be varied according to the particular system being analyzed
and according to the consequence of failure of the system. For example, shafts
may require higher factors of safety than boreholes to account for the more
direct, and possibly more extensive, potential leakage paths.

Requirements for Sealing Shafts

As noted previously, the function of a shaft seal will be to prevent water inflow
as well as to prevent radionuclide migration. During construction and waste
emplacement, water inflow will be reduced or prevented by freezing or grouting as
appropriate, and by the shaft lining and special water seals placed in aquifer
zones to prevent water flow behind the lining. Permanent seals placed after waste
emplacement must duplicate or support the function of the lining in preventing
water inflow. Consideration should be given in designing linings as to whether
the lining will eventually become part of the permanent seal system or whether it
will be removed to permit permanent seal emplacement.

An important factor to be considered with regard to shaft sealing is the requirement of shafts to be kept "open" for a specified period, initially to provide
access for waste emplacement and subsequently to provide the means for retrieval
of the waste, should that be necessary. It is a proposed NRC stipulation that it
must be possible to retrieve the waste at the same rate at which it is emplaced.
The degree to which shafts will be "open" during the retrieval period has yet to
be specified. At the completion of waste emplacement, at least three approaches
will be possible for the period during which retrieval capability is required.

o In the first approach, the shaft would be kept open with
   the provisions that the permanent seal design is completed
   and that the seal could be installed quickly and safely, if
   required. The premise of this approach is that the shaft
   seal is not the primary engineered barrier against radionuclide migration to the biosphere.

o In the second approach, the hoisting equipment would be
   removed and the shaft sealed in a partial manner in such a
   way that the seal could be removed quickly and safely
   without damage to the shaft. The seal installed might be
   part of the final permanent seal, or it might be a different
   design, intended specifically as a temporary measure.

o In the third approach, the shaft would be sealed permanently
   at the completion of waste emplacement. The seal design
   and seal materials used, however, would be such as to allow
   safe removal or penetration without damage to the integrity
   of the shaft, should this be necessary.

The first of the approaches is probably the simplest and has the apparent advantage
that the delay before the seal is emplaced should allow more advanced technology
to be incorporated. It is noted, however, that licensing requirements (or public
opinion) may not permit open shafts once the waste is emplaced and that complete
seal designs must be prepared and must be verifiable using existing technology,
even if the seals are not installed. A disadvantage of the third approach is that
it is doubtful that a "permanent" seal could be removed quickly enough to meet the
proposed NRC requirement that the waste should be retrieved within the same duration over which it was emplaced. At this stage, therefore, the second approach
appears to be preferable, and conceptual designs for shaft seals should take into
account the possibility of the seal being installed in two stages--a partial seal
installed for the retrieval period and a permanent seal completed when retrieval
is no longer required. An advantage of this method is that the performance of

the partial seal can be monitored during the retrieval period and any deficiencies in performance can be rectified by remedial action or by addition of further seals as part of the completed structure.

Requirements for Sealing Tunnels

The requirements for sealing tunnels are not readily apparent from examination of current waste isolation philosophy or existing repository designs. If it is assumed that tunnels will not be used as the primary means of access to a repository, tunnel seals will be required as the primary barrier against water inflow or the final barrier against radionuclide migration to the biosphere. Tunnel seals may be required within a repository to isolate parts of the repository in which waste emplacement is complete or they may be considered part of an overall 'final' seal placed partly in the shafts and partly in the tunnels connecting to the shafts. In either case the same comments regarding waste retrievability apply to tunnel seals as to shaft seals. Tunnel seals may indeed be the initial seals that are in place during the retrieval period while the shafts remain open or only partly sealed.

Review of Existing Repository and Seal Designs

In further considering shaft and tunnel sealing requirements it is useful to review existing repository conceptual designs, first to examine the location of shafts and tunnels relative to the repository, and second to evaluate the extent to which current designs take into account sealing requirements. Preconceptual designs for repositories in salt, granite, basalt and shale were developed for the Office of Waste Isolation (OWI) by Parsons Brinckerhoff Quade and Douglas (PBQD 1978a-j). Rather more detailed designs for salt domes and bedded salt were prepared by Stearns-Roger (1978) and Kaiser Engineers (1978) respectively. Shaft lining designs have been prepared also for the WIPP facility but are not yet available for detailed review.

It is not proposed in this paper to present a complete review of these conceptual designs. The following sections provide a brief summary of some aspects of the designs that have an impact on sealing, the main purpose being to highlight certain contrasts between the various designs and to illustrate that there is no unified approach to future sealing requirements.

Repository Layout: The PBQD conceptual design is the same for all host rock types with five shafts located in a central shaft pillar and connected to the storage rooms by five main corridors. These corridors are 30 ft. wide by 20 - 25 ft. high. Approximately 1000 linear ft. of each of the corridors between the shaft area and the storage rooms could be used for sealing if required. In the Kaiser design for bedded salt the shaft pillar is to one side of the repository and the four shafts are connected to the storage rooms by seven 2000 ft. long tunnels. In the Stearns-Roger design for dome salt, due to space restrictions, the shaft pillar is in the centre of an oval-shaped repository. Six shafts are connected to the storage rooms by as many as 18 short tunnels.

Shaft and Lining Design: All of the designs consider shaft excavation by conventional drill and blast techniques (one small diameter shaft in the PBQD design is to be drilled). The PBQD and Stearns-Roger designs specify pre-grouting from within the shaft for water bearing zones. In the PBQD design the shaft is concrete lined (12 in. thick) through the lowermost aquifer and a water seal and bearing ring are set at the bottom of the lining. In the Stearns-Roger design for dome salt the shaft is concrete lined (12 - 18 in. thick) 100 ft. into the salt; a chemical water seal is constructed behind the lining in competent anhydrite in the caprock above the salt. The shafts in the Kaiser design are fully lined with concrete -- no details of lining technique or water seals are given.

Sealing Considerations: The Kaiser design specifies that all storage rooms and the connections to the shafts will be backfilled with salt at decommissioning. The shafts will be sealed by stripping the lining and backfilling with a 'suitable material' to be defined by the borehole plugging program. The PBQD and Stearns-Roger designs give no consideration to backfilling or sealing techniques.

It is evident from this brief summary that existing repository designs do not place major emphasis on sealing requirements. If the designs are accepted and finalized future sealing efforts must be adapted to take into account existing repository layouts and existing shaft lining designs. A preferable approach is that repository designs should be modified to account for future sealing requirements. Areas in which such modifications might be appropriate include:

1. Repository layout
   Where feasible it is preferable to connect the repository and shafts with as few tunnels as possible. It appears advantageous to separate the shafts from the repository (as in the Kaiser design for bedded salt, for example) so as to provide the option for placing primary seals in tunnels as well as in vertical shafts.

2. Excavation techniques
   Excavation techniques in future seal zones should be chosen to minimize the potential for damage to the host rock. This applies to pretreatment methods, such as freezing or grouting, as well as to actual excavation techniques. In general, mechanical excavation methods are preferable to methods using explosives.

3. Lining designs
   Lining designs should specify whether the lining is to be removed at decommissioning or whether it is to be left in place. If practical, it may be preferable that shafts be unlined in sections where future seals are to be placed. Special consideration should be given to the longevity of seals in aquifer zones.

The comments above refer to relatively minor changes to existing repository designs. Future activity related to sealing requirements might point to deficiencies in repository designs that imply more radical changes in the designs. It is important that future repository and sealing design efforts should be closely coordinated.

Design Approach for Shaft and Tunnel Seals

A multiple-component seal concept has been proposed for repository openings (ONWI 1979b, Ellison and Shukla, 1980), primarily in order that different materials with varying seal properties can be matched to varying seal requirements. In the general case for repository seals, each component in the multiple-component seal will thus have a specific function, either to reduce permeability or to absorb radionuclides. Applied specifically to shafts other components may have additional functions, for example, to provide structural strength, or to provide a good bond with the host rock. A premise of the concept is that there is no single seal material with all the required physical and chemical properties for a complete seal. Additional reasons for pursuing a multiple-component conceptual design for shaft and tunnel seals are as follows:

   o Shaft seals will be constructed in more than one stage.
      For example, seals to prevent water inflow must be placed during construction, whereas seals provided to prevent radionuclide migration will not be required until a later stage. Seals provided to prevent water inflow during construction may be permanent or they may be replaced or modified at a later stage. Seal materials suitable for preventing water inflow may be different from those suitable for preventing radionuclide migration.

   o As noted previously, the seal to prevent radionuclide migration might be provided in more than one stage. This may provide the option for modifying the later stages to account for improved technology.

   o Shaft and tunnel seals will be required to have significant load-bearing capacity (much more so than borehole seals). Seals may be designed in such a way that some elements

carry the load, while other elements may be non-load-bearing
and composed of weaker or more flexible materials.

o Excavation of a shaft or tunnel may create a significant
   disturbed zone. Seal materials suitable for sealing a
   disturbed zone may be different than materials suitable for
   sealing within the opening.

Design studies act as a hub for the overall sealing study program in that they
determine to a large extent the goals for laboratory and field-testing and specialized study areas. Some sealing problems, or areas in which further detailed study
is required, may be identified only by detailed design effort. For this reason,
it is recommended that in addition to generic design studies, site-specific
detailed design studies should be initiated as soon as possible.

An implication of initiating detailed design at a relatively early stage in the
overall sealing design program is that it will be necessary to assume a particular
design concept as the basis of the detailed work. A suitable conceptual design
can be identified from the existing technology but this may not prove to be the
optimum concept once shaft operating requirements and other factors such as
materials properties are better understood. The recommended approach is to
initiate two parallel design efforts, one to develop detailed designs based on an
existing concept and one to modify and improve the basic concept.

Detailed designs for shaft and tunnel seals should be based on existing repository
designs and operating concepts as developed, for example, for the WIPP site. The
conceptual design for the seal should be derived from existing borehole and shaft
sealing studies. A suggested concept is based on a structural bulkhead combined
with impermeable and/or nuclide-absorbing but non-load-bearing fill materials.
The total multiple-component seal will utilize beneficial properties of different
materials. Detailed designs will include construction drawings and materials and
construction specifications.

Alternative conceptual designs will be developed for a wide range of shaft conditions, host rock characteristics and design requirements. The varying host rock
characteristics may include a variety of host rocks, stratigraphies, aquifer
locations, past loading conditions, etc. The objective will be to develop reference designs for individual seal components, each of which will be suitable for a
particular purpose (e.g., radionuclide absorption or prevention of water ingress)
and a particular geological situation. As appropriate, the final seal design
for a specific repository may be based on a multicomponent concept employing
several complementary materials. Alternative conceptual designs need not necessarily be adapted to existing shaft and lining designs. Conceptual design studies
may indicate that shaft and lining designs should be modified in order to facilitate sealing.

WORKPLAN FOR ADDITIONAL STUDIES RELATED TO SHAFT AND TUNNEL SEALING
Applicability of Existing Borehole Sealing Technology and Studies
to Shaft and Tunnel Sealing
It has been noted that much of existing repository sealing research has been
directed specifically towards sealing small-diameter boreholes. A review of this
research (ONWI 1979a) indicates many areas in which sealing studies are equally
applicable to shafts and tunnels as to boreholes. Areas of technological
overlap between borehole sealing and shaft sealing include definition of design
goals and basic design criteria related to repository operating modes, generic
geological studies, fundamental materials research, host rock characterization
and instrumentation.

In other areas there is a necessity to expand the scope of existing studies to
take into account the requirements for sealing shafts and tunnels as well as
boreholes. These areas include:

o Operational considerations: conceptual studies are required
   to define the operating requirements for shafts and tunnels
   and, specifically, the degree to which shafts and tunnels
   will remain open during the retrieval period.

o Excavation techniques: methods should be developed or
  chosen to minimize host rock disturbance.

o Conceptual design: shaft lining and seal designs must be
  coordinated. Conceptual designs must be developed for
  larger diameter openings and non-vertical inclinations.

o Materials: existing studies should be expanded to account
  for special requirements resulting from larger volumes,
  different placement techniques and different structural
  requirements applicable to shafts and tunnels.

o Characterization and treatment of disturbed zone.

A workplan developed to coordinate future shaft and tunnel sealing studies is
introduced in the following section.

Workplan for Additional Studies and Testing

The recommended workplan for the additional studies and testing required to advance
shaft and tunnel sealing technology to a level comparable with the existing technology for borehole plugging includes the following items:

o Definition of Shaft and Tunnel Operating Conditions and
  Requirements: Conceptual study is required to define
  operating modes for shafts during and after waste emplacement and to establish the requirements for sealing tunnels.
  Steps should be taken to integrate seal and shaft design
  programs in order to ensure design and construction compatibility.

o Continued Review of Borehole and Shaft Sealing Studies:
  Current efforts to review existing technology applicable to
  shaft and tunnel sealing should continue. Valuable input
  can be derived from borehole plugging studies and from
  shaft and tunnel sealing technology in other industries.

o Conceptual and Detailed Design: Design studies are necessary
  in order to fully identify design and construction problems
  that should be addressed in future study and testing
  programs. A concurrent detailed design and conceptual
  design approach is recommended:

  --Detailed designs based on existing repository and shaft
    designs and existing conceptual seal designs should be
    initiated as soon as possible.

  --Alternative conceptual designs should be developed in the
    form of a series of reference designs, each adapted to a
    specific sealing purpose (or material characteristic) and
    a particular host environment.

o Consequence Analyses: Consequence analyses should be
  performed to assess the necessity for permanently sealing
  shafts during the period when waste retrieval may be
  required and to assess the impacts of flaws within completed seals. For the latter application, consequence
  analyses should separately evaluate the impacts of pore-
  space at the seal/host rock interface, cracks within the
  seal and fractured rock within a disturbed zone around a
  seal. The analyses should address the impacts of water
  inflow as well as radionuclide migration.

o Testing Program: Additional testing is required in a
  number of specialized areas where existing borehole plugging
  technology is not directly transferable to meet shaft and
  tunnel sealing requirements. Laboratory testing should

examine materials properties taking into account the
special requirements of shafts such as volume, and consideration should be given to bench scale model tests to evaluate
seal behavior under a variety of conditions. Full-scale
testing at repository depths will have to be designed
carefully in view of the large costs involved. The recommended approach will be reduced-scale tests run in parallel
with other field tests for borehole plugs. In addition,
it will be necessary to conduct in situ tests to characterize the host rocks and the extent and nature of the
disturbed zone produced by excavation. A further possible
approach may be to conduct full-scale tests of seals or
bulkheads installed in facilities other than repositories
such as deep mines or within a specially constructed
shaft/tunnel test zone within the repository itself.

It will be noted that this workplan is based on two premises. First, it is proposed that site-specific as well as generic designs should be developed. Detailed
design work will reveal many of the specific problems that must be solved before a
finally acceptable design concept can be achieved. Generic studies and testing
programs should proceed in conjunction with design efforts and should be modified
periodically according to input from the design work. The second premise is the
importance of field testing to support laboratory and analytical studies. In this
regard, separate and perhaps somewhat different emphasis needs to be placed on the
testing of shafts and of tunnels. The sealing of tunnels connecting different
parts of a repository may be of greater significance in the emplacement and
retrieval phases, but shaft seals will have a more significant role in the permanent seal since tunnels will be in the repository host formation only, whereas
shafts will connect to the biosphere through aquifers above the repository host
formation. Tunnel seal tests are further complicated because of asymmetrical
stress and rock conditions around a tunnel and the fact that the roof-to-seal
interface will be difficult to duplicate in anything but a full-sized test. In
all cases, tests must be designed with specific objectives in view. These objectives may range from examination of a specific design problem to overall verification of seal integrity in a specific repository.

REFERENCES

Ellison, R. D. and D. K. Shukla, 1980, "Design Requirements and Concepts for
Penetration Seals at Deep Geleogic Waste Repositories -- A Progress Report,"
paper presented to Workshop on Borehole and Shaft Plugging, Columbus, Ohio, May
1980.

Kaiser Engineers, 1978, "NWTS Repository in a Bedded Salt Formation for Spent
Unreprocessed Fuel," Oakland, California.

Office of Nuclear Waste Isolation, 1980, "An Evaluation of Generic Similarities
and Differences in Requirements for Sealing Shafts, Tunnels and Boreholes," ONWI
90, draft report prepared by D'Appolonia Consulting Engineers, Inc.,for Battelle
Memorial Institute, Program Management Division, Columbus, Ohio.

Office of Nuclear Waste Isolation, Jan. 1979a, "The Status of Borehole Plugging
and Shaft Sealing for Geologic Isolation of Radioactive Waste," ONWI 15, prepared
by D'Appolonia Consulting Engineers, Inc., for Battelle Memorial Institute,
Program Management Division, Columbus, Ohio.

Office of Nuclear Waste Isolation, Aug. 1979b, "Repository Sealing Design--1979,"
ONWI 55, prepared by D'Appolonia Consulting Engineers, Inc., for Battelle Memorial
Institute, Program Management Division, Columbus, Ohio.

Parsons Brinckerhoff Quade and Douglas, Inc., 1978a, "Repository Preconceptual
Design Studies: Salt," Y/OWI/TM-36/8, Technical Support for GEIS: Radioactive
Waste Isolation in Geologic Formations, Office of Waste Isolation, Oak Ridge,
Tennessee.

REFERENCES
(Continued)

Parsons Brinckerhoff Quade and Douglas, Inc., 1978b, "Drawings for Repository Preconceptual Design Studies: Salt," Y/OWI/TM-36/9, Office of Waste Isolation, Oak Ridge, Tennessee.

Parsons Brinckerhoff Quade and Douglas, Inc., 1978c, "Repository Preconceptual Design Studies: Granite," Y/OWI/TM-36/10, Technical Support for GEIS: Radioactive Waste Isolation in Geologic Formations, Office of Waste Isolation, Oak Ridge, Tennessee.

Parsons Brinckerhoff Quade and Douglas, Inc., 1978d, "Drawings for Repository Preconceptual Design Studies: Granite," Y/OWI/TM-36/11, Office of Waste Isolation, Oak Ridge, Tennessee.

Parsons Brinckerhoff Quade and Douglas, Inc., 1978e, "Repository Preconceptual Design Studies: Shale," Y/OWI/TM-36/12, Technical Support for GEIS: Radioactive Waste Isolation in Geologic Formations, Office of Waste Isolation, Oak Ridge, Tennessee.

Parsons Brinckerhoff Quade and Douglas, Inc., 1978f, "Drawings for Repository Preconceptual Design Studies: Shale," Y/OWI/TM-36/13, Office of Waste Isolation, Oak Ridge, Tennessee.

Parsons Brinckerhoff Quade and Douglas, Inc., 1978g, "Repository Preconceptual Design Studies: Basalt," Y/OWI/TM-36/14, Technical Support for GEIS: Radioactive Waste Isolation in Geologic Formations, Office of Waste Isolation, Oak Ridge, Tennessee.

Parsons Brinckerhoff Quade and Douglas, Inc., 1978h, "Drawings for Repository Preconceptual Design Studies: Shale," Y/OWI/TM-36/15, Office of Waste Isolation, Oak Ridge, Tennessee.

Parsons Brinckerhoff Quade and Douglas, Inc., 1978i, "Repository Preconceptual Design Studies: BPNL Waste Forms in Salt," Y/OWI/TM-36/16, Technical Support for GEIS: Radioactive Waste Isolation in Geologic Formations, Office of Waste Isolation, Oak Ridge, Tennessee.

Parsons Brinckerhoff Quade and Douglas, Inc., 1978j, "Drawings for Repository Preconceptual Design Studies: BPNL Waste Forms in Salt," Y/OWI/TM-36/17, Office of Waste Isolation, Oak Ridge, Tennessee.

Stearns-Roger Engineering Corp., 1978, "NWTS Repository for Storing Reprocessed Wastes in a Dome Salt Formation," U. S. Department of Energy.

TABLE 1

COMPARISON OF BOREHOLES WITH SHAFTS AND TUNNELS AND
IMPLICATIONS FOR SEALING STUDIES

Characteristics	Differences	Implications for Shaft and Tunnel Sealing (Design and Construction)	Considerations for Shaft/Tunnel Sealing-Related Studies
Size	Boreholes range from 2" to 30" in diameter, whereas shafts/tunnels could typically be 6' to 30' in diameter.	Shafts/tunnels provide a larger pathway to biosphere than boreholes; hence more stringent demands on seal design and construction.	More significance in consequence analysis.
		Shafts/Tunnels may be accessible for direct observation and construction, allowing:	
		--direct access for host rock characterization,	
		--direct access for host rock preparation for seal emplacement; hence more careful treatment of disturbed zone is possible in shafts/tunnels than in boreholes,	
		--access for monitoring of seal behavior.	
		Shaft/tunnel seals will require one to three orders of magnitude greater volume of seal material than borehole seals; thus, seal material selection will require following considerations:	Consider cost, size effects and strength characteristics in materials study programs (in addition to other properties being evaluated for borehole sealing).
		--cost (large volume)	
		--property variability (or size effect)	
		--need to support greater seal weight, hence need for greater structural strength (than for borehole seals)	

TABLE 1

(Continued)

Characteristics	Differences	Implications for Shaft and Tunnel Sealing (Design and Construction)	Considerations for Shaft/Tunnel Sealing-Related Studies
		Greater surface area at host rock interface than for boreholes.	Greater emphasis needed on interface characterization and treatment than for boreholes.
Orientation and Shape	Tunnels are generally horizontal, shafts near vertical, boreholes could have various orientations, but predominantly vertical, with some horizontal. All of the penetrations will have distorted shapes between circular and rectangular with possibly irregular surfaces.	Slanted or horizontal boreholes and tunnels will make the placement of the seal near the tunnel roof difficult. Accordingly, design and construction procedures have to account for this difficulty by possibly placing the seal in lifts and under pressure.	Suitable placement techniques and materials should be developed to account for the tunnel/borehole seal orientation.
		Rock mass in tunnel roofs may be severely disturbed due to gravity and asymmetric stress relief. Also, because of asymmetric loading, tunnel seal will experience more complex stresses than a similar size shaft sealing.	Evaluate characteristics of tunnel roof disturbed zone and techniques for sealing (treating) it.
		Irregular shapes will require use of materials and placement techniques that can fill various irregular surface boundaries.	Materials that can seal irregular shapes should be evaluated. Consequences of not sealing all irregular corners should be evaluated.
Access Requirements for Waste Retrieval	Waste retrievability will depend on access	Shaft/tunnel seals should be constructed in stages such that the seal placed in the retrievability period,	Evaluate consequences of seal placed in stages.

TABLE 1

(Continued)

Characteristics	Differences	Implications for Shaft and Tunnel Sealing (Design and Construction)	Considerations for Shaft/Tunnel Sealing-Related Studies
	to waste through shafts and tunnels rather than through boreholes.	although providing adequate isolation, can be breached for waste retrieval. Permanent seal can be placed when retrieval is not a consideration.	Define requirements for retrieval as it may affect seal design.
Location and Depth	--Tunnels are usually within the repository host formation near repository depth. --Shafts penetrate from the ground surface to the repository host formation. Shafts probably do not penetrate below the host formation. --Boreholes can have varying depths, some even deeper than repository host formation.	Shafts and tunnels are unlikely to penetrate below the repository host formation, but connect to the ground surface. Tunnels are unlikely to intersect aquifers. Shafts and boreholes likely will intersect aquifers above the repository. Boreholes may intersect aquifers below the repository.	Develop sealing techniques for shafts intersecting aquifers compatible with permanent sealing requirements.

- 111 -

TABLE 1

(Continued)

Characteristics	Differences	Implications for Shaft and Tunnel Sealing (Design and Construction)	Considerations for Shaft/Tunnel Sealing-Related Studies
Excavation Technique	Boreholes are drilled, whereas shafts/tunnels could be constructed by either mechanical means or blasting.	Excavation and pretreatment method and quality control will affect characteristics of disturbed zone. This variation should be accounted for in shaft/tunnel seal design.	Develop construction techniques/specifications to minimize disturbed zone around shafts/tunnels.
Disturbed Zone	Will be generally larger for shafts/tunnels than for boreholes. Also, disturbed zone in an open penetration will be time-dependent due to stress relief.	Consideration of disturbed zone may be very significant in shaft/tunnel seal design and construction.	--Characterize disturbed zone. --Develop techniques to treat the disturbed zone.

Discussion

J. HAMSTRA, Netherlands

You mentioned a possible conflict between normal mining practice and the shaft sealing to be performed at a later stage. It is normal practice in rock salt mining and shaft sinking to have reinforced concrete insets at every connection to the shafts of the different mining levels. It is therefore not difficult to adapt these insets in such a way that they can be the support of the shaft sealing bulkheads. Any conflict can thus be easily eliminated.

D.K. SHUKLA, United States

Mine and shaft designs for nuclear waste repositories must take special care, in order not to create situations which may render future sealing of these penetrations either impossible or extremely expensive. For instance, any grouting of the rock around a shaft may be impossible to alter at a future date. Hence such grouting must be done so that it is either a part of or is at least compatible with the permanent seal to be installed. Another example may be the proposal to install steel lining in shafts. Removal of such lining may be necessary (although that has not been determined yet) and be very expensive if not impossible. If such conflicts do not exist, so much the better. But the point is that an evaluation of such potential conflicts before design and construction of tunnels and shafts appears to be necessary.

PENETRATION SEALING FOR THE ISOLATION OF
RADIOACTIVE WASTE IN GEOLOGICAL FORMATIONS IN EUROPE

J.R. Cole-Baker, C.A. Davenport, M.S.M. Leonard
D'Appolonia S.A.
Brussels (Belgium)

ABSTRACT

The countries in Europe which are developing the storage of radioactive waste in geological formations are concentrating on one or two potential host rocks, due to demographic considerations and the restricted choice of rock types. In order to optimise the studies necessary to construct the penetration seal the requirements for the studies need to be defined early in the overall programme. These requirements are summarised here, and a staged development of penetration sealing procedures appropriate for incorporation in typical European repository suitability studies is proposed. It is emphasized that international coordination of the studies will not only prevent needless duplication of effort, but will also assist in unifying licensing procedures.

RESUME

Les pays européens développant le stockage des déchets radioactifs dans des formations géologiques concentrent leurs efforts sur une ou deux formations d'accueil possibles à cause de facteurs demographiques et du choix limité de variétés de roches. Dans le but de rentabiliser au maximum les études nécessaires pour assurer le bon scellement de forages ou de puits, les critères d'études doivent être précisés dès le début du programme global. Dans cet article, on résume ces critères, et un développement par étapes des procédures de scellement des forages est proposé. Ces procédures sont particulièrement adaptées pour être incorporées dans des études approfondies typiques de dépôts en Europe. La coordination internationale de ces études est importante, non seulement parce qu'elle évitera la répétition inutile des efforts, mais aussi parce qu'elle facilitera la normalisation des procédures de certification.

1. INTRODUCTION

The countries in Europe which are performing studies on the isolation of radioactive waste in geological formations are generally each considering one or two rock types. This is because geological and demographic considerations leave little choice in the matter. Despite this, all the programmes to characterise the rocks into which it is proposed to bury the waste have certain points of similarity. For instance, the properties of thermal conductivity and permeability and the effects on the rock structure of high temperatures must be studied. These studies are necessitating the drilling of deep boreholes into the strata, or work underground in large-scale laboratories, such as in the Asse salt mine in Germany, or the shaft being constructed at Mol in Belgium.

It is recommended in this paper that a coordinated programme for developing penetration seal specifications should be initiated, which would start at the earliest stages of the repository suitability studies. The importance of having a coordinated European programme is that licensing issues in one country will affect adjacent countries, and also that the individual studies have many points in common.

To date, the aspect of borehole and shaft sealing has generally been given low priority in Europe, compared to other repository suitability aspects, although attention has been given to the backfilling of the underground tunnels and rooms of mined cavities [1, 2, 3]. In this paper aspects of sealing of penetrations which could be studied or perfected during the course of the work required to characterise the formations in which the future repository is to be situated are discussed. The aim is to provide an approach to a set of studies which can be integrated to the overall repository assessment. The work required can therefore be expected to be performed at a reasonable cost.

An essential element of the approach is that each country takes account of work being performed elsewhere in similar rock conditions, or on materials which may be used. However, the effects of the specific environment must be accounted for. Similarly each country should be prepared to share their knowledge with others working in the same field, preferably through a coordinating agency.

2. SEAL DESIGN

The basic function of a seal would be to prevent the penetration from becoming a preferred route to the biosphere of radionuclides from the repository. Studies directed towards the definition of seal performance have been performed [4], and a definition of seal performance most appropriate to the general case has been proposed. This can be stated as follows:

> "The radionuclide migration rate through the seal zone is always less by a specified factor of safety than an acceptable level determined by a risk and consequence analysis."

i.e.,
$$R_t < \frac{R_{acceptable}}{\text{Factor of Safety}}$$

where

R_t = potential transport of radionuclides through the seal zone,

$R_{acceptable}$ = acceptable potential transport of radionuclides through the seal zone.

The advantage of this definition is that it can be used in the context of the individual repository, and take account of the total environmental impact of possible radionuclide transport. In addition, the acceptable level of potential transport of radionuclides through the seal zone may be varied with time. This allows compatibility of the seal design with that of the repository, for which an acceptable transport of radionuclides through the seal zone will already have been defined.

Once the overall design requirement has been defined, and the appropriate factors of safety determined, the specific requirements of the seal must be considered. For instance, the seal may be a temporary one which will need to be removed and replaced at a later date, or it may be the final repository seal. In any case, it is advisable to consider the possibility of removing a seal since a seal which is subsequently found to be faulty may require to be replaced.

The design procedure for any seal will include consideration of:

- Waste Type and Containment
- Host Rock Environment
- Mode of Possible Radionuclide Transport
- Geological Stability

These aspects are reviewed in the following paragraphs.

2.1 Waste Type and Containment

Present-day concepts of the storage of high-level waste generally consider that the waste will be stored in canisters in a vitrified form. Corrosion resistant layers around the waste are provided to act as the first barriers to radionuclide migration. Depending on the waste type there may or may not be the possibility of significant gaseous products from the waste. In the case where transuranic nuclides are absent, gas production would be confined to corrosion and radiolysis.

Evidently, the seal design also depends upon whether the waste is low level, medium level or high level, which dictates the required longevity of the seal, and the type of repository. Thus, some countries may require that waste should be able to be recovered from the repository. In this case, the seal will need to be designed so that it can be removed, and subsequently replaced.

2.2 Host Rock Environment

Typical host rocks being considered for the disposal of radioactive waste in Europe include bedded and domed salt, granite and clay shafts. Consideration of the various drilling, stress/strain and chemical behaviour of these rocks suggests that the types of seal most appropriate for each of these rock types will be different, although multiple material and geometry concepts may yield similar solutions for different rock types.

2.3 Mode of Possible Radionuclide Transport

Radionuclide transport is an important aspect of current studies, as the effectiveness of future repositories is qualified at present by the number of radionuclides reaching the biosphere. Basically, transport is by convection and diffusion, and is modified by sorption of radionuclides by the host rock.

Transport by convection involves the mass movement of radionuclides with the permeant as a result of temperature gradient and depends on the rock mass permeability and the pressure gradient. If the permeant is gas, the transport by convection will be the dominant feature of radionuclide transport. As gases flow more quickly than liquids they would provide potentially faster permeant transport for radionuclides. The repositories should therefore be designed bearing in mind that heat dissipation should be rapid enough to ensure that the host rock remains stable.

In the context of borehole sealing, radionuclide transport by convection should be small, and the significant convection movements are likely to be at the seal boundaries and in any unsealed disturbed rock. Otherwise the main mode of radionuclide movement will be by diffusion. Even if the permeant is stationary, the radionuclides can move by diffusion. In free water, molecular diffusion is represented by Fick's Law [5].

Radionuclides can be absorbed by the host rock in a reversible phenomenon. The rates of absorption and dissorption are very dependent on the types
of radionuclide and host rocks. The process is selective and certain radionuclides
will be absorbed relatively quickly. However, it is likely that the presence of
a selectively absorbable molecule can be an obstacle to the absorption of others.
This phenomenon of sorption and dissorption will modify the concentration of
nuclides, and may affect diffusion rates.

2.4 Geological Stability

In all of the countries of Europe in which geological disposal of radioactive wastes is envisaged, there exists some risk of earthquakes. Situations may
arise where a repository would be situated close to a major fault, or in an area
where some small risk of ground movement might be present.

This aspect should be considered in the penetration seal design. The
most adverse impact for a borehole would be a side wall displacement and collapse
or damage to the integrity of a seal. Although specific risk analyses have not
been performed to date for borehole seals, the level of risk of significant damage
to an individual seal is considered to be extremely low.

Studies on the prediction of hazard to deep borehole seals which have
been carried out by the Savannah River Laboratory [6], have indicated that substantial earthquake damage to underground facilities is usually the result of displacements, primarily along pre-existing faults and fractures, or at the surface
entrance to these facilities. Evidence of this comes from both collection of
records of earthquake damage as a function of depth and calculation of displacement
fields resulting from earthquakes.

3. FACTORS AFFECTING CHOICE OF SEAL

Important aspects of penetration sealing that will require research and
development effort during the next few years are those of seal placement, and the
development of appropriate seal materials. Studies have been performed to identify
the locations of seals to perform different functions [4]. The purpose here is to
identify the aspects requiring further study and to suggest how this could be incorporated into the ongoing development projects.

3.1 Placement

Proven technology must be available to place the material into the
borehole. Many new seal materials are being developed in laboratories. However,
placement aspects are not receiving so much attention. Therefore, although several
new materials are showing promising capabilities regarding their physical properties, placement techniques required to use them in boreholes often need to be
developed. This development will be possible once deep exploratory boreholes are
available for field tests.

Research efforts should therefore use to the best advantage available
sites, to enable placement experiments to be performed. An essential part of these
experiments will be monitoring the workmanship of the sealing contractor. This
aspect is discussed in detail later in this paper.

3.2 Stability

Once in place, the change in physical properties of the seal material
with time must be acceptable. The acceptable condition of the seal is related to
the likelihood of radionuclide escape from the waste container and host rock.

The physical properties that should remain as stable as possible are
permeability, radionuclide sorption capability, deformation and conduction compatibility with the host rock, and chemical composition. The stability in the zone
of the seal-rock interface will be the most critical location for study.

Studies to define these properties, and to attempt accelerated aging experiments, may be made in the laboratory. However, the field conditions must be faithfully reproduced, and the model seal must closely resemble that which can be placed in-situ.

3.3 Permeability

Permeability is the most important factor in determining the likelihood for radionuclide movement through the seal zone. Permeability is very dependent on the seal zone structure. The permeability of the plug can be measured accurately in the laboratory, and is a relatively simple test to carry out. However, the overall seal zone permeability will be more dependent on the quality of the rock-seal interface, and treatment of any open fissures or joints. This aspect will need to be measured during in-situ tests, and also in the laboratory on recovered samples. Down-the-hole instrumentation can also monitor changes in the environment that depend upon the seal zone's permeability.

Permeability is one of the most obvious properties on which to base a present-day choice of seal material.

3.4 Radionuclide Sorption

The sorption properties of different materials are not as critical as the permeability, but a knowledge of the ability to filter out more active radio-nuclides would clearly be an important advantage at the time of seal material selection.

Studies on the sorption properties are being carried out on materials that are envisaged to act as binders for the waste materials.

3.5 Deformation Compatibility with Surrounding Rock

The properties of shrinkage, elasticity, brittleness, creep and strength, which are temperature- and time-dependent will require to be closely compatible with the host rock properties. The accuracy to which these properties need to be evaluated will be similar to that for the host rock.

Each of these properties can be evaluated in the laboratory, and their significance in quantitative terms is in their use in risk and consequence ana-lyses. Their qualitative assessment is more related to the rejection of certain materials as being obviously unsuitable.

3.6 Chemical Stability

The chemical stability will control the time-dependence of the different material properties. Chemical stability can be assessed in the laboratory, and is one of the first aspects of any new material that is studied. Chemical stability is therefore a useful parameter in qualitative assessment of a material's suitabi-lity.

The geochemical interactions at the seal-rock boundary will have an important role in the seal performance [7]. It is at this interface that the seal comes into contact with the chemical environment of the host rock. Selection of the seal material must take account of its behaviour in the groundwater environment of the seal-rock interface.

A particular feature in chemical stability assessment is the determina-tion of whether the material liberates gases in the down-the-hole conditions. The resistance to flow of materials to gases is far less than that for liquids, therefore potential for gas generation is an important aspect for consideration.

3.7 Other Considerations

A promising idea for the formation of penetration seals which could provide a complete barrier able to fulfill the total requirements of a seal is the multiple material/geometry concept [4]. As part of the seal placement techniques

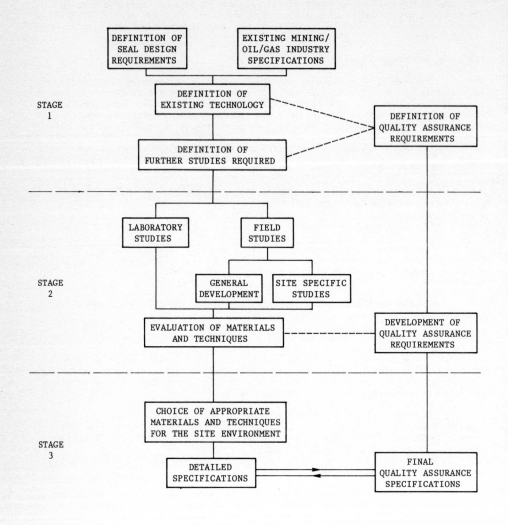

STAGE 1

STAGE 2

STAGE 3

Figure 1

STAGED DEVELOPMENT OF
PENETRATION SEALING
PROCEDURES

described above work would be required to develop tools able to produce the required shapes.

Once all the technical considerations have been studied and agreed, the final confidence in the seal will depend upon the workmanship of the sealing contractor. Therefore, detailed site and quality control procedures must be elaborated during the experimental phases of the programme.

4.   FURTHER STUDIES

Having defined the basic requirements of the seal, and discussed the aspects which affect seal performance and design, the areas in which further studies should be performed can be identified. These studies can be divided into those which are general for all environments and those which need to be performed specifically for one environment. For all these studies the project team should identify the performance required in the in-situ environment, so that specialists in each subject can work to a well-defined goal. An overall flow chart of the work required from the definition of seal requirements to the preparation of detailed specifications is given in Figure 1. This chart is given as an overall guide, as there will be general interaction between the various activities.

Studies will be required to refine the following aspects:

- Seal Loading Conditions
- Materials
- Seal Geometry
- Placement Techniques
- Instrumentation

All of these aspects are, of course, interrelated. It is intended here to emphasize the work which can be performed during the field characterisation studies.

4.1 Seal Loading Conditions

The studies to define seal loading conditions are site specific, and will form part of the overall repository design studies. The basic items to be defined are:

- Natural temperatures and gradients at depth.

- Time-dependent rock deformations associated with the penetration or repository excavation.

- Deformations associated with geological activity.

- Water pressure differences between aquifers intersected by the borehole.

- Thermal loading conditions resulting from the waste or from surface climatic conditions over long periods. Temperature fluctuation may be more important than actual temperatures if they cause relative displacements between dissimilar materials (i.e., host rock and seal or different seal materials).

- Radiation doses if the actual flow of a permeant occurs, regardless of how small.

- Gas pressures if gas is to be generated by the waste form or any waste-rock interactions.

Many of these items will be required to be confirmed by tests in exploratory boreholes. It is in these exploratory holes that the field studies discussed below should be performed.

4.2 Materials

Most of the available materials have been developed for the oil, gas and geothermal well industries. The properties and behaviour of materials may be specified in existing sealing specifications, either from the above industries or from the mining industry.

Although these existing materials may fall short of requirements for sealing boreholes in the context of radioactive waste disposal, they do provide a wide selection of available products, and have formed the basis of existing work on developing new materials [8].

In general, existing sealing technology has concentrated on the development of cements, which are relatively abundant and cheap materials, and relatively more information is available on this material than others in terms of longer term behaviour. The search for satisfactory seals for nuclear waste repositories has naturally turned to this direction, but other materials, such as bentonite clays possibly mixed with sands [1, 2, 9], are also being studied in detail. Compressed bentonite clays have been shown to be very impermeable and present indications are that they can act as buffer materials around certain types of waste repository. Of course, as is the case with all materials, the seal material should be stable in the expected host rock environment over the design period.

Many of the studies required to develop materials may be performed in the laboratory, or in large-scale experiments which model the in-situ environment.

4.3 Seal Geometry

Design approach studies have defined several basic geometric shapes for seal components [4]. A discussion of the application of these seal shapes is not within the scope of this paper. However, the practical feasibility of the shapes will require advances in downhole reaming and placement techniques.

4.4 Placement Techniques

The present-day placement techniques must be experimented to identify any difficulties not recognized at the planning stage. The time-dependent changes in the seal material are quickest immediately after mixing, and their behaviour during placement has to be confirmed. Furthermore, the procedures involved in placement to great depths have to be tested, irrespective of the seal material employed.

Test holes must be used for the field demonstration of the practicability of seal placement, and must determine the following:

- Confirm the condition of the borehole (verticality, size variations, stability and cleanliness).

- Ensure the seal materials are placeable.

- Verify the workmanship of the seal contractor.

- Test and install instrumentation to monitor seal and rock behaviour.

Pressure testing of the seal may not be useful as it may damage the seal and host rock in conditions that will not exist in practice.

This development of the existing downhole techniques must be envisaged at an early stage in the development of repository feasibility studies. Included in this development is the need to perfect methods of compacting possible seal materials, such as clays.

It must also be appreciated that seals may have to be replaced should waste repositories be located near to boreholes where placement tests have not given a satisfactory result. Therefore, allied to the development of placement techniques is the development of methods of seal removal and replacement. Methods of seal replacement to be studied and tested include:

- Overcoring
- Jetting
- Hole Re-Drilling

4.5 Instrumentation

Much work has been performed on developing materials, but relatively little work has been performed in Europe to develop deep hole instrumentation specific to the requirements of the geologic disposal of radioactive waste. Therefore, a brief summary of the characteristics which require measurement are discussed here.

The main objective of instrumentation is to define key parameters that will assist in determining the performance of the borehole seal in its environment. To accomplish this objective, the use of specially designed instrumentation will be necessary throughout each phase of the borehole sealing campaign.

The basic purpose of these instruments is to assess conditions within and around the borehole, and the instrumentation should function properly and reliably in the borehole environment. Environmental factors must be considered in the choice of existing instrumentation or in the development of new techniques, such as the following:

- Size and shape of borehole
- Temperature
- Natural radiation
- Stress level
- Fluids in the hole or along borehole face
- Type of rock
- Permeability

In addition, the instrumentation should not interfere with the integrity, longevity or performance of the borehole seal. Due to this requirement, two fundamental instrumentation features should be observed:

- Measurement techniques should be non-destructive to the seal material and host rock.

- Instrumentation to monitor seal performance within critical sealing areas should, if possible, be wireless, as this would eliminate possible deleterious results and any sealing problems associated with wired instrumentation.

Instrumentation requirements vary significantly depending on the state of development of borehole sealing techniques. Features of instrumentation are discussed for:

- Assessing the borehole environment before placement of the seal,

- evaluating plug quality during placement of the seal, and

- monitoring seal performance after placement of the seal.

Before seal placement, the characteristics of the host rock and borehole environment should be determined. Instrumentation should be able to monitor the following:

- Geometry of the borehole walls.
- Condition and aspect of the borehole wall surface.
- In-situ permeability of host rock and "disturbed" zone in the vicinity of the borehole.
- Degree and orientation of fracturing in host rock.
- Degree of fracturing within the "disturbed" zone.
- In-situ stress.
- Temperature.
- Natural radioactivity of host rock.

A detailed discussion of the instrumentation available to measure these aspects is not within the scope of this paper. However, whereas logging devices have been developed, principally for the petroleum industry [10, 11, 12, 13] devices to measure stress levels, for instance, probably require further development [14, 15].

When used to define the seal placement quality, requirements for instrumentation vary significantly, depending on the type of seal. In fact, depending on the material selected, instrumentation for this stage may not be necessary. However, some of the characteristics of the in-situ plug may be of interest, such as the following:

- Density of plug material
- Permeability (water and air)
- Temperature
- Strength
- Water content

After the plug has been placed, the integrity and performance of the seal may need to be monitored for long periods of time to detect any deterioration or changes in seal behaviour. Since plug deterioration could result in loss of mechanical properties and/or permeability, instrumentation to monitor these parameters is desirable.

In order to select appropriate instrumentation, key parameters must be defined. The definition of these parameters has been the subject of much discussion, and although no generally accepted list has been agreed, basic measurements defining borehole seal performance include [8]:

- Radionuclide concentration
- Water pressure
- Gas pressure
- Stress
- Permeability
- Fluid flow
- Temperature
- pH level

It should be noted that measurements of radionuclide concentration alone are not sufficient to enable predictions of in-situ seal performance to be made.

The elements listed above define key environmental conditions acting on the seal and provide a measure of how well the seal is functioning. Each element must be assessed and weighted to evaluate its contribution and necessity in borehole seal certification.

Finally, instrumentation to monitor plug performance should exhibit durability and a minimum of electronic component drift for the design life of the in-situ instrumentation.

If the integrity and longevity of borehole or shaft seal can be predicted with reasonable certainty, then long-term instrumentation of the borehole or shaft seal may not be required. However, instrumentation is required to monitor test seals in the experimentation stages, and to enable the effectiveness of materials and placement techniques in downhole or in-situ conditions to be evaluated. This is particularly true for borehole conditions, which always require remote placement techniques.

It is postulated that site and quality assurance procedures will not be finalised until successful seal placement has been demonstrated, and this includes monitoring and measurement of the seal during and after placement.

5.   DEVELOPMENT OF SEALING SPECIFICATIONS

The development of specifications for borehole or shaft plugs within a typical repository suitability study is described here with a view to putting into perspective the timetable required to develop the required technology.

European projects are typically concentrating either on tests in mined cavities - for example in the abandoned salt mine in Asse in West Germany or at Stripa in Sweden, or are at the stage of planning deep boreholes to characterise potential host rocks, for instance in the United Kingdom or Switzerland.

Some studies have already been performed to identify sealing materials and techniques appropriate for mined cavities [1], however less work has been performed to perfect borehole sealing techniques, and these will be emphasized here. The basic programme discussed here and outlined in Figure 1 will also apply to other studies.

5.1 Stage One Studies

These studies are essentially office studies of existing technologies, and definition of design aims and the requirements for further study or development. Ideally this stage should be started as early as possible in the overall repository suitability studies, so that the conclusions of this stage can be used to plan and integrate the further activities into the overall programme.

It is suggested that the initial steps of Stage One studies should concentrate on the one hand on the definition of seal design requirements, and on the other hand, on existing industry specifications for borehole plugging or mine shaft sealing.

The seal design requirements will include the definition of seal performance, such as the one suggested earlier in this paper. Each country will define an acceptable potential transport of nuclides through the seal zone depending on the overall potential transport through the repository host rock (the seal does not need to perform better than the host rock), and on the safety criteria adopted. Various scenarios will be considered, depending on whether disposal in deep boreholes or in mined cavities is envisaged, and on the potential repository locations. The type of waste to be stored will also influence this definition.

A review of existing specifications will assist in defining the routine methods used to seal penetrations.

Once these early studies are complete a thorough investigation of existing technology should be initiated. This study should include consideration of materials, placement technology and instrumentation, as discussed earlier. A cooperative effort between the countries of Europe would be advisable here, as many of the considerations will be common to each country. Specific technology applicable to proposed site conditions in individual countries can then be derived.

It is envisaged that the most appropriate way of defining the existing technology in each of the disciplines involved would be to have working parties comprised of specialists from the interested national bodies and representatives of industry and research organisations. It is envisaged that a final report would be prepared by a multi-disciplinary body to define existing technology. This final report should be a synthesis of the findings of the individual working parties and should discuss in concise terms existing technology as specifically applied to long-term penetration sealing.

This report could then form the basis for the development of further studies required to meet the seal design requirements defined previously.

In parallel to the work described above, working parties should study requirements of quality control or Quality Assurance related to sealing techniques. These early studies will start from existing specifications and codes of practice, and take account of the seal design requirements. Such studies are underway for the definition of overall repository Quality Assurance requirements and early consideration of the approach of the sealing specifications is essential. It is considered important that such requirements do not stop at written procedures, but emphasize the qualifications and experience of the persons involved. This direction must be given to the Quality Assurance requirements at the earliest stages.

It is envisaged that the Stage One studies could be completed within one year of establishing the appropriate working parties and coordinating groups.

5.2 Stage Two Studies

Stage Two studies will act on the recommendations made at the end of Stage One.

Laboratory studies will be aimed towards developing materials, and large scale model tests will be employed to model in-situ conditions. Instrumentation can also be experimented in the laboratory. However, the final test of laboratory findings will be full-scale field experiments. These will comprise general development studies, such as deep hole reaming and placement techniques, as well as instrumentation development. Site specific studies will concentrate on seal behaviour in proposed repository environments, including the field trials of proposed instrumentation technologies.

Since the field experiments would require a significant investment if carried out in isolation, they should be timed to coincide with the sinking of deep boreholes to characterise potential host rock formations or to take advantage of mined cavities or shafts. Hence the importance of defining early in the programme the field test requirements. Furthermore, it is likely that evidence from field tests will be better understood than results from laboratory tests by the non-specialised bodies associated with the making of decisions on the administrative aspects of waste disposal.

The result of the Stage Two studies, which will last several years, would be an evaluation of the various materials and techniques most appropriate to each country's site environment. Parallel and interacting development studies will be required to define further the Quality Assurance requirements, which will be close-ly allied to the licensing requirements of the country concerned. Again, the thrust of these requirements would be to ensure, and demonstrate, that all safety aspects of the sealing techniques would be adequately considered, both at the design and emplacement stages.

Although the studies in this stage will be both general and site, or country specific, a united approach to all the studies will be beneficial, since problems discovered or solved by one study group will often complement work performed by other groups. It is therefore suggested that the working groups and coordinating agency set up for the Stage One studies should continue their work in Stage Two.

5.3 Stage Three Studies

These studies need only be finalised during the commissioning of the repositories, since the Stage Two studies should have demonstrated the fact that the repository seals will function appropriately.

In this stage the materials and techniques which are most appropriate to the specific environment at each site can be finally chosen, and the detailed specifications for the seal emplacement elaborated. In parallel, the final Quality Assurance specifications can be prepared by the bodies responsible for the safety of the repository.

6. CONCLUDING REMARKS

This paper has attempted to review briefly the overall approach to penetration sealing which it is suggested should be taken by the countries of Europe. Some emphasis has been placed on performing large-scale tests in-situ, as it is postulated that in the end a major concern will be the workmanship, and assuring the workmanship in the field. To this end it is urged that the research bodies use to the best extent possible their proposed deep boreholes or underground mined cavities to perform these tests. This calls for early definition of the requirements of the various seals, and coordination of the effort by a multi-disciplinary international team to optimise each individual country's effort, and avoid unnecessary duplication.

The coordination should also include common agreement on licensing issues, particularly since proposed repositories may lie close to the borders of neighbouring countries.

REFERENCES:

[1] Bergstrom, A., A. Jacobsson, and R. Pusch: "Bentonite Based Buffer Substances for Isolating Radioactive Waste Products at Large Depths in Rock", International Symposium on the Underground Disposal of Radioactive Wastes, Paper No. IAEA-SM-243/22, Helsinki, Finland 1979.

[2] Griffin, J. R., H. Beale, W. R. Burton, and J. W. Davies: "Geological Disposal of High Level Radioactive Waste, Conceptual Repository Design in Hard Rock", International Symposium on the Underground Disposal of Radioactive Wastes, Paper No. IAEA-SM-243/93, Helsinki, Finland 1979.

[3] Manfroy, P., R. Heremans, M. Put, R. Vanhaelewijn, and M. Mayence: "Conception d'une Installation pour l'Enfouissement dans l'Argile de Déchets Radioactifs Conditionnés", International Symposium on the Underground Disposal of Radioactive Wastes, Paper No. IAEA-SM-243/3, Helsinki, Finland 1979.

[4] Office of Nuclear Waste Isolation, 1980 anticipated release: "Repository Sealing Design - 1979", Report No. ONWI 55, prepared by D'Appolonia Consulting Engineers, Inc., for Battelle Memorial Institute, Project Management Division, Columbus, Ohio, 1980 (in publication).

[5] Sousselier, Y., J. Pradel, and O. Cousin: "Le Stockage à Très Long Terme des Produits de Fission", Symposium on the Management of Radioactive Wastes from the Nuclear Cycle, Paper No. IAEA-SM-207/28, Vienna, Austria 1976.

[6] Pratt, H. R., D. E. Stephenson, G. Zandt, M. Bouchon, and W. A. Hustrulid: "Earthquake Damage to Underground Facilities", Conference on Rapid Excavation and Tunnelling, Atlanta, June 1979.

[7] Roy, D. M., M. W. Grutzeck, and P. H. Licastro: "Evaluation of Current Borehole Plug Longevity", ONWI Subcontract No. E512-00500, Office of Nuclear Waste Isolation, February 1979.

[8] Office of Nuclear Waste Isolation (ONWI): "The Status of Borehole Plugging and Shaft Sealing for Geologic Isolation of Radioactive Waste", Report No. ONWI-15, prepared by D'Appolonia Consulting Engineers, Inc., for Battelle Memorial Institute, Project Management Division, Columbus, Ohio, January 1979.

[9] Hatcher, S. R., S. A. Mayman, and M. Tomlinson: "Development of Deep Underground Disposal for Canadian Nuclear Fuel Wastes", International Symposium on the Underground Disposal of Radioactive Wastes, Paper No. IAEA-SM-243/167, Helsinki, Finland 1979.

[10] Schlumberger, Services Catalog, pp. 68, 1977.

[11] Dresser Atlas, Wireline Services Catalog, pp. 62, undated.

[12] Halliburton, Wellex Completion Logging Services, 1976.

[13] Harper, T. R. and D. U. Hinds: "The Impression Packer: A Tool for Recovery of Rock Mass Fracture Geometry", Rockstore 77, Stockholm, September 1977.

[14] Hoskins, E. R. and E. H. Oshier: "Development of a Deep Hole Stress Measurement Device", New Horizons in Rock Mechanics, Fourteenth Symposium on Rock Mechanics, June 1972.

[15] Russel, J. E.: "Data Reduction for a Deephole Device", Applications of Rock Mechanics, Fifteenth Symposium on Rock Mechanics, South Dakota School of Mines and Technology, September 1973.

Discussion

J. HAMSTRA, Netherlands

For the sake of those participants to the Workshop who are not familiar with the European situation it might be desirable to clarify on what authority you made your statements and to whom they are addressed.

Within the Commission of the European Communities (CEC) there is close cooperation on geologic disposal work performed in the different member countries. There will be a Symposium this month in Luxembourg at which status reports will be presented on the work performed during the last five years in the framework of the CEC contracts. Cooperative activities, specific to the three types of host rock studied, that is argillaceous sediments, cristalline rocks and rock salt, may also be expected in the field of borehole plugging and shaft sealing and borehole behaviour. I expect that the required research work in this specific area is partly already identified and underway within the scope of certain national programmes or will be discussed and formulated in due course to be incorporated in the overall geologic disposal efforts that will be accepted as CEC contract work for the next 5 year period.

J.R. COLE-BAKER, Belgium

The statements made, and the suggested development of penetration sealing procedures, are the opinions of the authors. The work which formed the basis of the paper is being performed as part of a study for the National Cooperative for the Storage of Radioactive Waste (NAGRA) of Switzerland, although the present statements are entirely the authors'opinions and do not necessarily reflect the ideas of the NAGRA staff. In view of the forthcoming plan of work for the next five years, to be funded by the CEC, the staged development procedure were suggested to stimulate discussion on the sealing issue.

BASALT WASTE ISOLATION PROJECT
BOREHOLE PLUGGING STUDIES - AN OVERVIEW

M. J. Smith
S. C. McCarel
Rockwell Hanford Operations
Richland, Washington (U.S.A.)

ABSTRACT

The objective of the Basalt Waste Isolation Project (BWIP) is to establish the feasibility of constructing a geologic repository for commercial high-level nuclear waste in the basalt flows within the Columbia Plateau and to provide the technology needed for design and construction of such a facility. The success of a repository located in basalt is dependent upon the ability to seal the facility at the end of the operational period. Therefore, a borehole plugging study is being conducted by the Engineered Barriers Group of the BWIP in order to develop, test, and demonstrate materials, equipment, and techniques for plugging the various man-made openings (boreholes, shafts, and tunnels) in and around the repository. These plugs are intended to severely restrict groundwater transport of toxic materials by both chemical and physical retardation. Verification of engineered plug system designs is accomplished through a proposed series of field tests and demonstrations.

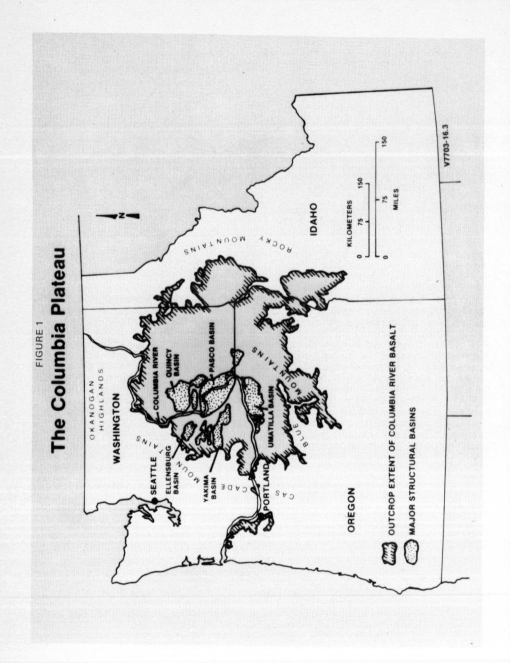

FIGURE 1

The Columbia Plateau

In February 1976, the U.S. Energy Research and Development Administration, precursor to the U.S. Department of Energy, established the National Waste Terminal Storage Program with the mission of providing storage facilities for commercial nuclear waste in various geologic formations within the United States. The objective of this program has been to investigate the properties of salt, granite, shale, and basalt in order to assess their suitability for terminal storage of high-level radioactive waste. As part of this program, the basalts underlying the Hanford Site in southeastern Washington were chosen as one of the candidate formations for evaluation. The Basalt Waste Isolation Project (BWIP) was initiated in 1976 with the objective of studying the feasibility of constructing a geologic repository for commercial high-level nuclear waste in the basalt flows within the Columbia Plateau and providing the technology needed for the design and construction of such a facility. The BWIP is administered by Rockwell International, Rockwell Hanford Operations, under contract to the U.S. Department of Energy.

The Columbia River Basalts are of particular interest as a viable medium for long-term nuclear waste disposal because of the degree of isolation afforded by this vast accumulation of continental basalt. As shown in Figure 1, the basalts underlying the Columbia Plateau occupy approximately 200,000 cubic kilometers of the Pacific Northwest and cover most of southeastern Washington as well as portions of the states of Idaho and Oregon. As such, they represent the second largest continental basalt province in the world. The Pasco Basin, a major structural feature near the center of the Columbia Plateau where the BWIP effort is concentrated, covers 2,000 square miles. The Columbia River basalt flows were formed between 9 and 15 million years ago, and are greater than 5,000 feet thick in the central part of the Pasco Basin where the Hanford Site is located.

The generalized stratigraphic column given in Figure 2 shows that the Grande Ronde Basalt is the most voluminous and deepest formation of the Columbia River Basalt Group. This formation which is more than 3,000 feet thick in the central part of the Pasco Basin, is comprised of from 30 to 40 basalt flows [1]. The known thickness of the Umtanum Flow, one flow of interest for location of a potential repository, ranges from 150 to 200 feet across the entire Basin and lies more than 3,000 feet below the surface in the central Pasco Basin.

One important feature that will determine the success of a repository located in basalt is the ability to seal the facility at the end of the operational period. The technology required to develop this capability must be explored in sufficient depth to ensure the safety and licensability of the repository as well as the isolation of waste components from man. In order to achieve this objective, a borehole plugging study is being conducted by the Engineered Barriers Group of the BWIP in order to develop, test, and demonstrate materials, equipment, and techniques for plugging the various man-made openings (boreholes, shafts, and tunnels) in and around the repository. These plugs are intended to:

- Severely restrict groundwater transport of toxic materials by both physical and chemical retardation;

- Isolate water bearing zones from the repository in order to maximize the travel time of migrating radionuclides;

- Provide structural strength to maintain the stability of the plug zone and prevent subsidence;

- Exhibit compatibility with the host geology and geologic environment;

- Exhibit longevity for the required period of isolation.

FIGURE 2

STRATIGRAPHY OF THE CENTRAL PASCO BASIN

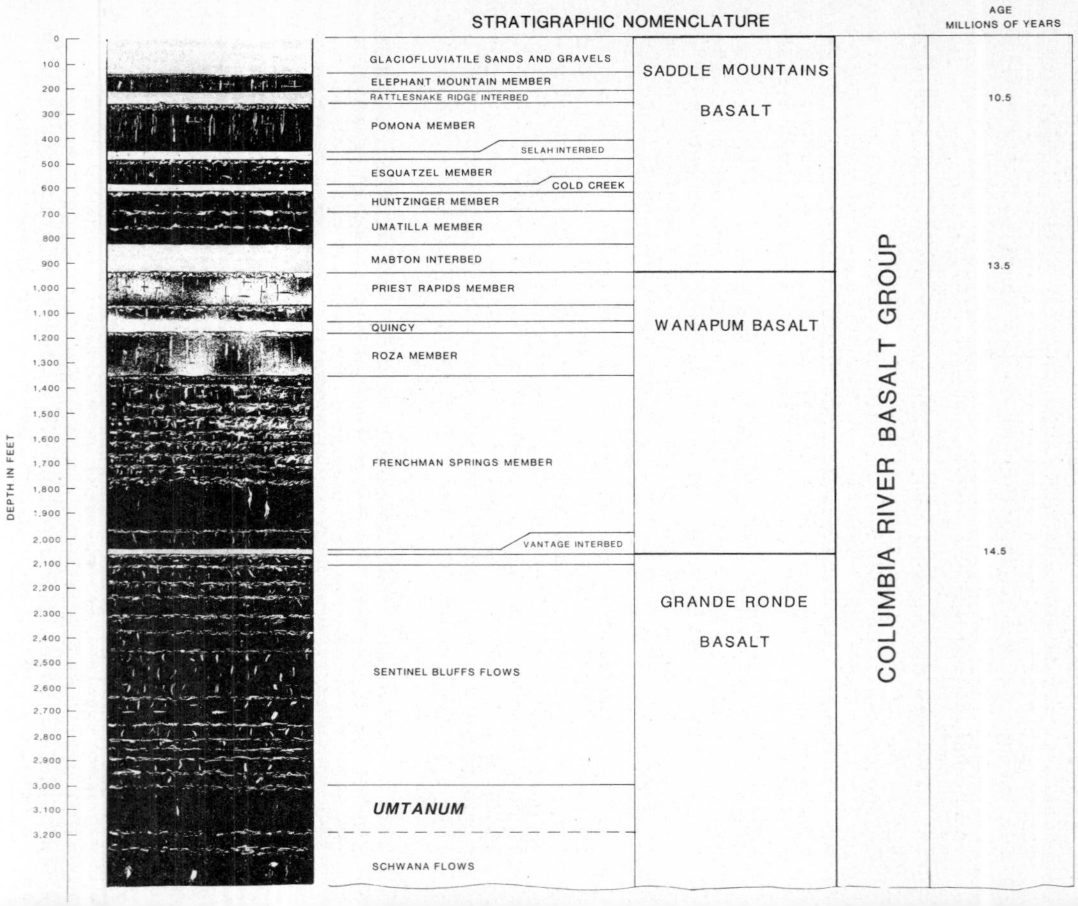

	STRATIGRAPHIC NOMENCLATURE	AGE MILLIONS OF YEARS

COLUMBIA RIVER BASALT GROUP

SADDLE MOUNTAINS BASALT — 10.5

GLACIOFLUVIATILE SANDS AND GRAVELS
ELEPHANT MOUNTAIN MEMBER
RATTLESNAKE RIDGE INTERBED
POMONA MEMBER
SELAH INTERBED
ESQUATZEL MEMBER
COLD CREEK
HUNTZINGER MEMBER
UMATILLA MEMBER
MABTON INTERBED

WANAPUM BASALT — 13.5

PRIEST RAPIDS MEMBER
QUINCY
ROZA MEMBER
FRENCHMAN SPRINGS MEMBER
VANTAGE INTERBED

GRANDE RONDE BASALT — 14.5

SENTINEL BLUFFS FLOWS
UMTANUM
SCHWANA FLOWS

DEPTH IN FEET

0, 100, 200, 300, 400, 500, 600, 700, 800, 900, 1,000, 1,100, 1,200, 1,300, 1,400, 1,500, 1,600, 1,700, 1,800, 1,900, 2,000, 2,100, 2,200, 2,300, 2,400, 2,500, 2,600, 2,700, 2,800, 2,900, 3,000, 3,100, 3,200

The U.S. Nuclear Regulatory Commission has recently proposed licensing procedures for disposal of high-level radioactive wastes at geologic repositories [2]. The proposed ruling sets forth requirements applicable to submitting a license application for such a facility. Several of these requirements relate to repository sealing. If it is assumed that an exploratory shaft is required for detailed in situ characterization of the potential repository site, assurance must be given in the Site Characterization Report that excavation of such a facility will not make the site unsealable. At the time that the License Application for construction of the repository is made, design considerations intended to facilitate decommissioning of the repository are to be identified. It is assumed that considerations made to facilitate sealing the repository at decommissioning will be a key part of this design. The Updated License Application for operation should resolve questions not answered during the construction authorization review process. It is assumed reasonable assurance that the repository can be sealed should be provided as part of this process. Finally, the License Application for closure should detail the results of backfilling and shaft sealing experiments.

The BWIP borehole plugging study will develop designs for plugging boreholes, shafts, and tunnels in or near a nuclear waste repository in basalt. The objectives of this study, which are compatible with Nuclear Regulatory Commission (NRC) requirements given above, are:

● To evaluate the natural environment where boreholes, shafts, or tunnels may penetrate the geologic units in or near a repository in basalt;

● To identify and test candidate plug materials which are compatible with existing and anticipated site-specific environmental conditions;

● To develop methods and equipment for plug emplacement; and

● To verify the design for the engineered plug system through a series of field tests and demonstrations.

In order to achieve these objectives, major emphasis in the BWIP borehole plugging program has been placed on the use of locally available natural materials that exhibit longevity in (ca., 10,000 years) and are compatible with the expected plug environment. The physical and chemical characteristics of this environment will greatly influence the choice of potential plug materials, which will, in turn, influence the selection of plug placement equipment and will ultimately determine plug system design. Thus, characterization of the plug environment and the choice of materials of known longevity in this expected environment are of critical importance to the success of the BWIP borehole plugging program.

The structural integrity of the plug designs developed will be provided by their ability to resist or accommodate deformation over time. While deformable, plastic materials will provide some physical continuity, engineering practice requires that more rigid plug components capable of transmitting the loads which may be imposed upon large plugs be employed. The BWIP recognizes that currently, the best technology for providing structural strength requires the use of man-made cements, and that the question of their longevity in a geologic environment will require extensive testing. On the other hand, natural self-cementing basalts being developed by the BWIP (discussed in the paper by F. N. Hodges, et al.) are easily demonstrated to possess chemical stability in a basalt environment, but the use of such materials will require a research and development effort to document their structural properties. As a result, both man-made and natural cements are being investigated in the BWIP borehole plugging program. Either one or both of these material types may be incorporated in final plug designs as a result of further material development and testing. A list of candidate borehole plugging materials is given in Table I.

TABLE I

PREFERRED CANDIDATE BOREHOLE PLUG MATERIALS*

Material	Desirable Attributes
Basalt	Relatively impermeable as a solid plug. Good strength for engineering purposes. Low (but some) ion exchange potential and moderate sorption capacity. Low cost, high availability.
Quartz (only as a solid plug for small boreholes)	Impermeable. Good strength. Chemically stable. Low cost.
Smectite clays	Very impermeable. High ion exchange potential and sorption capacity. Chemically stable. Low cost, readily available.
Clinoptilolite	High ion exchange potential and sorption capacity. Chemically stable. Low cost, readily available.
Steatite (only as a solid plug for small boreholes)	Low permeability. Good strength. Low cost.
Grouts Portland cements Hydrothermal cements	Low permeability. High adaptability. Low cost, readily available.
Concretes Portland cements and aggregate Hydrothermal cements and aggregate	Good strength. Low permeability. High adaptability. Low cost, readily available.

*Selected by extended dominance analysis.

Mixtures of the borehole plugging materials shown in Table I are being evaluated in a comprehensive laboratory physical and geochemical testing program. Mixtures selected for testing are based upon needed engineering properties in multiphase plugs and have application in plugging all sizes of man-made accesses (boreholes, shafts, and tunnels). Machinery selected to suit the environmental conditions, depths, diameters, and orientations of the accesses in a nuclear waste repository located in basalt are undergoing study. The most recently completed work has been the development of preconceptual designs for borehole tunnel and shaft-plugging systems. These designs have been aimed at fulfilling the following preliminary criteria:

• A design lifetime of 10,000 years;

- A maximum total seepage of 1 m^3/year through shaft and tunnel plugs after saturation;

- Reduction of waste leakage to below permissible levels assuming maximum credible amounts of radionuclide in solution in the repository;

- Sustaining a thermomechanical loading cycle from a 50°C temperature change;

- Producing a bond with tunnel or borehole walls to resist maximum credible axial forces.

In order to fulfill these criteria, plug designs must take into account the depth of disturbance of the wall rock created by borehole or tunnel excavation. This disturbance alters the permeability of the wall rock and affects the permeability of the plug-rock interface. Idealized models based on the conditions such plugs are expected to encounter were used to analyze the performance of monolithic plug elements for a variety of designs and conditions. Based on such calculations, combinations of plug elements (creating multiple-zone plug schemes) were devised to provide plugs which afforded superior performance over a wide range of repository environment conditions. A suite of "optimum" multiple-zone plug designs was chosen from those which were tested. Because many variables such as repository design, site location, exact hydrologic conditions, and so forth are yet to be determined, these designs are being treated as preconceptual for the present.

Preconceptual plug designs for sealing boreholes, tunnels, and shafts in a repository located in basalt are given in Figures 3, 4, and 5. The design for tunnels shows zones of concrete and mortared basalt blocks, interrupted at intervals along their length by seepage cut-off collars of clay/sand slurry. The cut-off collars extend across the entire plug cross section and into the wall rock extending through the disturbed rock zone and into the undisturbed rock zone. Basalt blocks, with mortared joints satisfy the requirements for structural support and long-term stability. The blocks would be hand masoned with cement mortar which is tentatively selected for use pending further evaluation of a ground basalt mortar. The design for shafts shows zones of concrete and of compacted clay/sand mixtures. Seepage cut-off collars used in the shaft are made of concrete. Preliminary calculations suggest a plug length of about 300 meters may be acceptable if the disturbed rock zone around the plugs is properly controlled during and after excavation. The scheme for plugging boreholes includes alternating zones of (1) gravel and unhydrated, compressed bentonite pellets in a hydrated bentonite originally introduced as a slurry and (2) zones of cement grout.

In the BWIP borehole plugging program, laboratory testing, materials selection, machinery development, and modeling of borehole plugs is coupled with a graduated series of field tests and demonstrations designed to test scale-up factors for materials and machines as well as elements of long-term performance in the in situ environment that cannot be simulated in the laboratory. The first of these tests, the Shallow Borehole Plugging Test, will involve the plugging of shallow boreholes under dry conditions. This test will yield engineering data on borehole plugging machines, materials, and techniques and on parameters such as plug/wall rock bonding and fracture filling around plugs. The Shallow Borehole Plugging Test is scheduled for completion during fiscal year 1983. A Second Plugging Field Test, which will probably take place within the Near Surface Test Facility (NSTF) at Hanford, will consist of two parts: (1) the plugging of small diameter horizontal and vertical boreholes under wet conditions, and (2) the plugging of a large diameter vertical shaft penetrating the NSTF from the surface. Data from this test will be used to define techniques and materials for fitting small diameter boreholes under adverse (wet) conditions and techniques to seal repository access shafts.

An Advanced Field Demonstration will be used to demonstrate the emplacement of prototype plugs and seals in the expected repository environment and for long-term observation of plug performance characteristics. This test has two distinct parts with parallel schedules. The first part is the deep remote plugging of a selected small diameter vertical core hole and is scheduled for completion in the 1980's. The second consists of plugging of tunnels and shafts in a subsurface test facility. The emplacement of plugs in the subsurface test area is scheduled to start in the early 1990's.

Based upon the results of the BWIP borehole plugging program to date, it appears feasible to design a composite plug system which employs natural and manufactured materials that will satisfactorily seal a nuclear waste repository in basalt.

ACKNOWLEDGEMENTS

The author wishes to thank Woodward-Clyde Consultants who under contract to Rockwell Hanford Operations provided the preconceptual systems and materials testing support necessary to complete the project reported in this paper.

REFERENCES

1. Basalt Waste Isolation, Geosciences Group, "Geologic Studies of the Columbia Plateau," RHO-BWI-ST-4, Rockwell Hanford Operations, Richland, Washington (1979).

2. U.S. Nuclear Regulatory Commission, "Disposal of High-Level Radioactive Wastes in Geologic Repositories: Proposed Licensing Procedures," Federal Register 44 (236), 70408-70420 (1979).

FIGURE 3

PRECONCEPTUAL DESIGN
FOR ACCESS TUNNEL PLUGS

(TYPICAL SECTION)

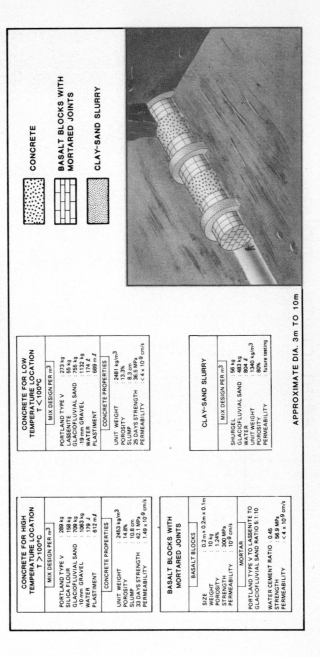

CONCRETE

BASALT BLOCKS WITH MORTARED JOINTS

CLAY-SAND SLURRY

CONCRETE FOR LOW TEMPERATURE LOCATION T < 100°C

MIX DESIGN PER m³

PORTLAND TYPE V	: 273 kg
LASSENITE	: 55 kg
GLACIOFLUVIAL SAND	: 755 kg
-19 mm GRAVEL	: 1132 kg
WATER	: 174 ℓ
PLASTIMENT	: 689 mℓ

CONCRETE PROPERTIES

UNIT WEIGHT	: 2481 kg/m³
POROSITY	: 13.3%
SLUMP	: 8.3 cm
25 DAYS STRENGTH	: 36.5 MPa
PERMEABILITY	: < 4 x 10⁻⁹ cm/s

CLAY-SAND SLURRY

MIX DESIGN PER m³

SHURGEL	: 56 kg
GLACIOFLUVIAL SAND	: 483 kg
WATER	: 804 ℓ
UNIT WEIGHT	: 1340 kg/m³
POROSITY	: 80%
PERMEABILITY	: future testing

CONCRETE FOR HIGH TEMPERATURE LOCATION T > 100°C

MIX DESIGN PER m³

PORTLAND TYPE V	: 289 kg
SILICA FLOUR	: 158 kg
GLACIOFLUVIAL SAND	: 709 kg
-10 mm GRAVEL	: 1063 kg
WATER	: 179 ℓ
PLASTIMENT	: 612 mℓ

CONCRETE PROPERTIES

UNIT WEIGHT	: 2453 kg/m³
POROSITY	: 14.8%
SLUMP	: 10.8 cm
33 DAYS STRENGTH	: 42.1 MPa
PERMEABILITY	: 1.49 x 10⁻⁹ cm/s

BASALT BLOCKS WITH MORTARED JOINTS

BASALT BLOCKS

SIZE	: 0.2m x 0.2m x 0.1m
WEIGHT	: 10 kg
POROSITY	: 1.24%
STRENGTH	: 300 MPa
PERMEABILITY	: 10⁻⁹ cm/s

MORTAR

PORTLAND TYPE V TO LASSENITE TO GLACIOFLUVIAL SAND RATIO 5:1:10

WATER CEMENT RATIO	: 0.45
STRENGTH	: 56.9 MPa
PERMEABILITY	: < 4 x 10⁻⁹ cm/s

APPROXIMATE DIA. 3m TO 10m

FIGURE 4

PRECONCEPTUAL DESIGN
FOR SHAFT PLUGS
(TYPICAL SECTION)

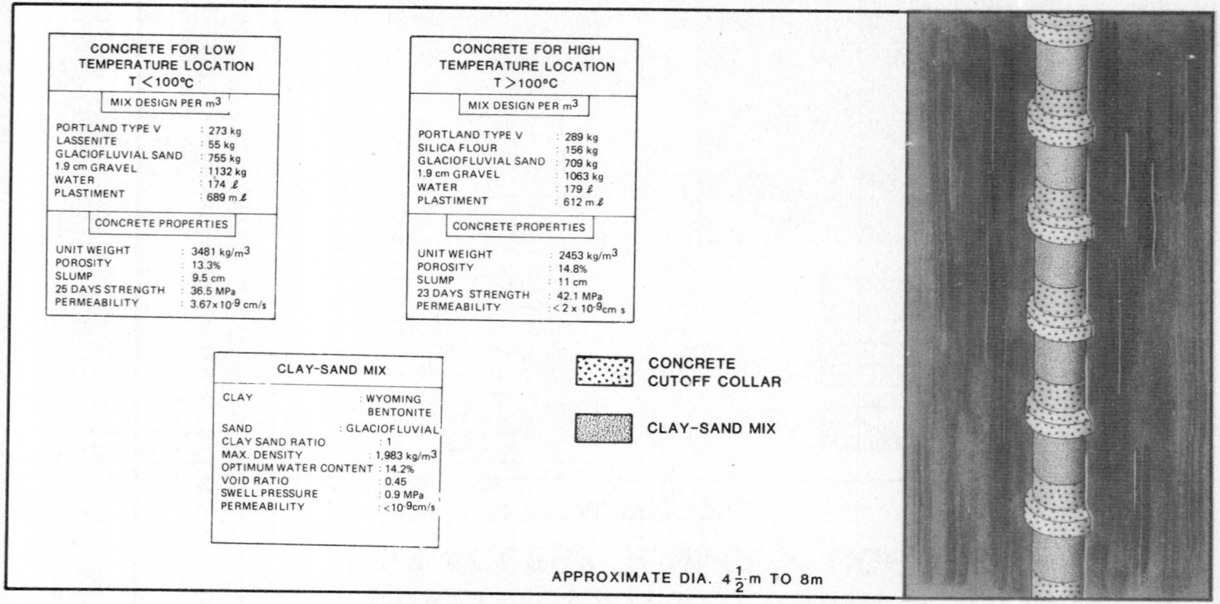

CONCRETE FOR LOW TEMPERATURE LOCATION T < 100°C

MIX DESIGN PER m³

PORTLAND TYPE V	: 348 kg
LASSENITE	: 55 kg
GLACIOFLUVIAL SAND	: 756 kg
1.9 cm GRAVEL	: 1132 kg
WATER	: 174 ℓ
PLASTIMENT	: 689 mℓ

CONCRETE PROPERTIES

UNIT WEIGHT	: 3481 kg/m³
POROSITY	: 13.3%
SLUMP	: 9.5 cm
25 DAYS STRENGTH	: 36.5 MPa
PERMEABILITY	: 3.67×10^{-9} cm/s

CONCRETE FOR HIGH TEMPERATURE LOCATION T > 100°C

MIX DESIGN PER m³

PORTLAND TYPE V	: 289 kg
SILICA FLOUR	: 156 kg
GLACIOFLUVIAL SAND	: 709 kg
1.9 cm GRAVEL	: 1063 kg
WATER	: 179 ℓ
PLASTIMENT	: 612 mℓ

CONCRETE PROPERTIES

UNIT WEIGHT	: 2453 kg/m³
POROSITY	: 14.8%
SLUMP	: 11 cm
23 DAYS STRENGTH	: 42.1 MPa
PERMEABILITY	: $< 2 \times 10^{-9}$ cm s

CLAY-SAND MIX

CLAY	: WYOMING BENTONITE
SAND	: GLACIOFLUVIAL
CLAY SAND RATIO	: 1
MAX. DENSITY	: 1.983 kg/m³
OPTIMUM WATER CONTENT	: 14.2%
VOID RATIO	: 0.45
SWELL PRESSURE	: 0.9 MPa
PERMEABILITY	: $< 10^{-9}$ cm s

CONCRETE CUTOFF COLLAR

CLAY-SAND MIX

APPROXIMATE DIA. $4\frac{1}{2}$ m TO 8m

- 140 -

FIGURE 5

PRECONCEPTUAL DESIGN FOR REMOTE BOREHOLE PLUGS

(TYPICAL SECTION)

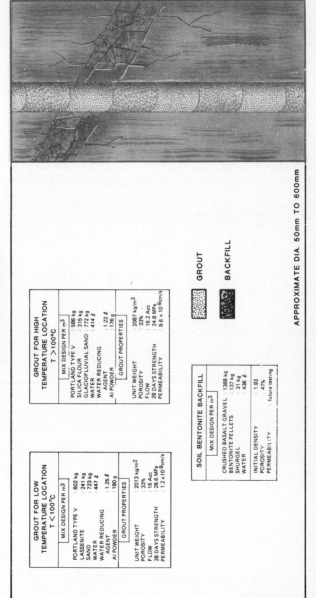

GROUT FOR LOW TEMPERATURE LOCATION T <100°C	
MIX DESIGN PER m³	
PORTLAND TYPE V	: 602 kg
LASSENITE	: 241 kg
SAND	: 723 kg
WATER	: 447 ℓ
WATER REDUCING AGENT	: 1.25 ℓ
Al POWDER	: 180 g
GROUT PROPERTIES	
UNIT WEIGHT	: 2013 kg/m³
POROSITY	: 33%
FLOW	: 16 Acc
28 DAYS STRENGTH	: 28.6 MPa
PERMEABILITY	: 1.2×10^{-9} cm/s

GROUT FOR HIGH TEMPERATURE LOCATION T >100°C	
MIX DESIGN PER m³	
PORTLAND TYPE V	: 586 kg
SILICA FLOUR	: 315 kg
GLACIOFLUVIAL SAND	: 772 kg
WATER	: 414 ℓ
WATER REDUCING AGENT	: 1.22 ℓ
Al POWDER	: 176 g
GROUT PROPERTIES	
UNIT WEIGHT	: 2087 kg/m³
POROSITY	: 33%
FLOW	: 16.2 Acc
28 DAYS STRENGTH	: 24.8 MPa
PERMEABILITY	: 9.6×10^{-9} cm/s

SOIL BENTONITE BACKFILL	
MIX DESIGN PER m³	
CRUSHED BASALT GRAVEL	: 1368 kg
BENTONITE PELLETS	: 137 kg
SHURGEL	: 31 kg
WATER	: 436 ℓ
INITIAL DENSITY	: 1.93
POROSITY	: 47%
PERMEABILITY	: future testing

GROUT

BACKFILL

APPROXIMATE DIA. 50mm TO 600mm

SEALING A NUCLEAR WASTE REPOSITORY
IN COLUMBIA RIVER BASALT: PRELIMINARY RESULTS

F. N. Hodges
Rockwell Hanford Operations
Richland, Washington (U.S.A.)

J. E. O'Rourke
G. J. Anttonen
Woodward-Clyde Consultants
San Francisco, California (U.S.A.)

ABSTRACT

The long containment time required of repositories for
nuclear waste (10^4 to 10^6 years?) requires that materials used for
repository seals be stable in the geologic environment of the repository and of proven longevity. A list of candidate materials for
sealing a repository in Columbia River basalts has been prepared and
refined through laboratory testing. The most feasible techniques
for emplacing preferred plug materials have been identified and the
resultant plugs have been evaluated on the basis of design functions.
Preconceptual designs for tunnel, shaft, and borehole seals consist
of multiple zone plugs with each zone fulfilling one or more design
functions. Zones of disturbed rock around tunnels and shafts,
resulting from excavation and subsequent stress release, are zones
of higher permeability and of possible fluid migration. In preliminary designs the disturbed zones are blocked by cut-off collars
filled with low permeability materials.

FIGURE 1

The Columbia Plateau

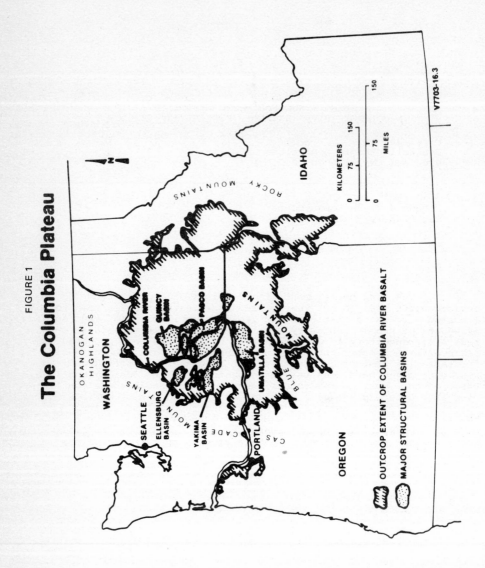

INTRODUCTION

The success of deep geologic disposal for high level nuclear waste is strongly dependent upon our ability to seal the many man-made openings that will exist in and around a repository. The most favorable geology will be nullified if the excavations necessary for emplacement of the waste provide rapid access to the biosphere. The plug systems for repository closure must prevent the transport of unacceptable quantities of radionuclides to the biosphere over the containment life of the repository. The containment times required, on the order of 10^4 to 10^6 years, are unique in engineering design experience and require materials of proven long term stability and a high degree of compatibility with the chemical and physical environment around the plugs. This latter requirement implies that material selection and plug design must be site specific.

The objective of the Basalt Waste Isolation Project (BWIP), operated for the U.S. Department of Energy by Rockwell Hanford Operations (Rockwell), is to assess the feasibility of placing a repository for the terminal storage of high level nuclear waste at depth in the basalts of the Columbia Plateau. As part of this effort, the BWIP borehole plugging program is assessing the feasibility of sealing a repository in Columbia River basalt after emplacement of waste and some undetermined period of retrievability. A preconceptual design study of plugging systems has been initiated (Woodward-Clyde Consultants, San Francisco, California) along with planning for a graduated series of site specific borehole plugging tests in the Columbia River basalts [1]. In a program of this type, where feasibility must be proven before repository operations can begin, emphasis is, of necessity, on existing and proven technology. However, according to current schedules repository closure will not take place until well into the twenty-first century. Our present schemes may be considered primitive at that time. Therefore, we should not, in our search for immediate solutions, fail to pursue less well known alternatives, that in the long run may prove far superior to our present schemes.

PLUG ENVIRONMENT

The very long containment times required of repository seals (plugs) requires a high degree of compatibility between plug materials and the physical and chemical environment in which the plug must exist. Thus, the physical and chemical environment will exert a strong influence upon the choice of plug materials, which will, in turn, influence the selection of emplacement machinery and plug design.

The Columbia River basalts, which form the Columbia Plateau of Washington, Oregon, and Idaho (Figure 1), are a thick sequence of thoeleiitic flood basalts erupted principally from fissures along the eastern margin of the province [2]. The basalts are relatively young, ranging in age from 6 to 16.5 million years [3,4] and in the upper part of the sequence basalt flows are interbedded with clastic sediments, particularly near the margins of the plateau. The Columbia River Basalt Group covers an area of approximately 200,000 km^2 and reaches a thickness of 1,500 m in the central part of the plateau (Pasco Basin). Individual basalt flows may reach thicknesses greater than 90 m (generally 22 to 36 m) and may represent volumes as great as 600 to 700 km^3 (generally 10 to 30 km^3). A generalized stratigraphic column for the central Columbia Plateau is presented in Figure 2. For a comprehensive treatment of the geology of the Columbia Plateau the interested reader is referred to Myers and others [5].

Several thick basalt flows within the Grande Ronde Basalt Formation of the Columbia River Basalt Group are considered prime candidates for repository horizons. In the central portion of the Pasco Basin the candidate flows occur between depths of 800 and 1,400 m.

FIGURE 2

PASCO BASIN STRATIGRAPHIC NOMENCLATURE

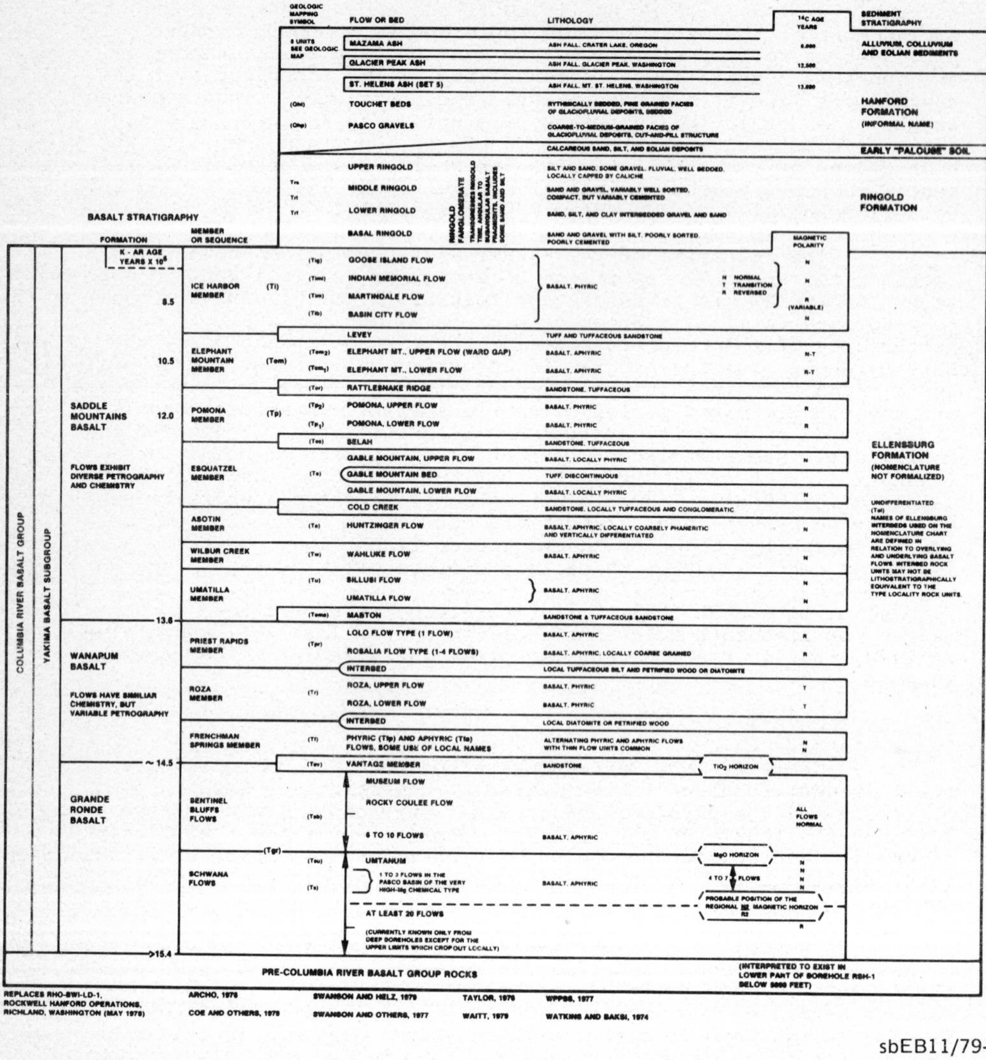

GEOLOGIC MAPPING SYMBOL	FLOW OR BED	LITHOLOGY	14C AGE YEARS	SEDIMENT STRATIGRAPHY
5 UNITS SEE GEOLOGIC MAP	MAZAMA ASH	ASH FALL, CRATER LAKE, OREGON	6,000	ALLUVIUM, COLLUVIUM AND EOLIAN SEDIMENTS
	GLACIER PEAK ASH	ASH FALL, GLACIER PEAK, WASHINGTON	12,000	
	ST. HELENS ASH (SET 5)	ASH FALL, MT. ST. HELENS, WASHINGTON	13,000	
(Qht)	TOUCHET BEDS	RYTHMICALLY BEDDED, FINE GRAINED FACIES OF GLACIOFLUVIAL DEPOSITS, BEDDED		HANFORD FORMATION (INFORMAL NAME)
(Qhp)	PASCO GRAVELS	COARSE-TO-MEDIUM-GRAINED FACIES OF GLACIOFLUVIAL DEPOSITS, CUT-AND-FILL STRUCTURE		
		CALCAREOUS SAND, SILT, AND EOLIAN DEPOSITS		EARLY "PALOUSE" SOIL
	UPPER RINGOLD	SILT AND SAND, SOME GRAVEL, FLUVIAL, WELL BEDDED, LOCALLY CAPPED BY CALICHE		
Tru	MIDDLE RINGOLD	SAND AND GRAVEL, VARIABLY WELL SORTED, COMPACT, BUT VARIABLY CEMENTED		RINGOLD FORMATION
Trl	LOWER RINGOLD	SAND, SILT, AND CLAY INTERBEDDED GRAVEL AND SAND		
	BASAL RINGOLD	SAND AND GRAVEL WITH SILT, POORLY SORTED, POORLY CEMENTED	MAGNETIC POLARITY	

BASALT STRATIGRAPHY

FORMATION	MEMBER OR SEQUENCE		FLOW OR BED	LITHOLOGY		MAGNETIC POLARITY	
	K - AR AGE YEARS X 10^6					N	
	ICE HARBOR MEMBER	(Tig)	GOOSE ISLAND FLOW			N	
		(Ti) (Tim)	INDIAN MEMORIAL FLOW	BASALT, PHYRIC	N NORMAL / T TRANSITION / R REVERSED		
	8.5	(Tim)	MARTINDALE FLOW			R (VARIABLE)	
		(Tib)	BASIN CITY FLOW			N	
			LEVEY	TUFF AND TUFFACEOUS SANDSTONE			
	ELEPHANT MOUNTAIN MEMBER	(Temg)	ELEPHANT MT., UPPER FLOW (WARD GAP)	BASALT, APHYRIC		N-T	
SADDLE MOUNTAINS BASALT	10.5	(Tem1)	ELEPHANT MT., LOWER FLOW	BASALT, APHYRIC		R-T	
		(Tor)	RATTLESNAKE RIDGE	SANDSTONE, TUFFACEOUS			
	POMONA MEMBER	(Tpu)	POMONA, UPPER FLOW	BASALT, PHYRIC		R	
FLOWS EXHIBIT DIVERSE PETROGRAPHY AND CHEMISTRY	12.0	(Tpl)	POMONA, LOWER FLOW	BASALT, PHYRIC		R	
		(Tos)	SELAH	SANDSTONE, TUFFACEOUS			
	ESQUATZEL MEMBER		GABLE MOUNTAIN, UPPER FLOW	BASALT, LOCALLY PHYRIC			
		(Tx)	GABLE MOUNTAIN BED	TUFF, DISCONTINUOUS			
			GABLE MOUNTAIN, LOWER FLOW	BASALT, LOCALLY PHYRIC		N	
	ASOTIN MEMBER		COLD CREEK	SANDSTONE, LOCALLY TUFFACEOUS AND CONGLOMERATIC			
		(Ta)	HUNTZINGER FLOW	BASALT, APHYRIC, LOCALLY COARSELY PHANERITIC AND VERTICALLY DIFFERENTIATED		N	
	WILBUR CREEK MEMBER	(Tw)	WAHLUKE FLOW	BASALT, APHYRIC		N	
	UMATILLA MEMBER	(Tu)	SILLUSI FLOW	BASALT, APHYRIC		N	
			UMATILLA FLOW			N	
		(Tems)	MABTON	SANDSTONE & TUFFACEOUS SANDSTONE			
WANAPUM BASALT	PRIEST RAPIDS MEMBER	(Tpr)	LOLO FLOW TYPE (1 FLOW)	BASALT, APHYRIC		N	
			ROSALIA FLOW TYPE (1-4 FLOWS)	BASALT, APHYRIC, LOCALLY COARSE GRAINED		N	
			INTERBED	LOCAL TUFFACEOUS SILT AND PETRIFIED WOOD OR DIATOMITE			
	ROZA MEMBER	(Tr)	ROZA, UPPER FLOW	BASALT, PHYRIC		T	
FLOWS HAVE SIMILIAR CHEMISTRY, BUT VARIABLE PETROGRAPHY			ROZA, LOWER FLOW	BASALT, PHYRIC		T	
			INTERBED	LOCAL DIATOMITE OR PETRIFIED WOOD			
	FRENCHMAN SPRINGS MEMBER	(Tf)	PHYRIC (Tfp) AND APHYRIC (Tfa) FLOWS, SOME USE OF LOCAL NAMES	ALTERNATING PHYRIC AND APHYRIC FLOWS WITH THIN FLOW UNITS COMMON		N N	
	~14.5	(Tsv)	VANTAGE MEMBER	SANDSTONE	TiO2 HORIZON		
GRANDE RONDE BASALT	SENTINEL BLUFFS FLOWS	(Tsb)	MUSEUM FLOW / ROCKY COULEE FLOW / 6 TO 10 FLOWS	BASALT, APHYRIC		ALL FLOWS NORMAL	
	(Tgr)		UMTANUM		MgO HORIZON	N	
	SCHWANA FLOWS	(Tsu) (Ts)	1 TO 2 FLOWS IN THE PASCO BASIN OF THE VERY HIGH-Mg CHEMICAL TYPE	BASALT, APHYRIC	4 TO 7 FLOWS	N N N	
			AT LEAST 20 FLOWS		PROBABLE POSITION OF THE REGIONAL R2 MAGNETIC HORIZON	N	
	>15.4		(CURRENTLY KNOWN ONLY FROM DEEP BOREHOLES EXCEPT FOR THE UPPER LIMITS WHICH CROP OUT LOCALLY)			R	

COLUMBIA RIVER BASALT GROUP — YAKIMA BASALT SUBGROUP

ELLENSBURG FORMATION (NOMENCLATURE NOT FORMALIZED)

UNDIFFERENTIATED (Tel) NAMES OF ELLENSBURG INTERBEDS USED ON THE NOMENCLATURE CHART ARE DEFINED IN RELATION TO OVERLYING AND UNDERLYING BASALT FLOWS. INTERBED ROCK UNITS MAY NOT BE LITHOSTRATIGRAPHICALLY EQUIVALENT TO THE TYPE LOCALITY ROCK UNITS.

RINGOLD FANGLOMERATE TRANSGRESSES RINGOLD TIME, SIMILAR TO BASAL RINGOLD PEDIMENTS, INCLUDES SOME SAND AND SILT.

PRE-COLUMBIA RIVER BASALT GROUP ROCKS

(INTERPRETED TO EXIST IN LOWER PART OF BOREHOLE RSH-1 BELOW 9000 FEET)

ARCHO, 1976   COE AND OTHERS, 1978   SWANSON AND HELZ, 1979   SWANSON AND OTHERS, 1977   TAYLOR, 1976   WAITT, 1979   WPPSS, 1977   WATKINS AND BAKSI, 1974

sbEB11/79-9

Vertical pressures at candidate repository levels should be essentially lithostatic and, assuming a density for basalt of 3.0 g/cm^3, should fall in the range of 24 to 42 MPa (240 to 420 bars). In situ horizontal stresses have not been measured at depth beneath the Columbia Plateau; however, the compilation of in situ stress measurements by Cook [6] indicates that the horizontal stresses at repository depths should fall in the range 0.5 to 1.5 times the vertical stress, or 12 to 65 MPa (120 to 650 bars). Temperature measurements in deep test wells indicate ambient host rock temperatures in the range 45° to 75°C. Modeling of a repository in basalt indicates that temperatures in the vicinity of canisters may be in excess of 200°C [7]; however, repository seals (exclusive of backfill) should be well removed from storage areas and maximum temperatures are expected to be in the vicinity of 100°C.

Groundwaters in the Grande Ronde basalts are essentially sodium-chloride-silicate-sulfate waters with very low dissolved solids content (Table I). Direct measurement of groundwater pH indicates values in the range 9.4 to 9.9. These high values for pH are consistent with groundwater isolated from the atmosphere and with dominant pH control resulting from the buffering action of silicic acid and carbonate

$$H_4SiO_4 \rightleftharpoons H^+ + H_3SiO_4^-, \quad Ka = 10^{-9.37} \text{ at } 65°C.$$

$$H_2CO_3 \rightleftharpoons H^+ + HCO_3^-, \quad Ka = 10^{-9.96} \text{ at } 65°C.$$

Oxidation potential (Eh) is an important parameter for modeling of mineral stabilities in aqueous environments; however, direct measurements of Eh are notoriously unreliable [8] and have not been made for groundwaters within the Columbia River basalts. Pyrite (FeS$_2$) exists as a secondary mineral within fractures and vesicles in the basalt [9] and allows estimation of the ambient Eh. Extrapolation of 25°C stability data for pyrite [10] indicates Eh values in the range -0.3 to -0.5 volts. These values for Eh are consistent with the low oxygen fugacities normally associated with basalts at higher temperatures [11] and indicates that Eh is controlled by the buffering capacity of the ferromagnesian minerals in the basalt.

TABLE I

AVERAGE CHEMICAL COMPOSITION OF
GRANDE RONDE GROUNDWATER AT 25°C

Ion	mg/l
Na^+	250
K^+	1.9
Ca^{2+}	1.3
Mg^{2+}	0.04
CO_3^{2-}	30
HCO_3^-	36
OH^-	3
$H_3SiO_4^-$	137
Cl^-	148
SO_4^{2-}	108
F^-	37

After Gephart and others [16].

Four different plug settings or plug subenvironments have been recognized for design purposes. These are (1) boreholes origi-

PRECONCEPTUAL BOREHOLE PLUGGING SITE ARRANGEMENT
(TYPICAL SECTION)

FIGURE 3

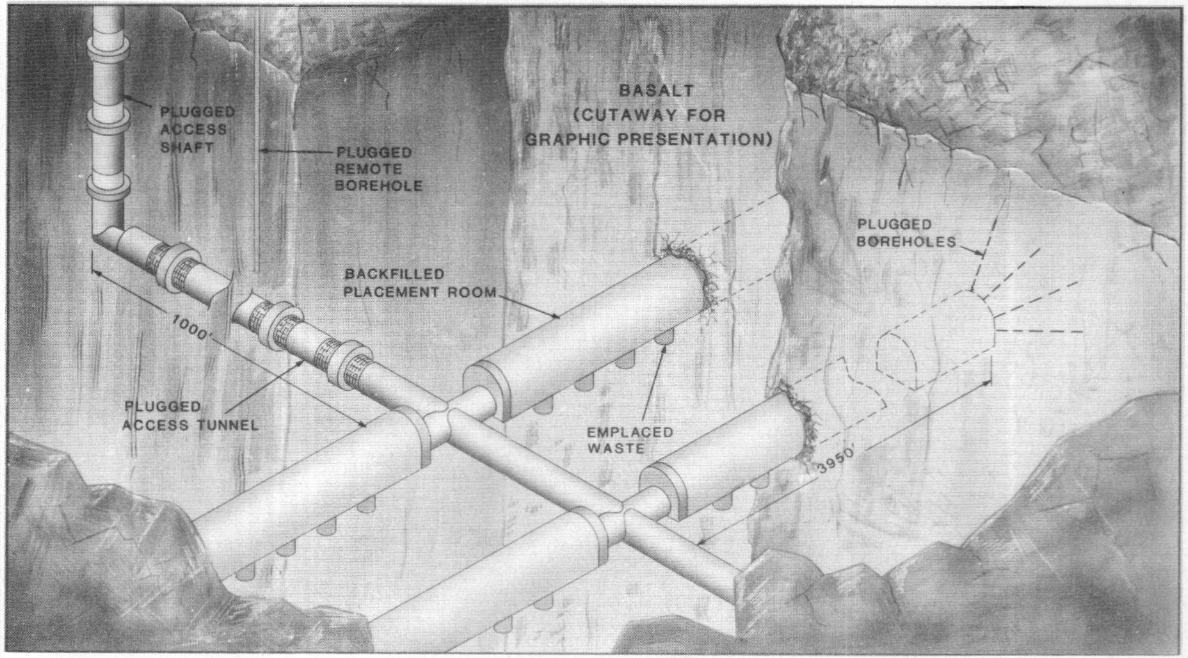

nating at the surface; (2) shafts; (3) boreholes originating under-ground; and (4) tunnels (Figure 3). Plugging of waste emplacement holes is not considered here. Boreholes originating at the surface are differentiated from shafts on the basis of human access. Shafts are defined as vertical holes large enough for a man to get down and work in. The minimum diameter for shafts has been arbitrarily set at one meter for design purposes; however, they are generally expected to be greater than four meters in diameter. Vertical holes less than one meter in diameter must be plugged through remote operations. Boreholes originating underground will generally be exploratory holes drilled ahead of excavation. Exploratory holes will normally be con-sumed by excavation; however, if substandard rock is encountered these holes must be sealed to protect the integrity of the repository. Boreholes originating underground are differentiated from tunnels on the basis of human access and the limit is arbitrarily set at one meter. Tunnels are expected to range between 3 and 10 meters in diameter.

MATERIAL SELECTION

The materials selected for use in plugs will strongly in-fluence the selection of plugging machinery and overall plug design. Thus, it is important to select a list of candidate plug materials early in the study.

An initial list of possible plugging materials was assem-bled (Table II) for evaluation. The list was not meant to be exhaus-tive, but rather to be a sampling of a wide range of materials. The initial list was then subjected to a dominance analysis [12,13] to select a list of candidate plug materials for preconceptual design and testing. During the decision analysis process the materials in the initial list were passed through a series of screening tests (Figure 4) to evaluate chemical stability in geologic environment within the Columbia River basalts, documented survival history and ability to fulfill one or more of three design functions; to inhibit fluid flow, to inhibit radionuclide migration, and to provide struc-tural integrity for the plug. Most man-made materials were dropped out at the first screen (step 2) because of lack of documented sur-vival history. Several ceramics and glass passed this screen because they have natural analogues with documented survival histories. Con-cretes were passed because of their relatively long documented sur-vival history (∿2,000 years); however, their documented history is considerably shorter than the required containment time and their stability under expected repository conditions has not been docu-mented. Thus, the acceptance of concretes as plug material is con-sidered tentative and, either the long term stability of concretes under repository conditions must be demonstrated or materials of proven longevity must be found to fill the role normally filled by concretes. The list of candidate plugging materials, those that passed all screens, is present in Table III, along with the major attributes of each material. It should be stressed that this is an initial list and will be modified as new data becomes available. In addition, Table III does not include additives. It is felt that se-lection of minor components, used to complement the major plug com-ponents, should be done only after specification and design of the major plug components.

LABORATORY TESTING

Both geochemical (hydrothermal) and mechanical testing of the candidate plug materials are in progress. Both testing programs have used the same materials and as far as possible have used materi-als readily available at the Hanford Site or in the northwestern por-tion of the United States. Materials used in the testing program to date include Columbia River basalt, aggregate and clay from the Han-ford Site, zeolite (clinoptilolite) from Oregon, and bentonites (smectite clays) from Oregon and Wyoming.

TABLE II

MATERIALS FOR CONSIDERATION

Igneous and Metamorphic Rocks
  Basalt
  Granite
  Natural glass
  Slate
  Serpentinite

Sedimentary Materials
  Shale
  Diatomaceous earth

Minerals
  Quartz
  Smectite clays
  Vermiculite
  Kaolinite
  Illite
  Talc
  Clinoptilolite
  Phillipsite
  Magnetite
  Hematite
  Barite
  Gypsum

Native Elements
  Gold
  Silver
  Platinum
  Carbon
  Copper

Natural Organic Materials
  Tar
  Amber

Processed Metals
  Lead
  Metal alloys

Ceramics
  Mullite
  Steatite

Manufactured Glasses

Quasi-Natural Materials
  Fly ash

Synthetic Resins
  Epoxy
  Polyacrylamide
  Phenoplast
  Aminoplast
  Polyester
  Polyvinyl
  Furnan compounds

Silicones

Polymers

Grouts
  Portland cements
    Type I
    Type II
    Type V
    Pozzolan
  Hydrothermal cements
  (calcium aluminate types)

Concretes
  Portland cements plug
    aggregate
  Hydrothermal cements
    plug aggregate

Asphalts

TABLE III

PREFERRED CANDIDATE PLUG MATERIALS

Material	Desirable Attributes
Basalt	Relatively impermeable as a solid plug. Good strength for engineering purposes. Low (but some) ion exchange potential and moderate sorption capacity. Low cost, high availability.
Quartz (only as a solid plug for small boreholes)	Impermeable. Good strength. Chemically stable. Low cost.
Smectite clays	Very impermeable. High ion exchange potential and sorption capacity. Chemically stable. Low cost, readily available.
Clinoptilolite	High ion exchange potential and sorption capacity. Chemically stable. Low cost, readily available.
Steatite Ceramic (only as a solid plug for small boreholes)	Low permeability. Good strength. Low cost.
Grouts portland cements hydrothermal cements	Low permeability. High permeability. Low cost, readily available.
Concretes portland cements and aggregate hydrothermal cements aggregate	Good strength. Low permeability. High adaptability. Low cost, readily available.

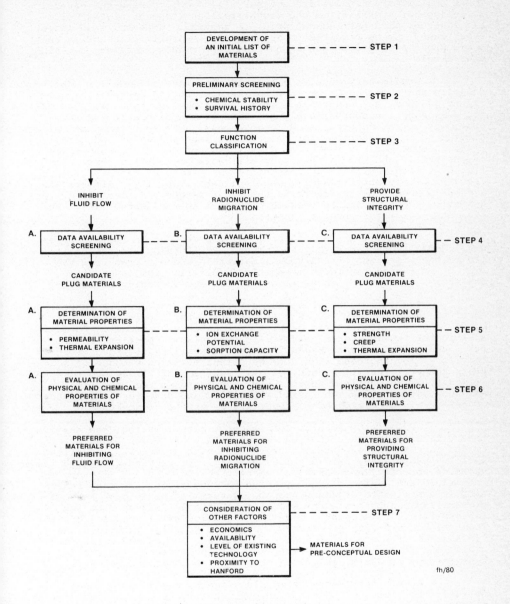

FIGURE 4

MATERIAL SELECTION PROCESS

Preliminary results from the geochemical testing (carried out by SRI International, Menlo Park, California), at temperatures up to 250°C and pressures up to 34.5 MPa, indicate that the candidate plug materials are compatible with each other and with the geochemical environment expected near a repository in Columbia River basalts. These preliminary results are not surprising considering the careful screening for stability in the geochemical environment of the Columbia River basalts carried out prior to testing. However, much longer experiments may be required to detect very low rates of reaction that would prove highly detrimental to plug performance over long containment times.

Preliminary runs in a stirred autoclave (250°C, 3.9 MPa) indicate that crushed Columbia River basalt, in the presence of simulated groundwater, undergoes a self-cementing reaction. In runs of two week duration approximately one half of the basalt in the autoclave was cemented into a hard durable mass. The self-cementation, similar to that described for reaction between sea water and basalt [14], is apparently a result of dissolution of the glassy phase within the basalt and precipitation of silica (amorphous?). This reaction offers the interesting possibility of producing natural cements from repository materials of known long term stability. Efforts are currently under way to determine the exact nature and kinetics of the cementing reaction and to produce pieces of self-cemented basalt sufficiently large for physical testing.

The physical testing program, carried out in the Woodward-Clyde Consultants laboratory in Oakland, California, is designed to aid in the selection of individual materials and mixtures from the list of preferred candidate materials (Table III) and to provide physical properties of these materials for use in design studies.

Preliminary results indicate that, of the cements tested, portland cement Type V is the most preferred for use in plugs, combining high sulfate resistance with acceptable mechanical performance. Addition of finely ground, high-silica pozzolan to portland cement was found to reduce shrinkage, increase workability, improve impermeability, and increase stability of cement mixtures exposed to moderate temperatures (less than 100°C). Addition of finely ground silica flour to portland cements [15] was found to reduce shrinkage, and substantially improve thermal stability and structural strengths of cement mixtures exposed to temperatures greater than 100°C.

Preliminary results for cohesive materials indicate that bentonite clay is the preferred material for use in clay/sand or skip graded clay/sand/aggregate mixtures. Its high plasticity, excellent swelling properties, and very low permeability, coupled with high ion exchange potential and sorption capacity should substantially decrease both fluid and radionuclide migration through plugs. Wyoming bentonite is superior to Oregon bentonite in terms of plasticity and compactability and both are far superior to Ringold "clays" from the Hanford Site. Both crushed basalt and glaciofluvial sand and gravel from the Hanford Site are strong, competent granular materials and are suitable for use as aggregate in concrete, compacted earth materials, and in premixed clay slurries. At the present time crushed zeolite appears useful only as a component in compacted backfill.

Preliminary tests using plug models, emplaced in holes drilled into blocks of basalt, indicate that it is possible to design mixtures of candidate plug materials which have permeabilities of less than 10^{-8} cm/sec and will form acceptable bond strengths with the host rock. Mud contamination of simulated borehole walls during model testing was found to substantially decrease the bond strength of miniature cement and soil plugs with the basalt. However, high bond strengths for compacted bentonite/sand mixtures cured at 100°C indicate the possibility of cementation between plug and basalt at high temperatures.

PLUG DESIGN

Preconceptual plug design has concentrated on plugging schemes that appear to be available on a near term basis (5 to 10 years). Numerous other plugging schemes have been identified and appear feasible with varying degrees of development. Schemes considered most feasible on the basis of the initial analysis were subsequently subjected to a technical analysis of their design performance under expected conditions. At this stage of the study only thermomechanical and hydraulic performances were felt to be capable of any useful prediction.

Numerical analysis of hydraulic performance indicates that the zone of construction disturbance around tunnels and shafts will have a profound influence upon plug design. This zone, resulting both from energy input during construction and subsequent stress relaxation, will be a zone of higher premeability and, if it extends out to one tunnel diameter, will have a cross-sectional area considerably larger than that of the tunnel. Grouting of the disturbed zone is contemplated; however, because of the uncertainties inherent in grouting, the proposed designs utilize cut-off collars through the disturbed zone. This technique involves excavating the disturbed rock-zone over a length of tunnel sufficiently short to promote over-burden load transfer (arching) to adjacent rock or temporary tunnel supports without further relaxation above the fresh excavation. The excavated collar would then be filled with low permeability plug material to slow fluid flow through the disturbed zone.

Following technical evaluation the plugging schemes were rated on their ability to fulfill five design functions: (1) core barrier performance; (2) plug-rock interface performance; (3) support performance; (4) disturbed rock zone performance; and (5) long term integrity. The ratings were carried out for each plug environment (boreholes, shafts, tunnels) and in each case no single scheme best performed all design functions. Thus, multiple zone plugs are proposed for each plug environment.

A preconceptual design for a tunnel plug consisting of alternating zones of concrete, clay/sand/silt slurry mixes and basalt blocks is presented in Figure 5. Concrete is considered appropriate on the basis of core barrier performance, plug-rock interface performance, and support performance. Clay/sand/silt slurry mixes are considered appropriate for disturbed rock zone treatment and basalt blocks, mortared with cement or bentonite are considered more appropriate for long term stability.

A preconceptual design for a shaft plug consisting of alternating zones of compacted earth material and concrete is presented in Figure 6. Compacted earth plugs are considered appropriate for core barrier performance, plug-rock interface performance, and to a lesser extent for long term integrity. Concrete is considered appropriate for support performance and is preferred for disturbed rock zone treatment both because it can provide support during the construction of cut-off collars and because it will limit axial movement of the plug.

A preconceptual design for a borehole plug consisting of gravel and clay slurry with bentonite pellets and cement grout is presented in Figure 7. Gravel and clay slurries containing pellets of compressed bentonite are preferred for core barrier performance and long term integrity, while cement grout is preferred for plug-rock interface performance.

SUMMARY

The geochemical environment within the basalts of the Columbia Plateau and a group of candidate plug materials believed to be

PRECONCEPTUAL DESIGN FOR ACCESS TUNNEL PLUGS

(TYPICAL SECTION)

FIGURE 5

CONCRETE

BASALT BLOCKS WITH MORTARED JOINTS

CLAY-SAND SLURRY

CONCRETE FOR LOW TEMPERATURE LOCATION T < 100°C

MIX DESIGN PER m³

PORTLAND TYPE V	: 273 kg
LASSENITE	: 55 kg
GLACIOFLUVIAL SAND	: 755 kg
19 mm GRAVEL	: 1132 kg
WATER	: 174 ℓ
PLASTIMENT	: 689 mℓ

CONCRETE PROPERTIES

UNIT WEIGHT	: 2481 kg/m³
POROSITY	: 13.3%
SLUMP	: 8.3 cm
25 DAYS STRENGTH	: 36.5 MPa
PERMEABILITY	: < 4 x 10⁻⁹ cm/s

CLAY-SAND SLURRY

MIX DESIGN PER m³

SHURGEL	: 56 kg
GLACIOFLUVIAL SAND	: 483 kg
WATER	: 804 ℓ
UNIT WEIGHT	: 1340 kg/m³
POROSITY	: 80%
PERMEABILITY	: future testing

CONCRETE FOR HIGH TEMPERATURE LOCATION T > 100°C

MIX DESIGN PER m³

PORTLAND TYPE V	: 289 kg
SILICA FLOUR	: 158 kg
GLACIOFLUVIAL SAND	: 709 kg
10 mm GRAVEL	: 1063 kg
WATER	: 179 ℓ
PLASTIMENT	: 612 mℓ

CONCRETE PROPERTIES

UNIT WEIGHT	: 2453 kg/m³
POROSITY	: 14.8%
SLUMP	: 10.8 cm
33 DAYS STRENGTH	: 42.1 MPa
PERMEABILITY	: 1.49 x 10⁻⁹ cm/s

BASALT BLOCKS WITH MORTARED JOINTS

BASALT BLOCKS

SIZE	: 0.2 m x 0.2 m x 0.1 m
WEIGHT	: 10 kg
POROSITY	: 1.24%
STRENGTH	: 300 MPa
PERMEABILITY	: 10⁻⁹ cm/s

MORTAR

PORTLAND TYPE V TO LASSENITE TO GLACIOFLUVIAL SAND RATIO 5:1:10

WATER CEMENT RATIO	: 0.45
STRENGTH	: 56.9 MPa
PERMEABILITY	: < 4 x 10⁻⁹ cm/s

APPROXIMATE DIA. 3m TO 10m

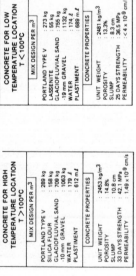

stable within that environment have been identified. The initial
laboratory testing program has refined the list of candidate plug
materials and has provided physical parameters for preconceptual
design. Preconceptual design schemes have been proposed for tunnels,
shafts, and boreholes; however, these are very preliminary and an
extensive program of development and validation, including in situ
field test, will be required before licensable plug designs are
available.

REFERENCES

1. Smith, M. J.: "Basalt Waste Isolation Project Borehole Plugging
    Studies - An Overview," This Volume.

2. Waters, A. C.: "Stratigraphic and Lithologic Variations in the
    Columbia River Basalt," American Journal of Science, 259, 581-611,
    1961.

3. Watkins, N. D. and Baksi, A. K.: "Magnetostratigraphy and Oro-
    clinal Folding of the Columbia River, Steens and Owyhee Basalts
    in Oregon, Washington and Idaho," American Journal of Science,
    274, 148-   , 1974.

4. McKee, E. H., et al.: "Duration and Volume of Columbia River
    Basalt Volcanism; Washington, Oregon and Idaho," Geological
    Society of America Abstracts with Programs, 9, 463, 1977.

5. Myers, C. W., et al.: "Geological Studies of the Columbia Pla-
    teau: A Status Report," RHO-BWI-ST-4, Rockwell Hanford Opera-
    tions, Richland, Washington, 1979.

6. Cook, N. W.: "An Appraisal of Hard Rock for Potential Under-
    ground Repositories of Radioactive Waste," LBL-7004, Lawrence
    Berkeley Laboratory, University of California, 15 p., 1977.

7. Hardy, M. P. and Hocking, G.: "Numerical Modeling of Rock
    Stresses Within a Basaltic Nuclear Waste Repository - Phase II -
    Parametric Design Studies," RHO-BWI-C-23, 303 p., 1978.

8. Morris, J. C. and Stumm, W.: "Redox Equilibria and Measurements
    of Potentials in the Aquatic Environment," Equilibrium Concepts
    in Natural Water Systems, American Chemical Society, Washington,
    D.C., 270-285, 1967.

9. Benson, L. V.: "Secondary Minerals, Oxidation Potentials, Pres-
    sure and Temperature Gradients in the Pasco Basin of Washington
    State," RHO-BWI-C-34, Rockwell Hanford Operations, Richland,
    Washington, 21 p., 1978.

10. Garrels, R. M. and Christ, C. L.: Solutions, Minerals, and
    Equilibria, Freeman, Cooper & Company, San Francisco, 419, 1965.

11. Carmichael, I.S.E., et al.: Igneous Petrology, McGraw-Hill,
    New York, 739 p., 1974.

12. Holloway, C. A.: Decision Making Under Uncertainty, Prentice-
    Hall, Englewood Cliffs, New Jersey, 552 p., 1979.

13. Keeney, R. L. and Raiffa, H.: Decisions with Multiple Objectives,
    Wiley and Sons, New York, 569 p., 1976.

14. Bischoff, J. L. and Dickson, F. W.: "Seawater-Basalt Interactions
    at 200°C and 500 bars: Implications for Origin of Sea-Floor
    Heavy-Metal Deposits and Regulation of Sea Water Chemistry,"
    Earth and Planetary Science Letters, 25, 385-397, 1975.

15. Roy, D. M., et al.: "Borehole Plugging by Hydrothermal Transport - Final Report," ORNL/SUB-4091/3, Materials Research Laboratory, The Pennsylvania State University, University Park, Pennsylvania, 78 p., 1976.

16. Gephart, R. E., et al., "Hydrologic Studies Within the Columbia Plateau, Washington," RHO-BWI-ST-5, Rockwell Hanford Operations, Richland, Washington, 1979.

FIGURE 6

PRECONCEPTUAL DESIGN FOR SHAFT PLUGS
(TYPICAL SECTION)

CONCRETE FOR LOW TEMPERATURE LOCATION
T < 100°C

MIX DESIGN PER m³

PORTLAND TYPE V	: 273 kg
LASSENITE	: 55 kg
GLACIOFLUVIAL SAND	: 755 kg
1.9 cm GRAVEL	: 1132 kg
WATER	: 174 ℓ
PLASTIMENT	: 689 m ℓ

CONCRETE PROPERTIES

UNIT WEIGHT	: 3481 kg/m³
POROSITY	: 13.3%
SLUMP	: 9.5 cm
25 DAYS STRENGTH	: 36.5 MPa
PERMEABILITY	: 3.67×10⁻⁹ cm/s

CONCRETE FOR HIGH TEMPERATURE LOCATION
T > 100°C

MIX DESIGN PER m³

PORTLAND TYPE V	: 289 kg
SILICA FLOUR	: 156 kg
GLACIOFLUVIAL SAND	: 709 kg
1.9 cm GRAVEL	: 1063 kg
WATER	: 179 ℓ
PLASTIMENT	: 612 m ℓ

CONCRETE PROPERTIES

UNIT WEIGHT	: 2453 kg/m³
POROSITY	: 14.8%
SLUMP	: 11 cm
23 DAYS STRENGTH	: 42.1 MPa
PERMEABILITY	: 2×10⁻⁹ cm s

CLAY-SAND MIX

CLAY	: WYOMING BENTONITE
SAND	: GLACIOFLUVIAL
CLAY-SAND RATIO	: 1
MAX. DENSITY	: 1983 kg/m³
OPTIMUM WATER CONTENT	: 14.2%
VOID RATIO	: 0.45
SWELL PRESSURE	: 0.9 MPa
PERMEABILITY	: <10⁻⁹cm/s

▦ CONCRETE CUTOFF COLLAR

▨ CLAY-SAND MIX

APPROXIMATE DIA. 4½ m TO 8m

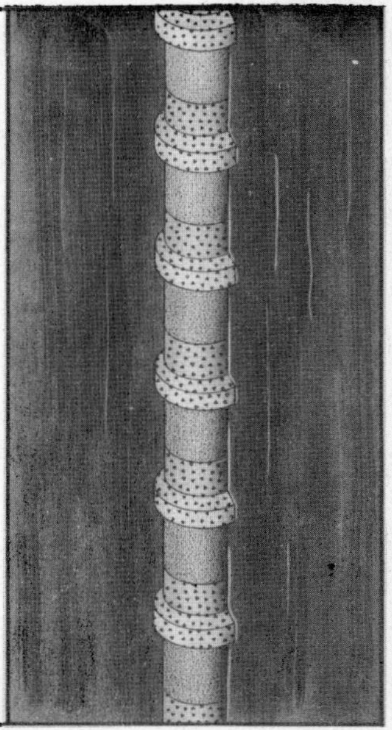

PRECONCEPTUAL DESIGN FOR REMOTE BOREHOLE PLUGS
(TYPICAL SECTION)

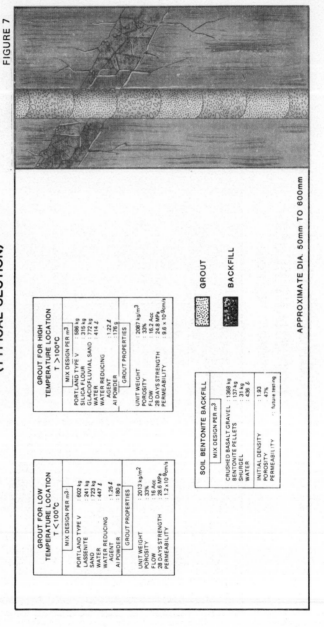

FIGURE 7

GROUT FOR LOW TEMPERATURE LOCATION
T < 100°C

MIX DESIGN PER m³

PORTLAND TYPE V	: 602 kg
LASSENITE	: 241 kg
SAND	: 723 kg
WATER	: 447 ℓ
WATER REDUCING AGENT	: 1.25 ℓ
Al POWDER	: 180 g

GROUT PROPERTIES

UNIT WEIGHT	: 2013 kg/m²
POROSITY	: 33%
FLOW	: 16 Acc
28 DAYS STRENGTH	: 28.6 MPa
PERMEABILITY	: 1.2 x 10⁻⁹ cm/s

GROUT FOR HIGH TEMPERATURE LOCATION
T > 100°C

MIX DESIGN PER m³

PORTLAND TYPE V	: 586 kg
SILICA FLOUR	: 315 kg
GLACIOFLUVIAL SAND	: 772 kg
WATER	: 414 ℓ
WATER REDUCING AGENT	: 1.22 ℓ
Al POWDER	: 176 g

GROUT PROPERTIES

UNIT WEIGHT	: 2087 kg/m³
POROSITY	: 33%
FLOW	: 16.2 Acc
28 DAYS STRENGTH	: 24.8 MPa
PERMEABILITY	: 9.6 x 10⁻⁹ cm/s

SOIL BENTONITE BACKFILL

MIX DESIGN PER m³

CRUSHED BASALT GRAVEL	: 1368 kg
BENTONITE PELLETS	: 137 kg
SHURGEL	: 31 kg
WATER	: 436 ℓ
INITIAL DENSITY	: 1.93
POROSITY	: 47%
PERMEABILITY	: future testing

GROUT

BACKFILL

APPROXIMATE DIA. 50mm TO 600mm

Discussion

S.J. LAMBERT, United States

I am aware of the Stanford experiments on the interaction between seawater at various temperatures and basaltic rocks. In such interactions, it is almost inevitable to get amorphous hydrated silica. Considering the reactivity of amorphous hydrated silica in geologic systems and the transitions it can undergo I would question the suitability of the basalt-cement, without further evaluation. Perhaps enthusiasm over its use is premature.

F.N. HODGES, United States

We are not proposing the use of self-cemented basalt without further evaluation. We think that it is an interesting possibility that we plan to explore fully. We are presently evaluating the solids produced in the experiments and hope to know, in the near future, whether the cement is crystalline or amorphous. We plan to investigate the range of conditions under which the cementation occurs, the physical properties of the materials, and the effects of possible recrystallization on strength.

D.M. ROY, United States

I believe that one must be cautiously optimistic about "cementing" properties of a material such as finely powdered basalt which requires (on the basis of early experiments) stirring of dilute suspensions at elevated temperatures (250°C) and pressures for prolonged periods of time. The crystallization of quartz as a cementing matrix also seems to us highly unlikely within finite time at any except experimentally unfeasible temperatures and pressures (based upon our previous feasibility studies). The material, however, should be an acceptable fill.

F.N. HODGES, United States

I agree that caution is needed ; however, we feel the need to find a back up or an alternative to concrete. We plan to fully investigate the conditions and kinetics of the reactions involved and possible techniques for speeding or duplicating the processes. I wish to point out that self-cementation might take place naturally over reasonable time periods in plugs located near a repository where temperatures will be relatively high. Incidentally that is exactly the kind of environment where the stability and properties of portland type cements are most suspect.

A.M.L. BOULANGER, France

Dans votre présentation vous avez fait allusion à certains essais devant vous permettre de juger de la pérennité des matériaux de remplissage présélectionnés.

Pouvez-vous nous donner des détails sur la nature de ces essais et nous préciser quel degré d'extrapolation vous pouvez en espérer.

F.N. HODGES, United States

The hydrothermal tests were not intended to prove longevity. The materials were carefully selected for documented longevity prior to testing. The tests, carried out between 150° and 250°C under saturated steam pressure, were intended as reconnaissance experiments to test for reactions between various plug materials and between plug materials and basalt with simulated Hanford ground water.

HIGHLY COMPACTED BENTONITE FOR BORE-HOLE AND SHAFT PLUGGING

R. Pusch, Professor in soil mechanics,
          University of Luleå, Sweden
A. Bergström, Civ.Eng., Nuclear Fuel Safety
          Project, Stockholm, Sweden

ABSTRACT

Dense bentonite is a very efficient sealing substance for plugging
bore holes and shafts. Blocks of highly compacted bentonite will
take up water, swell and fill up slots and irregularities and create
a perfect contact with the confining rock. This can provide a homogeneous plug with sufficient mechanical strength and an extremely
low permeability.

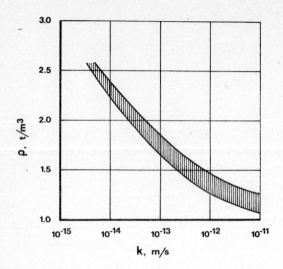

Fig. 1. Average relationship between coefficient of permeability (k) and bulk density (ρ) of Na bentonite. The hatched area accounts for scattering and for variations in the hydraulic gradient.

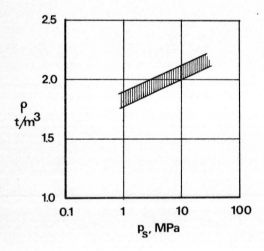

Fig. 2. Swelling pressure of water saturated Na bentonite. The hatched area accounts for scattering and variations in electrolytes in the pore water.

1.   INTRODUCTION

Bentonite is the term for montmorillonite-rich clay formed by devitrification of the natural glass component of volcanic ash deposited in prehistoric lakes and estuaries. Dense bentonite takes up water and swells to several times its original volume. This property, which is primarily due to the affinity to water of the montmorillonite crystal lattice and its adsorbed cations, means that density gradients produce water redistribution so that a homogeneous state is eventually produced in confined bentonite samples. This self-healing ability suggests the use of highly compacted bentonite for bore hole and shaft plugging.

In this article a possible technique and various criteria for plugging are discussed and a pilot test of bore hole sealing is reported.

2.   SEALING PROPERTIES OF HIGHLY COMPACTED BENTONITE

Effective sealing of bore holes and shafts requires a swelling potential of the bentonite to guarantee that the contact between the bentonite and confining rock is always perfect. Also, the bentonite should have a lower permeability than that of the surrounding rock. Its plastic consistency and swelling potential must be retained for the entire period during which sealing is required, which is several thousand years for a nuclear waste repository. Chemical stability is a necessary prerequisite for this and it can be disturbed by a number of factors, mainly temperature and pH. They have been considered in a comprehensive study made by the Swedish Nuclear Safety Project (KBS) [1], from which it can be concluded that bentonite is chemically stable for geological ages in the pH range found in granite and gneiss bedrock in Sweden and for temperatures lower than 100°C.

The permeability is an essential sealing parameter [2]. Examples from the literature and own test results suggest the relationship between bulk density (water saturated condition) and permeability shown in Fig. 1. Complete ion exchange from Na to Ca, which can take place under natural conditions, increases the permeability 2 to 3 times.

For a bulk density of about 2 t/m^3 of water saturated Na bentonite k will be 10^{-13} to 10^{-14} m/s which is lower than the value usually found for ordinary crystalline rock. Densities of this order of magnitude can be obtained by using high compaction pressures. The greater part of the KBS study concerned commercial, granulated bentonite powder "Volclay MX-80", which was compacted by applying a pressure of 50-100 MPa. The powder has an average water content of about 10%, which corresponds to a degree of water saturation of about 50% at a bulk density of 2-2.1 t/m^3. This density is obtained when a pressure of 50 MPa is used for the compaction.

If a partly water saturated, compacted bentonite sample has access to water but is confined so that its total volume remains constant, it will exert a swelling pressure on the confinement. The water uptake proceeds until an equilibrium is obtained at which the swelling pressure reaches a maximum value. Its magnitude is indicated by the diagram in Fig. 2.

3.   BORE HOLE PLUGGING

3.1 Criteria

A possible technique should be to fill the bore holes with cylindrical samples of dense bentonite, which then take up water, swell and form practically impervious plugs. The insertion of bentonite samples into deep bore holes involves a number of difficulties, the main ones being:

A. The filling must be fairly complete, i.e. the slot between the rock and the samples must be narrow in order to yield a sufficiently high final density.

B. The filling operation must run without any trouble, i.e. samples must not get stuck.

C. The swelling of the samples must not take place so rapidly that the filling cannot be completed.

The first difficulty has to do with the bearing capacity of the samples during the swelling, and with the required permeability of the bentonite core. For practical purpose the bulk density of the expanded, fully swollen sample should be at least 1.7 t/m^3, which corresponds to a k-value of about $10^{-12} - 10^{-13}$ m/s.

As to the bearing capacity it is clear that if it is too low, the bentonite column may be compressed under its own weight and leave an open space in the upper part of the bore hole. This can happen if the swelling pressure is lower than the pressure produced by the overlying bentonite column in steeply oriented holes. In practice, the bentonite will adhere to the bore hole walls ("silo effect") which largely reduces the vertical pressure. Thus, a rough estimation suggests that a water saturated, homogeneous, bentonite with an average bulk density of 1.8 - 2.0 t/m^3 can form a mechanically stable column in bore holes of any length. Considerable initial inhomogeneities or density variations which may appear in course of the swelling, seem to be completely eliminated by the strong self-healing ability of the bentonite [3].

The second difficulty suggests the use of a pipe device which protects the bentonite samples from coming into contact with the rock wall during the filling operation. The pipe must be richly perforated to permit the bentonite to migrate through its walls and to form, eventually, a homogeneous medium embedding it.

The third difficulty arises if the filling operation takes more than 2-3 days. It is therefore required that 1000 m long or even longer holes be filled in not more than one day. This suggests the use of a pipe device which is successively filled and lowered into the hole and which is then left there. The pipe should preferably be made of copper as concluded from the KBS study. Vertical, as well as inclined or horizontal bore holes can be plugged by means of this technique. The outer end is preferably sealed with cement.

3.2 Pilot test

A pilot test was made to investigate the character and homogeneity of the bentonite in a simulated ϕ 56 mm bore hole approximately 10 days after filling it with Na bentonite cylinders and distilled water. The actual inner diameter was ϕ 53 mm, and the outer diameter of the perforated pipe and the bentonite cylinders ϕ 51 mm and 45 mm, respectively. The bentonite had an initial water content of 10% by weight and a bulk density of 1,85 t/m^3. The wall thickness of the perforated pipe was 2 mm and its total volume 60

cm^3. Theoretically, the bulk density and water content after complete redistribution of soild matter and water should thus become 1.5
t/m^3 and 75% respectively. The device was equipped with a local
water inlet for the uptake of additional water. The non-uniform
conditions for water take-up in a real bore hole in rock are therefore imitated. Fig. 3 shows the application of the bentonite in the
perforated pipe, and Fig. 4 the test device before applying the lid.

The electronic pressure gauge showed that the swelling
pressure was built up fairly rapidly. 24 hours after filling up the
system with water the recorded pressure was about 60 kPa. 48 hours
after the start of the test it was 80 kPa, and after 1 week it had
reached a value of 180 kPa. This indicates a rather fast migration
of the bentonite. After 10 days the device was dismantled and it
was confirmed that bentonite had formed practically homogeneous
media outside as well as inside the perforated pipe. Fig. 5, which
gives the average water content distribution in the bentonite after
10 days, confirms that the degree of homogeneity is high already
at this early stage. Systematic laboratory and field tests are in
preparation.

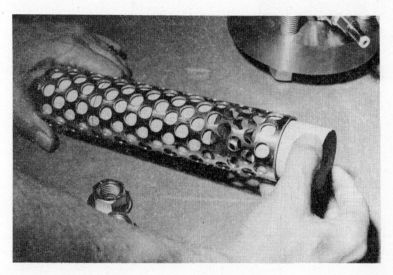

Fig. 3.   Filling of the perforated pipe with "air-dry" bentonite.

4.   SHAFT PLUGGING

Highly compacted bentonite has been suggested as a sealing
substance for shaft plugging in the KBS concepts [1]. Very regularly
shaped cubical or parallelepiped-blocks can easily be produced
by iso-static compaction on an industrial scale and this offers a
possibility of applying highly compacted bentonite in a dense arrangement also in large voids (Fig. 6). This means that the bulk
density after water uptake can be very high and the swelling pressure correspondingly high. Highly compacted bentonite is preferably
applied in stacks separated by less dense fill. This fill will be
somewhat compressed by the swelling pressure exerted by the bentonite stacks but the displacements can be controlled by choosing
appropriate densities of the various components. If the stacks of
highly compacted bentonite blocks are combined with slots in the
rock made by seam drilling and by grouting in drill holes extending
from these slots, a very efficient shaft plugging can probably be
obtained.

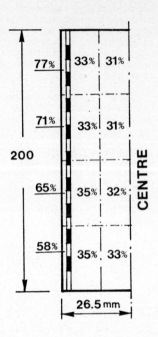

Fig. 4. Pipe device with water inlet seen at lower end and pore pressure gauge at mid-height. The device is turned upside down here.

Fig. 5. Water content distribution in percent in the pilot test core. Vertical scale 1:2, horizontal scale 1:1.

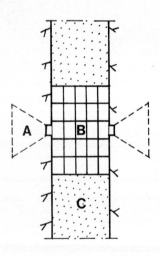

Fig. 6. Shaft or tunnel plugging. A) Grouting with bentonite suspension. B) Blocks of highly compacted bentonite. C) Sand-bentonite.

5.　REFERENCES

[1] "Handling and final storage of unreprocessed spent nuclear fuel", Vol II (Technical) 1978, Swedish Nuclear Safety Project, Fack, S-102 48 Stockholm, Sweden.

[2] Pusch, R.: "Highly compacted Na bentonite as buffer substance", KBS Technical Report No. 74, 1978.

[3] Pusch, R.: "Self-injection of highly compacted bentonite into rock joints", KBS Technical Report No. 73, 1978.

Discussion

G.M. IDORN, Denmark

Is there any relationship between the viscosity of swelling bentonite and temperature ?

R. PUSCH, Sweden

Yes, we know roughly the influence of temperature. Further investigations are going on, but in order to avoid trouble we stay below 75°C.

K. MATHER, United States

You have made your experiments with Wyoming bentonite which is predominantly a sodium bentonite. If, in the environment of the repository, the ground water contained divalent cations such as Ca^{++} which would displace the sodium from exchange sites and replace it, the swelling pressure of sodium (or monovalent cation) bentonite would be somewhat reduced to the limited swelling, although still extensive, of a bentonite with a divalent cation. Would this have any effect or consequence on the scenario that you envision ?

R. PUSCH, Sweden

This is correct. An ion exchange from sodium to calcium will reduce the swelling pressure and increase the permeability. The changes would be considerable in the case of change from a pure sodium-bentonite to a pure calcium-bentonite. For the Wyoming Mx80 bentonite the changes would be moderate since this bentonite already holds a considerable amount of calcium in exchange positions in its natural state. The expected changes in physical and mechanical properties have been taken into consideration in the KBS analyses. I believe that, in practice, natural calcium-bentonites might very well be usable.

EBAUCHE D'UN PROGRAMME EXPERIMENTAL EN VUE DU COLMATAGE D'UNE INSTALLATION
D'ENFOUISSEMENT POUR DECHETS RADIOACTIFS CONDITIONNES DANS UNE FORMATION D'AR-
GILE PLASTIQUE EN BELGIQUE

par

R. Heremans   P. Manfroy
Centre d'Etude de l'Energie Nucléaire, Mol, Belgique

et

R. Funcken   M. Mayence
Société Tractionel Engineering, Bruxelles, Belgique

RESUME

        Le Centre d'Etude de l'Energie Nucléaire de Mol a entrepris depuis six
ans un programme de recherches sur l'enfouissement de déchets radioactifs conditionnés dans une formation argileuse oligocène présente à profondeur moyenne sous son
site propre.
        Après des travaux en laboratoire et sur terrain destinés à caractériser
la formation et suite à l'étude conceptuelle d'une installation complète d'enfouissement à l'échelle industrielle, il a été décidé d'entamer la construction d'un
puits et d'une galerie expérimentale dans la formation considérée.
        Ce laboratoire souterrain permettra de conduire un vaste programme d'ex-
périences portant sur les caractéristiques de l'argile "in situ" ainsi que sur les
revêtements. Il permettra également de tester différents matériaux de colmatage
ainsi que les techniques de mise en oeuvre qui s'y rapportent.

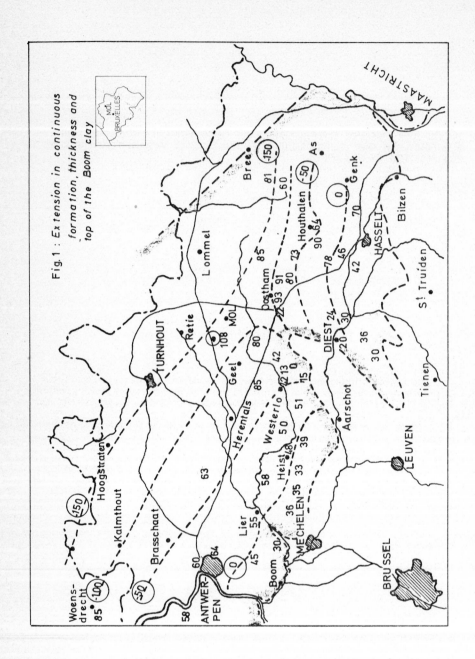

Fig.1 : Extension in continuous formation, thickness and top of the Boom clay

0. Introduction

Depuis fin 1973 le Centre d'Etude de l'Energie Nucléaire (C.E.N./S.C.K.) à Mol, Belgique, a entrepris un programme de R & D sur le rejet en formation géologique de certains déchets radioactifs insolubilisés. (1)

Dans le cadre de ce programme, un inventaire des formations géologiques du sous-sol belge susceptibles de convenir à cet effet a été élaboré avec l'aide du Service Géologique National.

L'une des formations selectionnées dans cet inventaire est constituée d'argile et est présente à profondeur moyenne en dessous des installations du C.E.N. /S.C.K..

Une campagne de sondages et des mesures géophysiques de haute résolution ont permis de préciser la structure de la formation tandisqu'un programme d'expérimentations intensif en laboratoire sur les échantillons prélevés "in situ" lors des sondages ont conduit à caractériser les propriétés physiques, chimiques, minéralogiques et géomécaniques de l'argile.

Une étude conceptuelle et de faisabilité d'une installation d'enfouissement de déchets conditionnés dans l'argile a été réalisée avec l'aide d'un bureau d'ingénierie spécialisé. Cette étude a permis entre autre de dégager un certain nombre de techniques ayant trait au colmatage et au remblayage des installations de rejet. Elle a permis aussi de se rendre compte qu'il existait de nombreuses questions très importantes qui resteraient sans réponse tant qu'une installation expérimentale n'aurait pas été implantée dans la couche sélectionnée.

C'est pourquoi le C.E.N./S.C.K. a entrepris au début de cette année les travaux préliminaires de creusement de ces installations qui devraient être opérationnelles vers la fin de 1982. Parmi les nombreuses expériences prévues durant les prochains années, quelques unes d'entre elles relatives à la nature et à la mise en place de matériaux de colmatage pourront être developpés.

1. Description générale de la formation argileuse sélectionnée

L'argile sélectionnée est une formation sédimentaire datant de la période Oligocène ($\simeq 30 \times 10^6$ années) connue en Belgique sous le nom "d'argile de Boom" du nom de sa zone d'affleurement dans la région d'Antwerpen. Elle s'enfonce vers le N-E sous un recouvrement de terrains miocènes et pliocènes essentiellement sableux.

1.1. Cadre géologique de la région de Mol

Le site du C.E.N./S.C.K. est situé dans la partie N-NE de la Belgique (voir Fig. 1). L'altitude moyenne par rapport au niveau de la mer y est de 25 mètres. Différents sondages effectués en 1975 ont permis de dresser la stratigraphie des couches sous jacentes (voir Fig. 2). Celles-ci se présentent, depuis le toit du Crétacé jusqu'à la surface du sol, suivant une succession de formations argileuses et de formations sableuses d'épaisseurs variables. Parmi les assises argileuses seules l'argile d'Ieper (argile des Flandres) et l'argile de Boom se distinguent par leur épaisseur de l'ordre d'une centaine de mètres.

De ces deux assises, seule l'argile de Boom présente une grande homogénéité sur toute son épaisseur Deux formations aquifères sableuses encadrent la couche d'argile de Boom, l'une, en dessous, est constituée des sables fins à faible perméabilité, l'autre au dessus, est constituée d'une succession de couches de sables plus grossiers contenant une nappe aquifère exploitée dans la région.

Une campagne de sismique réflexion de haute résolution a montré que la couche d'argile était, dans cette région, d'une très grande régularité, dépourvue d'accidents tectoniques tels que des failles ou des flexures. Cette partie du territoire a été jusqu'à présent reconnue comme asismique.

(1) Programme réalisé à frais partagés avec la Commission des Communautés Européennes de Bruxelles.

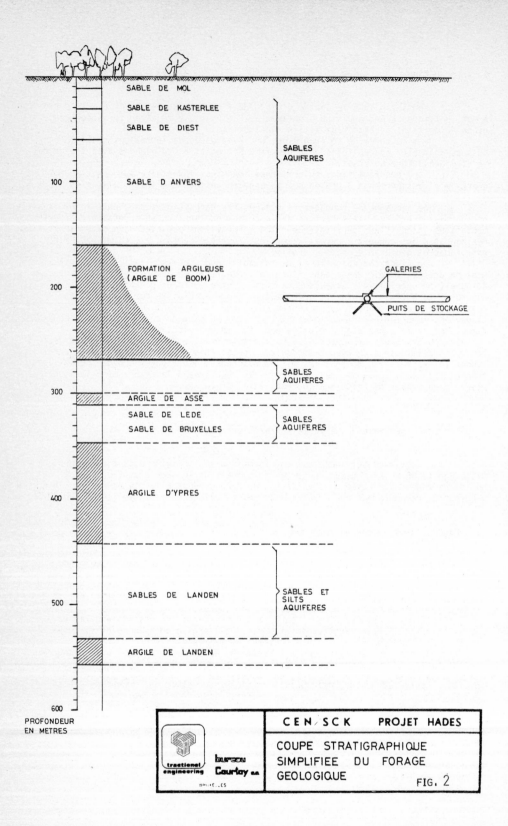

SABLE DE MOL

SABLE DE KASTERLEE

SABLE DE DIEST

SABLES
AQUIFERES

SABLE D ANVERS

100

FORMATION ARGILEUSE
(ARGILE DE BOOM)

GALERIES

200

PUITS DE STOCKAGE

SABLES
AQUIFERES

300

ARGILE DE ASSE

SABLE DE LEDE

SABLE DE BRUXELLES

SABLES
AQUIFERES

ARGILE D'YPRES

400

SABLES DE LANDEN

SABLES ET
SILTS
AQUIFERES

500

ARGILE DE LANDEN

600

PROFONDEUR
EN METRES

CEN/SCK   PROJET HADES

COUPE STRATIGRAPHIQUE
SIMPLIFIEE DU FORAGE
GEOLOGIQUE   FIG. 2

tractionel
engineering

bureau
Courtoy s.a

BRUXELLES

1.2. Propriétés de l'argile de Boom

Les sondages ont permis de prélever de nombreux échantillons d'argile tant remaniés que non remaniés qui ont permis la détermination en laboratoire des caractéristiques physiques, chimiques et géotechniques qui figurent au tableau I.

De nombreuses propriétés de l'argile, qui ne peuvent être détaillées ici, ont pu être également étudiées notamment concernant les propriétés de corrosion sur les métaux ainsi que le comportement aux rayonnements.

2. Résultats de l'étude conceptuelle concernant les possibilités de colmatage et de remblayage.

2.1. Cadre de l'étude

2.1.1. But de l'étude et quantités de déchets à stocker (hypothèses de travail)

Un des buts de l'étude était d'évaluer les techniques à mettre en oeuvre pour la réalisation éventuelle d'une unité complète d'enfouissement de déchets conditionnés dans l'argile. Les déchets pris en considération étant ceux de moyenne et haute éctivité découlant de l'exploitation pendant 30 ans d'une puissance installée de 10.000 MW électriques. Ces déchets se répartissent comme suit :

- Déchets de haute activité, de longue demie-vie et fortement générateurs de chaleur : 9.000 fûts de 1,5 m de hauteur et 0,3 m de diamètre.

- Déchets de haute activité, de longue demie-vie et faiblement générateurs de chaleur : conditionnés dans les mêmes fûts et suivant les mêmes quantités que les précédents.

- Déchets moyennement actifs de longue demie-vie et non générateurs de chaleur : 150.000 fûts de $\simeq$ 0,9 m de hauteur et $\simeq$ 0,6 m de diamètre.

2.1.2. Configuration retenue pour les installations souterraines d'enfouissement.

- Les installations se composent des puits d'accès et de ventilation et d'un réseau de galeries parallèles (voir Fig. 3).

- Les puits sont au nombre de quatre : deux puits d'accès d'un diamètre utile de 4,5 m et deux puits de ventilation d'un diamètre utile de 2 m. Tous ces puits seront de section circulaire et auront une hauteur de 220 m. Ils seront foncés par le procédé de congélation.

- Le réseau de galerie se composera (voir Fig. 4) d'une galerie principale d'accès et de galeries secondaires réservées à l'évacuation des déchets. Certaines d'entre elles seront reservées aux déchets de haute activité qui seront introduits dans des puits radiaux obliques, d'autres seront reservées aux déchets de moyenne activité qui représentent un volume beaucoup plus important et qui seront empilés de manière à remplir la quasi-totalité de la section utile. Toutes ces galeries seront de sections circulaires d'un diamètre utile de 3,5 m pour les galeries secondaires et de 4,5 m pour la galerie principale. La méthode de creusement retenue pour les galeries secondaires est celle du tunnelier avec pause simultanée, du revêtement lequel sera constitué de voussoirs nervurés en fonte nodulaire. La longueur totale des galeries à prévoir pour le dépôt de la totalité des déchets est de l'ordre de 16 Km.

2.2. Implication de la configuration choisie sur les opérations de colmatage des cavités

2.2.1. Qualités générales attendues des matériaux de colmatage

Les matériaux de colmatage doivent présenter les qualités suivantes :

- réduire les effets de l'effondrement des ouvrages souterrains sur la déformation plastique de l'argile qui les surmonte.(les voussoirs en fonte ne cesseront d'assurer leur rôle de soutènement que très progressivement et vraissemblablement qu'après plusieurs siècles),

TABLEAU 1

COMPOSITION AND MAIN PROPERTIES OF BOOM CLAY
SAMPLED AT THE MOL SITE

- Natural water content (weight %)	: $\sim$26
- Chemical composition of dry material (%)	: $\sim$64 SiO_2,14 Al_2O_3 5.9 Fe_2O_3,2.2 K_2O 1.4 Na_2O ,0.6 Cao 0.5 TiO_2 ,0.7 MgO weight loss at 1000 °C 9.8
- Mineralogical composition of the fraction <20μ (%)	: $\sim$illite 25, smectite 20 vermiculite 30, illite- montmorillonite interstratified 15, chlorite + chlorite-vermi- culite interstratified 10
- Granulometric composition (%)	: d < 2 μ : 49 2μ< d < 60 μ : 47 60μ < d< 200μ : 3.5 d > 200 μ : 0.5
- Bulk density ($t.m^{-3}$)	: $\sim$1.93
- Dry density ($t.m^{-3}$)	: $\sim$1.53
- Plasticity limit (%)	: $\sim$33
- Liquidity limit (%)	: $\sim$82
- Index of plasticity (%)	: $\sim$49
- Permeability ($cm\ s^{-1}$)	: between $1.4.10^{-8}$ and $4.7.10^{-10}$
- Porosity (%)	: between 34.6 and 44
- Saturation degree (%)	: between 88.4 and 100
- Elasticity modulus at the origin ($kg\ cm^{-2}$)	: between 1000 and 3500
- Undrained shear strength -Cu ($kg\ cm^{-2}$)	: between 3.3 and 8.6
- Apparent cohesion C' ($kg\ cm^{-2}$)	: 1.1 average value
- Angle of shearing resistance φ'_u (°)	: 19
- Thermal conductivity ($Wm^{-1}C^{-1}$)	: $\sim$1.3 at 20 °C, 0.3 at 100 °C $\sim$0.45 at 300 °C
- Specific heat ($Wh.kg^{-1}\ °C^{-1}$)	: $\sim$0.26 at 25 °C $\sim$0.41 at 275 °C
- Natural radio-activity ($\mu Ci.100\ g^{-1}$ dry sample)	: ^{40}K $\sim$2.10^{-3} $^{226}Ra\sim$2.1 10^{-4} $^{232}Th\sim$1.2 10^{-4}
- Ion exchange capacity	: $\sim$20 meq 100 g^{-1}

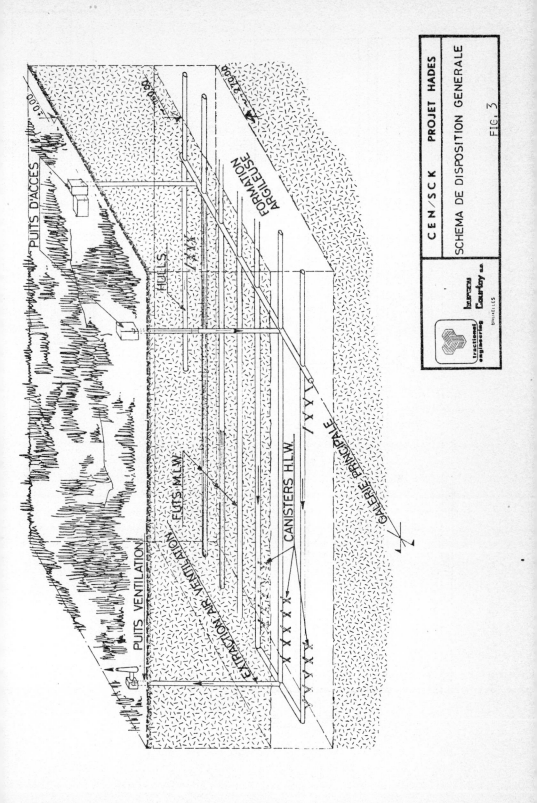

PUITS D'ACCES

PUITS VENTILATION

FORMATION ARGILEUSE

HULLS

FUTS M.L.W.

EXTRACTION AIR VENTILATION

CANISTERS H.L.W.

GALERIE PRINCIPALE

+0.00

-190.00

-230.00

CEN/SCK   PROJET HADES

SCHEMA DE DISPOSITION GENERALE

FIG. 3

tractionel engineering

bureau Bontay s.a.
BRUXELLES

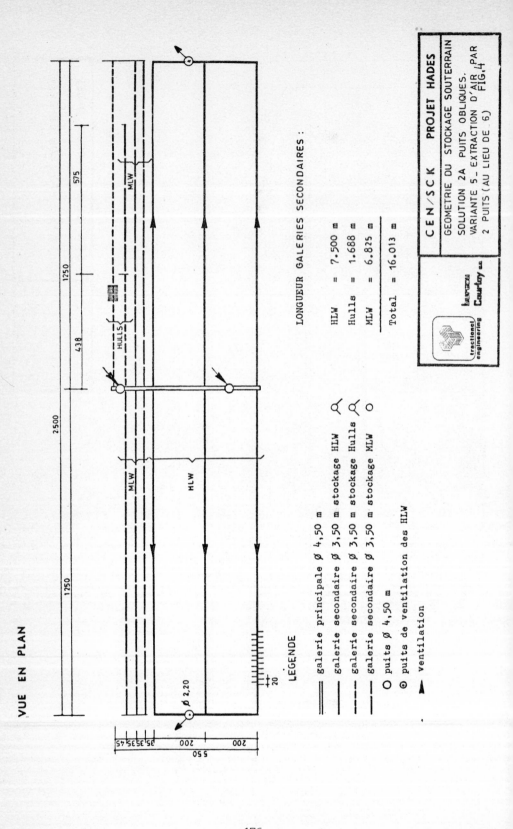

VUE EN PLAN

LÉGENDE

galerie principale ∅ 4,50 m

galerie secondaire ∅ 3,50 m stockage HLW

galerie secondaire ∅ 3,50 m stockage Hulls

galerie secondaire ∅ 3,50 m stockage MLW

○ puits ∅ 4,50 m

⊙ puits de ventilation des HLW

➤ ventilation

LONGUEUR GALERIES SECONDAIRES :

HLW   = 7.500 m

Hulls = 1.688 m

MLW   = 6.825 m

Total = 16.013 m

CEN/SCK   PROJET HADES

GEOMETRIE DU STOCKAGE SOUTERRAIN
SOLUTION 2A PUITS OBLIQUES.
VARIANTE 5 _ EXTRACTION D'AIR PAR
2 PUITS (AU LIEU DE 6)   FIG,4

tractionel
engineering Courtry s.a

- être très peu perméables,

- présenter un certain degré de plasticité leur permettant de s'adapter à des déformations éventuelles,

- éviter d'offrir entre le terrain et les revêtements un chemin préférentiel pour la migration des radionucléides,

- présenter la propriété de fixer les radionucléides afin de freiner leur migration éventuelle de manière similaire à celle de l'argile,

- avoir un comportement à la chaleur favorable au maintien des qualités précitées pour une augmentation modérée de température,

- présenter une bonne compatibilité physico-chimique avec l'argile et les matériaux de structure

2.2.2. Les différents types de colmatage et les matériaux mis en oeuvre.

Différents types de colmatage devront être mis en oeuvre durant les travaux d'excavation, pendant la phase d'enfouissement des déchets et enfin après celle-ci une fois qu'il aura été décidé de refermer le site de façon définitive.

- Colmatage par injection simultanément au creusement des galeries (voir Fig. 5). Le creusement d'une galerie circulaire par tunnelier laisse subsister un vide annulaire entre l'extrados des voussoirs constituant le revêtement et le massif en place. Afin de limiter les déformations de l'argile en place à proximité immédiate du revêtement une injection devra se faire simultanément à l'avancement de la machine de creusement. Le matériau injecté sous pression devra présenter une bonne résistance à la compression tout en ayant de bonnes propriétés d'étanchéité.

- Colmatage des puits radiaux obliques d'évacuation de déchets de haute activité (voir Fig. 6).

x colmatage d'intrados : L'espace compris entre le terrain et le revêtement métallique des puits radiaux devra être comblé par un matériau possédant toutes les qualités citées en 2.2.1., mais qui de plus sera à même de supporter l'élévation de température résultant de la proximité des déchets sans donner lieu à un retrait susceptible d'entraîner la formation d'un interface ouvert.

x colmatage d'intrados (voir Fig. 6) : L'espace annulaire compris entre les cylindres de déchets et l'intrados du revêtement devra être rempli par un matériau présentant des qualités différentes de celles énoncées en 2.2.1.. Sa fonction est en effet d'assurer une répartition uniforme de la chaleur vers le massif argileux, tout en permettant une récupération éventuelle des cylindres pendant une certaine période. Il s'agira donc d'un matériau non cohérent, isogranulaire et chimiquement amorphe pour les températures envisagées, pouvant être aspiré à l'aide d'un dispositif de type suceuse. Un sable siliceux de granulométrie adéquate semble actuellement le matériau le mieux adapté compte tenu des problèmes de mise en oeuvre et de réversibilité du dépôt. Afin d'augmenter les capacités d'échanges ioniques, ce sable pourra éventuellement être additionné de glauconite, très abondante dans la région.

- Remblayage des galeries

x remblayage des galeries d'évacuation des déchets de moyenne activité (voir Fig. 5) : Celui-ci se fera au fur et à mesure de l'empilement des fûts de déchets lesquels rempliront la quasi-totalité de la section utile de la galerie. Il est clair qu'une telle opération menée à son terme ne serait en aucun cas reversible. C'est la raison pour laquelle le matériau de remblayage doit présenter, outre une excellente résistance à la compression permettant d'obvier à une défection à terme de la fonction de soutènement des voussoirs, une très faible tendance à la fissuration sous contrainte (bonne imperméabilité) jointe à d'excellentes capacités de sorptions.

x remblayage des galeries d'évacuation pour déchets de haute activité : Une fois les déchets de haute activités logés dans leurs puits radiaux et la période de surveillance écoulée ces galeries devront être remblayées à leur tour après avoir été préalablement remplies par les déchets encore disponibles à ce moment (par exemple les pièces provenant d'opérations de démantèlement d'installations). Ce remblayage marquera l'irréversibilité du dépôt.

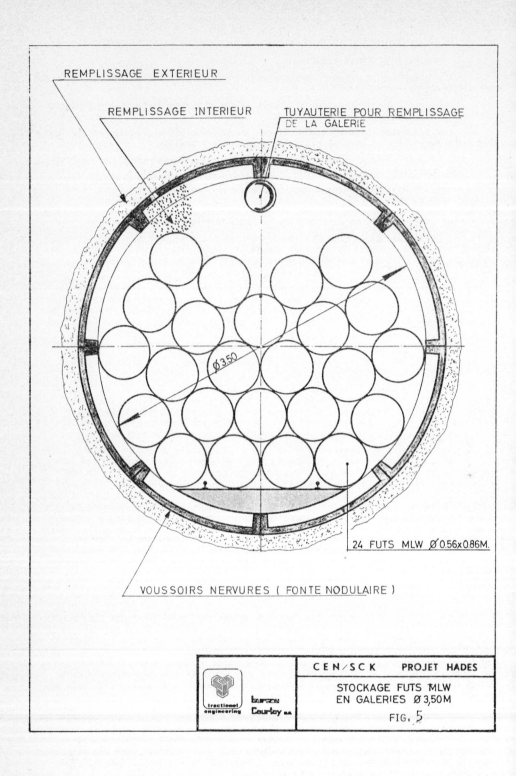

REMPLISSAGE EXTERIEUR

REMPLISSAGE INTERIEUR

TUYAUTERIE POUR REMPLISSAGE DE LA GALERIE

Ø 3.50

24 FUTS MLW Ø 0.56x0.86M.

VOUSSOIRS NERVURES (FONTE NODULAIRE)

	CEN/SCK PROJET HADES
tractionel engineering burgeon Courtoy s.a.	STOCKAGE FUTS MLW EN GALERIES Ø 3,50M FIG. 5

COUPE VERTICALE

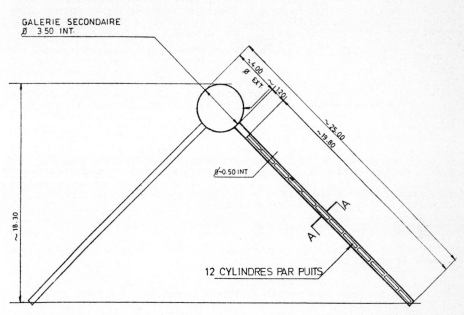

GALERIE SECONDAIRE
Ø 3.50 INT.

~18.30

~4.00
Ø EXT. ~1.20

~25.00
~19.80

Ø~0.50 INT

A

A

12 CYLINDRES PAR PUITS

DISTANCE ENTRE PUITS :

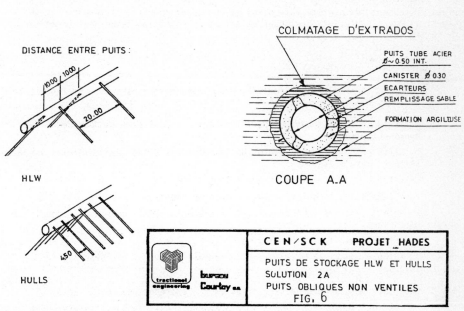

10.00 10.00
20.00

HLW

4.50

HULLS

COLMATAGE D'EXTRADOS

PUITS TUBE ACIER
Ø~0.50 INT.

CANISTER Ø 0.30

ECARTEURS

REMPLISSAGE SABLE

FORMATION ARGILEUSE

COUPE A.A

tractionel engineering
bureau Courtoy s.a.

C E N / S C K   PROJET HADES

PUITS DE STOCKAGE HLW ET HULLS
SOLUTION 2A
PUITS OBLIQUES NON VENTILES
FIG. 6

- Colmatage des forages de congélation. Ces forages de congélation sont disposés en couronnes concentriques autour des puits et constituent un système de drains vers la formation hôte, drains qu'il est indispensable de colmater de façon parfaite afin d'éviter tout risque de migration préférentielle vers les nappes a-quifères et la surface.

Pour cela il conviendra de retirer d'abord les tubes congélateurs, d'ob-struer ensuite les sondages au moyen de bouchons d'argile tassés par battage ou de bentonite injectée sous pression sur toute la hauteur de la couche d'argile tra-versée et compléter enfin le colmatage avec du sable sur la hauteur restante.

- Remblayage des puits. La dernière phase du remblayage consistera à com-bler les puits d'accès de façon à rendre à la formation les qualités d'isolement qu'elle présentait avant le début des travaux. Pour cela il est envisagé de recon-stituer les couches géologiques traversées de façon à les rendre pratiquement simi-laire à leur état initial, tout au moins dans et à proximité de la couche argileuse.

Cette contrainte impose de démolir la paroi des puits sur une hauteur plus grande que l'épaisseur d'argile traversée après recongélation préalable d'un anneau de terrain d'épaisseur correspondante. La partie supérieure du revêtement du puits demeurant dans le terrain sera comblée à l'aide de sable compacté et sa stabilité sera assurée à sa base par une fondation adéquate.

Les sondages de congélation ayant permis l'opération seront ensuite col-matés. Outre le fait que le coût d'une telle opération serait du même ordre de grandeur que celui du fonçage des puits, il conviendra de tenir compte des problè-mes de stabilité qui risquent de se poser, dus en grande partie aux phénomènes de compaction différée des couches reconstituées.

3. Programme expérimental futur concernant les matériaux de colmatage et leur tech-nique de mise en oeuvre.

Le programme expérimental qui sera développé dans les années futures au C.E.N./S.C.K. s'attachera à résoudre les problèmes spécifiques mis en lumière lors de l'étude conceptuelle. Ce programme devra conduire au choix de différents maté-riaux présentant les qualités adéquates et convenant pour les différents cas envi-sagés, il devra également développer et mettre au point les techniques permettant de mettre en oeuvre ces matériaux.

3.1. Matériaux de colmatage.

Ces matériaux présenteront des propriétés différentes suivant leur desti-nation dans les différentes cavités des installations souterraines.

3.1.1. Matériaux de colmatages des puits radiaux d'enfouissements pour déchets à haute activité.

Différentes sortes de sables de compositions et de granulométries variées seront testées afin d'étudier :

- leur comportement thermique à long terme

- leur comportement vis-à-vis des rayonnements

- leur capacité de corroder les revêtements métalliques avec lesquels il seront en contact

- leur capacité a être retirés

- leurs capacités de sorption (notamment en ce qui concerne les sables glauconitiques)

3.1.2. Matériaux de colmatage des galeries et à l'extrados des revêtements.

Ces matériaux seront constitués de mélanges dans lesquels interviendra toujours en plus ou moins grande proportion un ciment permettant d'assurer les pro-priétés de stabilité mécanique et différents adjuvants tels que la bentonite ou l'argile traitée ou tous autres minéraux assurant au mortier de bonnes capa-cités de sorptions.

Les expériences sur ces matériaux devront étudier :

- leur résistance mécanique

- leur capacité d'échanges ioniques

- leur mise en oeuvre par injection ou projection

- leurs caractéristiques de vieillissement

- leur capacité de corrosion des revêtements métalliques

- leur lixiviabilité.

3.1.3. Matériaux de colmatage des puits et des forages.

Les puits et les forages étant les voies d'accès préférentielles vers le site il conviendra de les boucher de la façon la plus parfaite possible. Le matériau qui semble le plus adéquat à cet égard semble être l'argile que l'on aura extraite lors de la phase de creusement.

La question se pose de savoir si cette argile doit être reconditionnée avant d'être mise en oeuvre. Il apparait que, pour minimiser les phénomènes d'interface, l'argile de colmatage devra se rapprocher les plus possible de l'argile en place. Or l'argile extraite lors de la phase de creusement aura été remaniée mécaniquement (perte de sa texture orientée), aura été exposée à l'air (phénomènes d'oxydation), aura perdu une partie de son eau intersticielle (phénomènes de retrait et de variation de composition chimique globale). Il peut paraitre donc illusoir de vouloir colmater avec une argile identique à celle de la couche en place.

Une autre approche est de reconditionner complètement l'argile avant de la mettre en oeuvre comme matériau de colmatage. Dans ce cas on viserait à obtenir un matériau de composition voulue aussi régulière que possible et aussi proche que possible du matériau idéal (dont la composition peut être différente de celle de l'argile du massif en place). Une telle approche nécessite un traitement mécanique et chimique de l'argile avant sa mise en oeuvre :

- traitement mécanique (séchage, suivi d'un broyage poussé) permettant d'homogénéiser les granulométries de la masse argileuse et de répartir les différentes fractions minéralogiques de façon la plus régulière possible,

- traitement chimique (addition de produits basiques comme des carbonates, addition de produits à grande capacité d'échanges ioniques comme des zéolites, éliminations de certains composés gênants comme la pyrite), destiné a fournir un matériau se rapprochant le plus du matériau idéal déterminé par approches successives en laboratoire.

3.2. Techniques de colmatage.

En l'état actuel du programme de recherche et développement du C.E.N./ S.C.K., il est prématuré de vouloir tester toutes les techniques de mise en oeuvre qui seront nécessaires au colmatage de toutes les cavités. Néanmoins du fait que la construction d'un puits et d'une galerie expérimentale vient de débuter, certaines techniques particulières concernant les colmatages des vides situés entre l'extrados des revêtements et le massif d'argile devront être développées.

- traitements des vides d'extrados de la galerie.

Ceux-ci pourront être colmatés avec différents matériaux types au cours de l'édification de la galerie expérimentale et leur propriétés testées "in situ" dans des conditions réelles. De nombreux essais portant sur le comportement global de ces matériaux pourront être entrepris (contraintes-déformations, perméabilité, pression interstitielle, interfaces, corrosion).

- traitements des vides d'extrados des puits radiaux.

Il est prévu de foncer à partir de la galerie expérimentale de nombreux forages radiaux nécessaires aux diverses expériences à entreprendre dans le massif d'argile. Certains de ces forages d'un diamètre pouvant aller jusqu'à 50 cm seront mis à profit pour tester les techniques de forage, de placement, des revêtements et d'injection à l'extrados de ceux-ci.

- colmatage des forages de congélation.

Le puits d'accès expérimental en cours de fonçage au C.E.N./S.C.K. se réalise suivant la méthode de congélation. Deux couronnes concentriques de 16 forages de congélation chacune seront donc disponibles après complétion du puits pour divers essais d'enlèvement des tubes et de colmatage au moyen de divers matériaux.

3.3. Expérimentations futures pour le remblayage de la galerie et de son puits d'accès.

A long terme, une fois que tous les essais "in situ" auront été réalisés , la galerie et son puits d'accès pourraient eux mêmes faire l'objet d'une expérience de remblayage à grande échelle. Les enseignements suceptibles d'être retirés d'une telle entreprise doivent justifier le sacrifice des installations souterraines.

Ces enseignements devraient donc concerner les matériaux de remblayage les plus adéquats et leurs techniques de mise en oeuvre mais aussi le comportement à long terme des installations ainsi rebouchées. Ce dernier point implique l'existence d'un réseau de surveillance permettant de transmettre des données depuis le sous sol jusqu'à la surface, sans apporter de perturbations aux paramètres mesurés.

Le développement d'un tel réseau n'ira pas sans poser des problèmes conceptuels et méthodologiques de difficulté considérable. En tout état de cause les expériences de remblayage à grande échelle ne doivent pas être attendues avant au minimum une dizaine d'années.

DRAFT PLUGGING PROCEDURE FOR DEEP HLW DISPOSAL BOREHOLES IN ROCK SALT

J. Hamstra
Netherlands Energy Foundation (ECN)
PETTEN N.H., The Netherlands.

ABSTRACT

A draft plugging procedure is presented for sealing deep vertical
boreholes in a salt dome burial facility, immediately after they are filled
with HLW canisters.
In addition to the normal plugging and sealing procedure attention is also
paid to emergency plugging provisions in case an unexpected flooding of the
burial facility might arise.
The plugging and sealing system is discussed and a number of "What-If"
questions are tentatively answered.

RESUMÉ

Un système est presenté pour la fermeture hermétique des orifices de
forage verticaux et profonds pratiqués dans un stockage de déchets
radioactifs dans un dome de sel, immédiatement après y avoir déposé des
containers avec des déchets de haute activité.
En plus du processus normal de fermeture hermétique, une attention toute
particulière est prêté aux dispositifs de bouchage d'urgence au cas
où une inondation se produirait au site de stockage sous-terrain.
Le système de fermeture hermétique est discuté en tentant de répondre à
un certain nombre de questions sur des événements qui pourraient survenir.

1. INTRODUCTION

In the Netherlands Energy Research Foundation (ECN) conceptual design for a high-level waste repository to be excavated in a medium size salt dome the disposal of canisters of vitrified waste is envisaged in deep vertical boreholes |1|. An optimal use of the vertical dimension in which the rocksalt is available for disposal purposes in a salt dome aims at the realization of low maximum rock salt temperatures in the HLW-disposal areas and of a high rate of spreading of the canisters over a large rock salt volume.

A recent drilling experiment performed in the Asse II-salt mine has shown that vertical disposal boreholes of 1 foot diameter can be realized with a dry drilling technique up to a length of 300 m and down to a depth of more than 1000 m.

Thus a HLW disposal geometry has become feasible in which several hundred canisters can be stacked one on top of each other in deep vertical bore-holes that are spaced about 50 m apart.

The upper part of each boring will be provided with a concreted steel casing that will have its top flange at disposal roadway floor level arranged as a centric flange for positioning the filling station, to be used for unloading the HLW canisters from their transport flasks.

Borehole convergence measurements, now underway in the experimental deep borehole, indicate for the Asse II rock salt at 1042 m depth a limited closure rate of about 0.1 mm per day. Notwithstanding this indication that the borehole diameter could have limited oversize compared to the outer diameter of the canisters, the borehole still is planned to have an oversize in diameter of about 50 mm.

It is assumed that the HLW canisters will be lowered by wire rope from the filling station into its position in the disposal borehole. Directly above the grab mechanism a bottom unloading barrel will be provided for transporting granulated material with which the gap between the borehole and the canisters is to be filled in-situ, once the canister is in position.

This transport barrel can also be used for bringing sufficient filling material down for making a first radiation shielding plug on top of the stack of canisters, prior to removing the filling station from above the open borehole. The succeeding plugging and sealing steps can then be performed while working above and around an open upper part of the borehole.

Whereas the containment shield of at least 200 m thickness, that will remain undisturbed around the burial facility, is considered to be the dominant barrier against any future radionuclide release, the plugging and sealing of the HLW disposal boreholes only plays a secondary role in providing adequate long-term isolation of the HLW.

The draft plugging procedure described in this paper was mainly worked out as a preventive provision, in case an unexpected flooding of the disposal mine during the operational period would be assumed feasible.

Because of the great number of canisters disposed of in one deep borehole, the length of borehole that has to remain available for plugging and sealing a filled borehole can be provided and the plugging and sealing procedure itself executed elaborately at relative low costs per unit.

The point to be stressed is that both the disposal in deep boreholes and the following draft plugging and sealing procedure for these deep boreholes have been worked out without any intention for retrievability.

Finally it should be kept in mind that the temperature conditions for a deep borehole disposal geometry can be defined by stating that the maximum rock salt temperatures at the disposal borehole wall next to a HLW canister will remain below 150°C, if heat production of the canisters will be limited to about 360 Watt per meter length.

2. REQUIREMENTS FOR FILLING, PLUGGING AND SEALING OF HLW DISPOSAL BOREHOLES

2.1. Radiation shielding requirements

Adequate radiation shielding and radiation protection is to be provided in the disposal roadway area directly surrounding the disposal borehole openings in the floor of the disposal roadway during the filling operation and the subsequent plugging and sealing operations.
Maximum radiation levels are still to be defined that will enable a continuous exposure of disposal miners, working in two shifts, at a rate of 30 underground exposure hours per week, at a frequency of one borehole filling every six weeks, a time span of about 90 hours for filling one borehole and a subsequent 30 hours for completing the plugging and sealing of that borehole.

Tentative radiation levels are

0.2 mr/h as a general level at roadway floor level;
2   mr/h at 1 m above floor level for a very limited area directly around
        the borehole under filling operation; and
50   mr/h at the surface of the HLW canister transport flasks and at the
        surface of any component of the filling station.

2.2. Alignment requirements

For alignment purposes during the first stage of drilling the deep borehole and later during the filling operation, the top part of the boring will be provided with a concreted steel casing of 3 m length.
For centring the filling station on top of the borehole the top flange of the casing will be executed with a centric flange.

2.3. Transport cask and filling station requirements

Even if an unshielded vertical transport of HLW canisters would be feasible from the surface down into the disposal mine, the horizontal transport at the disposal level still will have to be in transport casks.
The transport cask is to be provided with a sliding lid at the top and at the bottom. On top of the disposal borehole a filling station is to be situated, in which the filled transport cask can be situated and unloaded.
The filling station is to be provided with a sliding bottom lid centred on top of the disposal borehole, a construction to receive the transport cask in a vertical position, a shielded top part to receive the grab mechanism and the transport barrel for filling material and a hoist-winch.
The transport barrel may be assumed to be lowered from the shielded top part of the filling station for filling with granulated material during the time that the emptied transport casks is changed for a filled one.
Each HLW canister may be assumed to be kept in a central position in the borehole by the guide strips of the transport barrel during the in-situ filling of the gap between the canister and the borehole wall.
The transport barrel may also be used for bringing the plugging material down for making a first radiation shielding on top of the stack of canister prior to removing the filling station from above the open borehole.
The filling station is to be constructed in such a way that it need not much dismantling for a horizontal transport from one borehole to the next.

2.4. Plugging and sealing requirements

The complete plugging and sealing procedure has to provide for a reliable watertight seal against brine instrusion as a consequence of an assumed accidental flooding of the disposal mine during the operational period.
The watertightness is to be maintained under changing conditions such as

- the thermal expansion movement of the roadway floor and a local fracturing
  of this floor due to this movement,
- the borehole convergence that will compress the air in the voids still
  present once the gap around the canisters is filled with granulated material,
- a vertical movement of the canisters relative to the borehole plugging due
  to the density difference between the stack of canisters and the surrounding
  rock salt.

3. PLUGGING AND SEALING PROCEDURE

The plugging and sealing operation will start immediately following the
positioning of the last HLW canister in a borehole on top of the stack of
canisters. As shown in fig. 1 the first step in this procedure will be the
filling of the borehole over a length of at least 3 m with a mixture of clay
dust, crushed rock salt and possibly also fly ash.
This column of granulated material is assumed to provide sufficient radiation
shielding to allow for the filling station to be removed.
Once the filling station is removed, or with that station still in place but
with its sliding bottom lid permanently open, the plug mixture will be
vibrated and finished off with its top at exactly 1 m below the underside
of the borehole casing.
Apart from its initial radiation shielding function this first plug of rock-
salt-clay dust-fly ash mixture also will form a last barrier against unwished
brine intrusion down the borehole. The clay dust is assumed to absorb the
brine leak coming down and close the passage-way by its swelling reaction.
The fly ash is also assumed to absorb the brine it contacts and cement the
seepage ways.
Should however, in case of a flooding of the disposal mine, the assumption
be made that also this last barrier could fail and that brine could come
in contact with the HLW canisters than the clay and the fly ash will still
have an ad- and absorptive function for certain radionuclides, should they
become dissolved in the brine and start their pathway out of their confinement.
The initial plug of about 3 m length will be followed by a second plug, this
time of 2 m length and of clay dust, that will be vibrated in sections.
This second plug is purposely situated at the transition zone concreted steel
casing-rock salt. Its function is primarily to act as a barrier against brine
instrusion, both through the overlying borehole plug or along the outside of
the concreted casing, by its swelling reaction, when contacting brine.
Should the assumption be made that its sealing function could fail and should
also the underlying last barrier be assumed to fail than the potential pathway for the buried radionuclides out of their confinement still will be through
this clay material. Thus the clay can also be allotted an ad- and absorptive
function for certain radionuclides in case their migration upward through the
borehole would start.
The upper 2 m of the borehole casing on top of the clay plug is to be filled
with a good quality salt concrete.
In case the borehole plugging is to provide for an overhead protection against
a direct vertical penetration from a hypothetical future reconnaissance drilling into the stack of canisters, punch caps might be added to this concrete
to seriously obstruct a future drilling intrusion.
A cover plate provided with packing and bolted onto the centric flange is to
complete the closure of the disposal borehole opening.
If an obstruction against a future drilling intrusion is really judged
desirable then the cover plate can be replaced by a sugarbread shaped cast
steel plug which stem will replace the upper 60 cm of the concrete filling
of the casing. Prior to putting this cast steel plug in place a stopper of
cold bitumen will be put on top of the concrete. The bitumen will be pushed
out by the weight of the plug thus filling the gap between the plug stem and
the borehole lining.

Just for completeness sake an area of approximately 3 x 3 m of the roadway
floor in way of the borehole will be covered with two layers of reinforced
bitumen the first fixed and rolled onto a clean rock salt surface, the
second crossways fixed and rolled on top of the first one.

4. EMERGENCY PLUGGING PROVISIONS

In case of an unexpected accidental flooding of the disposal mine the
emergency situation might arise that the miners evacuate their stations on
such a short notice that normal procedures can remain unfinished at any step
in the procedure.

It should however be recognized that all the salt mine floodings that happened
the last 50 years, announced themselves far ahead of the moment of final
evacuation.

It therefore is assumed feasible to have emergency procedures that are to be
followed prior to any snap evacuation preceding the ultimate flooding of the
disposal mine.

An emergency plugging provision is envisaged to provide for an emergency plug
of granulated material followed by a stopper at any stage of the borehole
filling and plugging procedure.

The filling station will have to be provided with a special filling opening
in its bottom lid that can be unplugged in case of emergency.

Through this filling opening plugging material is to be supplied into the
borehole at any stage of the filling procedure. For certain there is filling
material readily available, that was to be used for filling the gap between
the canisters and the borehole wall. That can be used first to provide a plug
of 3 m height by emptying some five 50 liter bags through a funnel into the
filling opening.

More important will be the final filling with clay dust, that should also be
available at the filling station because it is needed anyhow during the normal
plugging procedure. Sufficient clay dust may be assumed to be available for
emergency use to provide an additionnal clay column of 2 m height above the
first plug.

This clay dust will, once it is wetted by the brine entering the borehole,
swell and thus accomplish a stopper function.

Should the emergency plugging procedure has to be performed immediately after
the last canister was emplaced and prior to any step of the normal plugging
procedure, than the special filling opening is also to be used to realize a
filling with the same materials that were to be used for the normal plugging
procedure. It must even be feasible to provide a salt concrete filling of the
top part of the casing.

It should be realized that part of the 30 hours a normal plugging procedure is
estimated to last is required for the hardening of the concrete, the subsequent
closure of the borehole casing and the final sealing with reinforced bitumen.
Under urgency conditions a rather adequate emergency plugging should be
realizable in a matter of hours.

5. DISCUSSION AND JUSTIFICATION OF THE PLUGGING PROCEDURE

This draft plugging procedure derived from a deliberate attempt to incorporate
a number of answers to different "What-If"-questions, related to the most
extreme operational period hazard, an unexpected accidental flooding of the
disposal mine.

It may well be that model calculations on radionuclide release rates following
a flooding accident will indicate that certain barriers incorporated in this
draft proposal are redundant. Yet the procedures as proposed in anticipation
to credible and uncredible events should not be weighed by rational arguments
alone. The regulatory procedures may well require some superabundance on
radionuclide containment measures.

The proposed plugging procedure is relative unexpensive because

- the barriers are provided at the source and in the relative smallest
  dimensions: that of the borehole;
- due to the great length of the disposal boreholes the total number of
  boreholes required for disposal of a given quantity of HLW will be relative
  small; and
- the additional length of borehole required to provide space for the multi-
  barrier plug construction can be provided at relative low additional
  drilling costs.

Reviewing the successive man-made barriers from the roadway floor level down into the disposal borehole, the following argumentation can be given for each successive barrier.

The two layers of reinforced bitumen fixed and rolled onto the rock salt surface of the roadway floor directly surrounding the borehole will have sufficient flexibility to follow the bending and elongation of the roadway floor that will be caused by the thermal loading of the rock salt surrounding the HLW canisters. Local fracturing of the surface of the roadway floor may occur due to the thermal expansion movements. These fractures will remain covered and protected against brine intrusion by the bitumen covering.

The concrete steel casing will follow the rock salt in which it is embedded in its upward movement. Its length is assumed to be an order of magnitude longer than the length of the fractures that might occur in the floor surface.

The salt concrete filling of the casing may also be assumed to follow the steel casing in its movement and to maintain its watertight attachment to the steel casing.

The cover plate provided with packing and bolted onto the centric flange of the steel casing is also assumed to remain protected by the reinforced bitumen covering.

The plug of clay dust, partly in and partly below the steel borehole casing, is a first back-up in case the bitumen covering would fail to keep the brine from penetrating towards the borehole casing, the brine would find its way along the cover plate and the attachment of the salt concrete filling to the steel casing should not be fully watertight.

Either a brine seepage down the inside of the steel casing or a seepage along the outside surface of the concreted casing will end up in the plug of clay dust. Through its swelling reaction and its subsequent plastic behaviour this plug of clay dust may be assumed to close the brine passage way and stop further intrusion of the brine down the borehole.

The last barrier consisting of a column filled with a mixture of crushed rock salt clay dust and possibly fly ash will by its two last components also have a certain capacity to stop an eventual brine seepage from penetrating towards the emplaced HLW canisters.

In case it is assumed that this local multi-barrier system still could fail, then part of its material composition will effect that the only transport mechanism for dissolved radionuclides to migrate out of the borehole will be diffusion. Diffusion through material columns that, especially for the long-lived actinides, have proven ad- and absortive qualities.

These qualities will not be deteriorated too much by relative high ambient temperatures, because the clay dust part of the column will be 3 m away from the top of the line source.

Is there any chance that certain developments may in time have an unfavourable effect on the plugging system?

For instance will the air, originally present in the voids of the granulated filling material around the emplaced canisters, be pressed out of the borehole by the inevitable borehole convergence?

This air, gradually compressed up to the rock pressure inforce around the borehole, will be forced upward along the stack of canisters or may also remain partly trapped between the granules. It may be expected that some 10 liters of air, compressed to some 100 bar, will present itself at the top of the stack of canisters.

Our tentative assumption is that whether the air will remain trapped or gradually diffuse away towards the roadway opening, it will do no functional harm to the plug system.

Is there any harm to be expected to the plug system from a vertical movement of the stack of canisters relative to the borehole wall and the borehole plugging?

It may be assumed that the canisters that are unloaded from their transport casks will have a somewhat higher temperature, than they will have once they are emplaced. This may result in an initial shrinkage of the stack length during the first time period after the plugging took place.

The rock salt supporting the bottom canister may well react under the stack load by displacement of some rock salt into the gap area around bottom part of that canister.

Our tentative assumption for this development is that all movements will be demonstrable small and negligible in respect to possible consequences for the borehole plugging.

Is there any consequence to be expected for the plug system from possible brine migration from the surrounding rock salt towards the emplaced canisters?
It was mentioned before that with the emplacement of the canisters a filling of the gap between the canister and the borehole wall with granulated material is envisaged.
It is assumed that this material will act primarily as a brine absorber.
It is also assumed that borehole convergence will cause a timely and sufficient pressure rise that no boiling will occur at the prevailing rock temperatures of less than 150°C. An eventual corrosion process on the canister walls will be sufficiently slow to allow the surrounding rock salt to build up its full rock pressure around and against the glass blocks.
The tentative assumption for this development is that once the canister have become fully embedded in the rock salt there are no consequences to be envisaged for the plug system from an eventual brine migration towards or corrosion along the stack of canisters.

Finally the question can be raised whether radiation damage to the filling material around the canisters and the rock salt that surrounds the borehole can have subsequent consequences for the plug system.
The answer to this question will partly depend on the choiche of filling material. In and around the stack of canisters solids in the form of canister wall material and filling material are readily available to react with products of radiolysis. For the present it is assumed that these reactions will have no consequences for the plug system.

If no other short term or long term changes in the direct rock salt environment around the borehole plugging system are envisaged that can detrimentally affect this system, than the final question can be raised whether some intruder from outside could break the plugging system part of the containment of the HLW canisters.
There is no answer to an assumption that a future generation might start a solution mining operation in the salt dome once used for waste disposal.
It is recognized that such an operation could result in the return to the biosphere of radionuclides. Because of the rather large spacing of the boreholes the amount of canisters that may be exposed to dissolution will than be limited.
It should however also be recognized that the subsequent chemistry required to treat the brine solution will be performed by a limited number of persons, who may be assumed by their profession to have sufficient analysing capacity to discover in time what materials they are handling.
There may be an answer to the other assumption that a future generation might perform reconnaissance drillings into the salt dome, once used for waste disposal.
The answer under consideration is to provide a sugarbread shaped cone of cast steel, as shown in fig. 1, on top of each borehole. This cone will make any direct hit through the borehole casing impossible because it will definitively deviate the reconnaissance drilling away from the borehole.
Whether this cast steel cone is really to be includes in the plugging procedure is however very questionable.

This paper is submitted for presentation at the OECD-NEA Workshop on Borehole and Shaft Plugging, as a draft proposal, open for discussion, possibly also on the questions in this last paragraph.

REFERENCE

|1| J. Hamstra and P.Th. Velzeboer, "Design study of a radioactive waste repository to be mined in a medium size salt dome". ECN-78-023, January 1978.

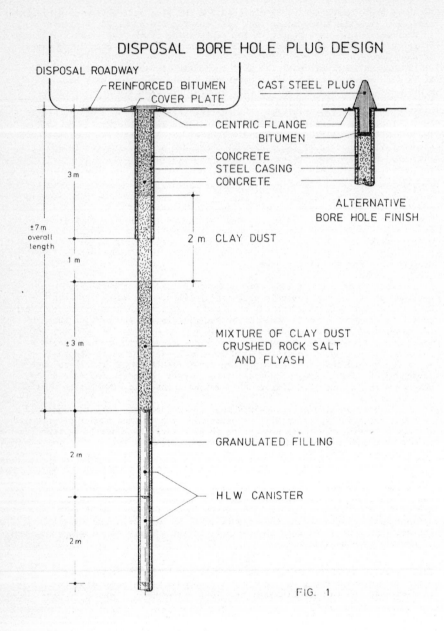

DISPOSAL BORE HOLE PLUG DESIGN

DISPOSAL ROADWAY

REINFORCED BITUMEN
COVER PLATE

CAST STEEL PLUG

CENTRIC FLANGE
BITUMEN

CONCRETE
STEEL CASING
CONCRETE

ALTERNATIVE
BORE HOLE FINISH

3 m

±7 m
overall
length

1 m

2 m   CLAY DUST

± 3 m

MIXTURE OF CLAY DUST
CRUSHED ROCK SALT
AND FLYASH

2 m

GRANULATED FILLING

HLW CANISTER

2 m

FIG. 1

Discussion

W. FISCHLE, Federal Republic of Germany

Why is fly ash used in the borehole plugging mixture ?

J. HAMSTRA, Netherlands

Fly ash is considered because it is expected to give more body to the plug column, once it is wetted with brine, and also because of its ion-exchange properties.

F.N. HODGES, United States

My question is in two parts : first, have you made calculations that would indicate the temperature distribution resulting from your canister configuration, and, second, have you estimated how long the steel components in the system would survive in the chemical environment of a salt dome ?

J. HAMSTRA, Netherlands

We have made temperature rise calculations for rock salt surrounding the emplaced canisters of high-level waste for different disposal geometries (Report ECN 42).

We have not estimated the longevity of the container material after emplacement. In our hazard analysis work no value is attributed to the container wall as a barrier.

R.D. ELLISON, United States

Will the mined entries above the plugs be backfilled ? If yes, what material will be used ?

J. HAMSTRA, Netherlands

The mine roadways will be backfilled with crushed salt and possibly fly ash once the boreholes are filled with waste canisters.

R. PUSCH, Sweden

Is the present rate of uplift due to diapirism known sufficiently well to guarantee that we do not get a rise of the canisters and maybe surface exposure in case of erosion of the caprock and the underlying salt ?

What will be the risk of large canister movements due to density differences and heat ? Could it be that a sufficient amount of waste is concentrated to generate the risk of criticality ? Such concentration might be produced by the presence of internal dome structures, which govern the rate and direction of movement.

J. HAMSTRA, Netherlands

Geologists can study the diapiric history of a specific salt dome and estimate the rate of uplift in the recent past.

Predictions on future upward movement can be based on these data and on the information whether or not there is still supply of rock salt in the mother layer surrounding the basis of the dome. There is a recent report by Dr. Werner Jaritz on upward movement of rock salt in the Zechstein basin in which the maximum upward flow rate during the diapiric stage is stated to be 0.2 to 0.4 mm per year and in which the rate of uplift in the post-diapiric stage is estimated to be two orders of magnitude lower.

With vitrified high-level reprocessing waste there is no future criticality risk ; this could only be the case with spent fuel if it were considered a waste.

Any movements of emplaced canisters relative to the surrounding rock salt would be very small, because the density difference is relatively small.

M.J. BARAINCA, United States

In answer to the question relative to the uplift of salt domes and the movement of canisters within a salt repository I wish to add that the uplift of domes is being studied in the United States. Our experience indicates that uplift rates are very small (less than 0.01 mm/year). An experimental program carried out by Louisiana State University (LSU) on the inland salt domes which utilizes a precise leveling network and tilt meters has not detected significant dome movement. Other natural phenomena such as changes in the water table do affect the surface instruments ; these natural variations are so great that any dome uplift is masked. Further information could be obtain from the LSU reports.

In relation to the question on movement of canisters the movement of canisters was insignificant during the period of study at the Avery Island salt dome. Dick Robinson of ONWI could provide more details on this particular aspect.

ACTIVITIES IN THE FIELD OF BOREHOLE AND
GALLERY PLUGGING IN THE ASSE SALT MINE

W. Fischle
Gesellschaft für Strahlen- und Umweltforschung mbH München
Institut für Tieflagerung - Wissenschaftliche Abteilung
Schachtanlage Asse
3346 Remlingen

ABSTRACT

Am Ende der Einlagerungsphase von radioaktiven Abfällen muß der Kontakt mit der Biosphäre verhindert werden. Untersuchungen sollen zeigen, mit welchen Materialien Bohrlöcher, Kavernen und Strecken langzeitig verschlossen werden können. Im folgenden werden in situ-Versuche in der Asse beschrieben.
1. Ein Test zur Versiegelung von Bohrlöchern.
2. Ein Versuch zum Verschließen von Kammern und Verfüllen von
   Strecken.
Es sollen sowohl die stofflichen Eigenschaften als auch die Verfülltechniken untersucht werden.

SUMMARY

After the emplacement of radioactive waste, contact with the biosphere must be prevented. Tests shall show with what kinds of materials boreholes, caverns and galleries can be locked for a long time. The following describes two in situ-tests in the Asse:
1. A borehole plugging test.
2. A full scale chamber entrance sealing test.
The quality of materials shall be tested as well as the technology of filling.

In the deep geological disposal of radioactive waste various types of free valium have to be mined. From the surface exploratory-boreholes and shafts will be sunk. Underground galleries, chambers and caverns will be mined and disposal boreholes drilled at mine level.

After the emplacement of radioactive waste contact with the biosphere must be prevented.

Therefore borehole-plugs and gallery-dams are being developed as multiple barriers to insure isolation of waste. Since European salt repository mine workings will intersect several types of mineral deposites with dips from $\emptyset$ - 180 degrees. A broad selection of sealing materials and technologies must be developed.

An important material consideration is the reaction of the fill with various host rocks, particularly during the emplacement incuring phase. For example, wet pumpable materials are generally not compatible with carnalite. In regard to emplacement technologies the requirements for vertical boreholes, galleries and large drum filled caverns are greatly different.

Material selection for the borehole plugs must consider contact with different minerals and brines under elevated temperatures and pressures.

The following describes two in situ-tests in the Asse:

A borehole plugging test and
a full scale chamber entrance sealing test.

For the material suitability testing of borehole sealing materials for middle and high active waste a test gallery was prepared on the 775 m-level in the Asse.

It is situated in the Younger Halite (third Zechstein sequence). At this point it slopes about 45 degrees to the southwest. Here the salt cristals are clear till milky with grain-sizes from 0,2 to 2,0 cm. Apparently, folding during diapirism has stressed the salt and the cores can be crumbled with the fingers. Clay and sulphite flakes of 0,5 to 4 cm are dispersed throughout the unit and contribute less than 1 % of the weight.

In this experimental drift 20 boreholes of 10 m depth and a 170 mm diameter were bored at a distance of about 2,5 m. There are three groups of these boreholes, parallel perpendicular and diagonal to the original bedding plains.

A gas permeability probe with mechanical packers is being used to determine if the borehole intersects permeable cracks.

The packer is set 1 m deep into the borehole and the pressure drop from 7,5 bar is monitored.

The lithostatic pressure at this depth is about 17,4 N/mm^2. The ratio of borehole pressure to the host rock pressure is 1:23. According to the literature, the host rock cannot be fractured at this pressure. However fractures and zones of loose salt can be detected and taken into account. Should a pressure drop occur the packer will be set deeper in order to determine the location of the permeable zone.

Laboratory pretesting has selected the various seal materials. A previous in situ borehole plugging demonstration in the Asse used a multimaterial layered seal, which was stable for about 1 year. The present test will determin the performance and thickness requirements of the single components of a multi-seal. The workability will be tested considering the special handling precautions, necessary when the hole is filled with high active waste.

At that point the durability under convergence pressure of the host rock is monitored by gas-permeability measurements under the previously mentioned conditions.

After the pressure remains constant for a long time a core hole will be drilled into the plug until a pressure drop is seen. the point at which the pressure drops will indicate the thickness of the drop necessary for a hermetic seal.

The cores from these plugs will be examined in the laboratory testing for brine resistance, heat transfer and mechanical properties.

We planned to use the following materials: rock salt, cementious products with and without additives, bitumen, melamine resin, prefabricated parts of steel or ceramics.

We have selected prefabricated plugs because:

1. They can be produced to specific tolerances.
2. The properties can be reproduced.
3. They can be stored.
4. They can be compatible with the waste handling system.

Since the emplacement of these prefabricated-plugs requires a gap between the borehole wall and the plug which will close quickly at the temperature expected with high level waste. These conditions will be simulated in one or two of the boreholes using heaters.

The second in situ-experiment is the backfilling of a chamber entrance.

When the disposal in a particular gallery or a section of the mine is complete then the sealing is required.

Special materials and techniques are necessary. The goal is to construct dams with a high density strengh and homogeneity.

On the 750 m-level in the Asse a gallery of 40 m length, 1,6 m width and 1,7 m height is prepared to be filled. This gallery is an entrance way to a chamber and walled on the chamber end.

The axis is from northeast to southwest with a slide incline. It intersects in the major carnalite, polyhalite, kiserite and older halite.

About 50 % of the gallery is in the carnalite. Carnalite is a very soluble potassium mineral and very hygroscopic up to six waters of hydration.

This test imposes the following conditions:

1. Filling can only be done from one end.
2. Materials requiring water to cure must not allow free water to escape.
3. The filling material must tightly bind to the top of the gallery and swell to fill all open spaces.

In the laboratory a series of materials have been tested. A fly-ash based material from Poland has been selected for this in situ-test. It is a dry powder requiring 25 % of water for curing. It reaches a relatively high compressive strengh of 10 N/mm² after a swelling little more than 10 %. Katacit will be filled into the gallery according to a Polish patented process.

To monitor the sealed gallery the following instruments have been installed:

1. Oriented flat jacks to measure the swelling pressure.
2. Hydraulic pore pressure probe.
3. 5 gas permeability probes mounted along the gallery.
4. 4 guide tubes for ultrasonic measurements along the complete length of the gallery.
5. 2 hose leveling instruments to watch the movements of the floor and the sealing.

After a period core samples of the Katacit will be taken and tested in the laboratory. The properties of the in situ-cured material will be compared to the previous laboratory consults.

These two tests shall determine the feasibility of various materials for the sealing of boreholes and galleries.

Discussion

R.H. GOODWIN, United States

Could you tell us the material components and thicknesses of the in situ plug discussed in your talk which remained stable for one year ? Do you know what caused the plug to become unstable ?

W. FISCHLE, Federal Republic of Germany

I do not know the composition of the plug by heart.

Stable in my report means that the gas pressure remained constant.

D.M. ROY, United States

You have described the use of a fly ash material, katasite, as a component of a borehole plugging mixture. Have you analysed it ? Is it a proprietary material, and are there any additional components beyond the fly ash in the initial material ? The expansion described is greater than that observed in usual fly ashes characterized by high calcium and sulfate.

W. FISCHLE, Federal Republic of Germany

Katasite shall not be used as a component of the borehole plugging material ; we want to test it for sealing horizontal paths in a salt mine as there are rooms, galleries and so on. I do not know the chemical composition of katasite.

T.O. HUNTER, United States

When will the horizontal backfill experiments by started ?

W. FISCHLE, Federal Republic of Germany

The instrumentation has been emplaced. We hope that back-filling will start in the first part of this year.

R.D. ELLISON, United States

How is the gap between the borehole wall and prefabricated plugs filled ?

W. FISCHLE, Federal Republic of Germany

We expect that the prefabricated plugs can be made very exactly. The unavoidable small gap will be closed by the salt convergence, accelerated by the temperature rise due to the high-level waste.

G.M. IDORN, Denmark

What is the mechanism by which fly ash expands in the boreholes ?

<u>W. FISCHLE</u>, Federal Republic of Germany

        This particular fly ash is a Polish product. I do not know the reasons of the swelling properties.

<u>K. MATHER</u>, United States

        Fly ashes from lignites can acquire swelling properties.

ROLE OF BOREHOLE PLUGGING IN THE EVALUATION OF THE
WASTE ISOLATION PILOT PLANT

T. O. Hunter
Experimental Programs Division 4512
Sandia National Laboratories
Albuquerque, New Mexico 87185
United States of America

ABSTRACT

        Research on borehole plugging (BHP) is part of an integ-
rated strategy to develop technology that can assure successful
nuclear waste isolation. The application of this strategy to the
Waste Isolation Pilot Plant (WIPP) in southeastern New Mexico has
included an assessment of the role BHP plays in the development of a
repository at that site. This paper presents a description of the
WIPP site, repository design, and the current research and development program. The status of drill holes--those drilled for petroleum
and potash exploration and those drilled for site characterization--
within the proposed site boundaries is presented. Sixty-six holes
are present on the 7700 hectare (19,000 acre) site, yet only 8 penetrate as deep as the proposed repository location. The assumptions
made about shaft and borehole sealing in consequence assessment
studies are presented. The results of these studies indicate that
borehole seals with effective permeabilities greater than tens of
darcies would result in doses to maximally exposed individuals of
less than 0.01% of natural background.

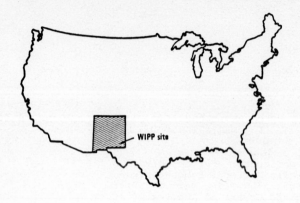

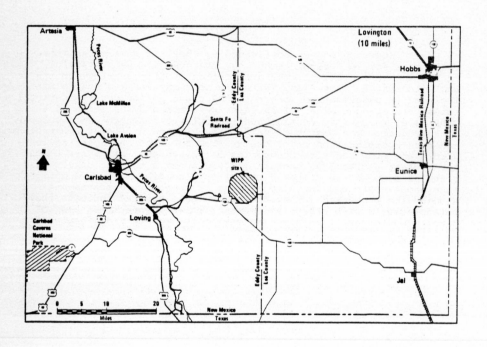

Figure 1.   GENERAL LOCATION OF WIPP SITE

INTRODUCTION

The U.S. Department of Energy (DOE) proposed the development of a repository for disposal of nuclear waste from the defense programs at a site in New Mexico. This project, called the Waste Isolation Pilot Plant (WIPP), has a twofold mission: a pilot repository for retrievable storage of defense transuranic (TRU) wastes, and a facility for experimentation on the interaction of defense high-level wastes and bedded salt.[1] Current recommendations by the DOE include the termination of project activities which support the unlicensed disposal of TRU wastes and research and development on defense waste.[2] The site in southeastern New Mexico would continue to be evaluated along with other sites in the United States for potential consideration as a licensed repository for both defense and commercial high-level wastes.

SITE AND REPOSITORY DESCRIPTION

Studies relating to the design, operation, and safety assessment of the proposed WIPP repository are based on assumptions about the characteristics of the waste to be received there. The long-term (post-operation) assessments that were performed are based on TRU wastes, whose general characteristics were conservatively presumed to be those presented in Table I.[1] These wastes are primarily contact-handled and produce insignificant quantities of heat and penetrating radiation. A detailed study of the potential interaction between these wastes and the environment imposed by the WIPP repository has been performed for development of waste acceptance criteria.[3]

TABLE I

WIPP WASTE FORMS
(Transuranic Wastes)

o   TRU Content $>$ 10 NCi/gm

CONTACT HANDLED

o   210-liter drum $<$ 5-10 gm Pu
o   0.04 Watt/drum
o   Surface Dose Rate (SDR)   200 mrem/hr

REMOTE HANDLED

o   200 mrem/hr $<$ SDR $<$ 100 rem/hr
o   4 watts/(0.6 x 4.6 m can) (2 x 15 ft can)

Site Characterization

Geologic and hydrologic investigations have been in progress in southeastern New Mexico since 1972. The site presently under investigation is called the "Los Medanos" site and is located in the Delaware Basin, which is one of a series of sedimentary basins within the Permian Basin that extends into Texas, Oklahoma, Colorado, and Kansas. The general location of the WIPP site is shown in Figure 1. Extensive geologic and hydrologic investigations have resulted in a detailed characterization of the site under consideration.[4]

The stratigraphy at the site is illustrated by the cross-section shown in Figure 2. The intervals of particular

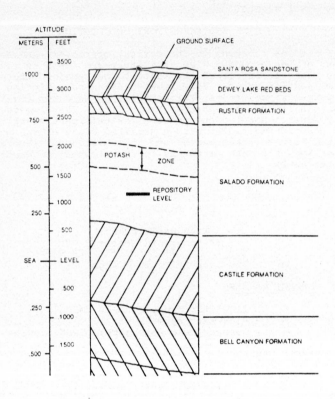

Figure 2.   STRATIGRAPHY AT WIPP SITE

interest in the evaluation of the consequences of radionuclide release and consequently the integrity of penetrations are: (a) Rustler Formation, which, at the center of the site, extends from 168 m (550 ft) to 247 m (810 ft) below the surface; (b) the Salado, which extends from the Rustler to about 850 m (2800 ft) below the surface and (c) the Delaware Mountain Group, which is about 1220 m (4000 ft) thick with its top at about 1220 m (4000 ft) below the surface.

The Rustler is primarily anhydrite and siltstone, but does contain two dolomite beds--the Culebra and Magenta, the most signifi- cant aquifers in the area--each of which are about 7.6 m, (25 ft) thick at depths of about 220 m and 186 m (720 and 610 ft), respec- tively. The Culebra and Magenta are considered confined aquifers with an average porosity of about 10% and a calculated transmissivity ranging from 10^{-10} to 1.5×10^{-4} m^2/s (10^{-4} to 140 ft^2/day), depend- ing on the locality and degree of fracturing. The total dissolved solids range from 3000 to 60,000 ppm.[4] The Salado Formation con- tains the location of the proposed repository horizon (660 m, 2160 ft) within its lower member. At the repository location, the Salado is primarily halite but also has thin beds of anhydrite and poly- halite. Thin clay zones are also present. The Salado does not con- tain circulating groundwater.

The Delaware Mountain Group consists primarily of sand- stone, limestone, and shale and has been characterized as having an average porosity of 10% and an average conductivity of 7×10^{-8} m/s (0.02 ft/day). The upper formation in this group is the Bell Canyon Formation. Groundwater yields from wells in Bell Canyon Formation are approximately 3.8×10^{-5} to 9.5×10^{-5} m^3/s (0.6 to 1.5 gal/minute).[4]

Repository Design

The proposed repository for defense wastes at the WIPP site is depicted in Figure 3. It consists of a single-level excavation using conventional room and pillar mining techniques. Surface facilities used during operation include a waste handling building, an underground personnel building for support of under- ground operations, a storage-exhaust-filtration building, an administration building, and various support buildings. The sub- surface development includes four shafts to the underground, storage rooms for TRU wastes, and areas for experiments.[5]

EXPERIMENTAL PROGRAM

The development of the WIPP has been supported by an exper- imental program that addresses the technical concerns identified in the design and safety assessment of the facility. This program in- cludes eight program areas that are oriented towards specific issues.[6] A brief description of these areas follows:

Tru Waste Characterization Studies--Studies of the interaction of transuranic wastes with the salt environ- ment, including assessment of potential degradation mechanisms and the impact on the repository and radio- nuclide isolation.

High-level Waste Interaction--Studies of the interaction of thermal and radiation fields from heat-producing wastes with the salt environment and the impact on the waste form encapsulating materials.

Thermal-Structural Interaction--Development and verifi- cation of methods to evaluate the response of the host

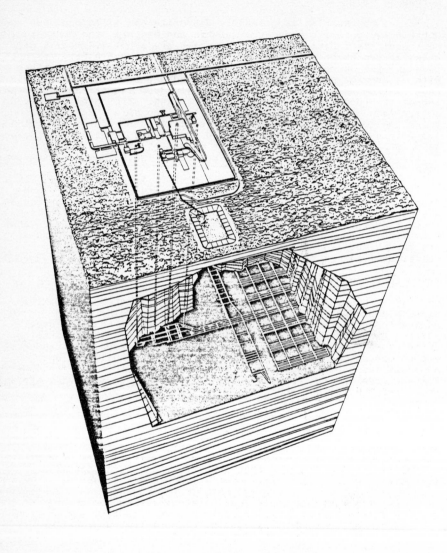

Figure 3.   THE PROPOSED WIPP REPOSITORY

rock to both the ambient conditions upon excavation and the enhanced deformation anticipated with heat-producing waste forms.

Permeability--The characterization of the properties of the host rock for permeation of gases or liquids.

Brine Migration--Assessment of the potential for mobilization of natural fluids in the salt and the subsequent interaction with waste containers.

Borehole Plugging--Quantification of the technology for sealing man-made penetrations into or near the storage horizons.

Nuclide Migration--Characterization of the potential for radionuclide migration in the WIPP environment.

Operation and Design--The demonstration and certification of safe operational techniques and appropriate design assumptions.

Each of these areas is being methodically evaluated to accumulate additional data and to correlate all available knowledge into an assessment of the potential consequences to the integrity of the waste isolation mechanism.

Each area may consist of several development phases:

Modeling	Analytical formulations that can represent or predict phenomena of interest
Laboratory Analyses	Specific laboratory investigations to develop important parameters and accumulate relevant data
Bench Scale	Intermediate-scale investigations employing larger samples than typically encountered in a laboratory scale
In Situ Testing	Non-radioactive experiments and demonstrations performed in undergound salt environments that can incorporate in situ parameters
WIPP	The collection of requirements from each program area into a description of and comprehensive plan for experiments to be performed in the portion of the WIPP facility devoted to that purpose

Borehole Plugging (BHP) is only one component in the development program. The overall assessment of the integrity of the waste isolation system requires the integration of each of these components. The appropriate role of BHP is best evaluated by considering the relative influence of it and other factors, e.g., characterization of the waste stored, interactions with the host rock, response of rocks to repository development and post-operational effects, migration of radionuclides in the local and regional geosphere, in providing acceptable containment and isolation of wastes.

The generic nature of the BHP program has resulted in a cooperative effort between the WIPP program and the Office of Nuclear Waste Isoation (ONWI). This cooperative effort will allow transfer of technology to repository development at other sites.

Figure 4

IMMEDIATE AREA
WIPP SITE DRILL HOLE STATUS

Sandia Laboratories

SEPTEMBER 1978

--- LEGEND ---

⊕ DEEP PRODUCING GAS
╋ ABANDONED WELL
⊕ DEEP & ABANDONED
⊕ POTASH DRILL HOLES
○ GEOLOGICAL HOLES
● HYDROLOGICAL HOLES
⊕ ERDA POTASH DRILL HOLES
⬨ (LAND WITHDRAWAL BOUNDARY)

V-U S POTASH CO (MISSISSIPPI POTASH)
D-DIS DUVAL CORP
I-INTERNATIONAL MINERALS
PP-PERMIAN POTASH
F DM-FC-KERR MCGEE

--- NOTES ---

1. ONLY THOSE HOLES WITHIN A 5 MILE GRID NORTH
   AND SOUTH AND EAST AND WEST OF ERDA 9 ARE
   SHOWN WITH THE EXCEPTION OF THE ERDA/WIPP
   SPONSORED HOLES INDICATED

2. DRILL HOLES NOT SHOWN TO SCALE

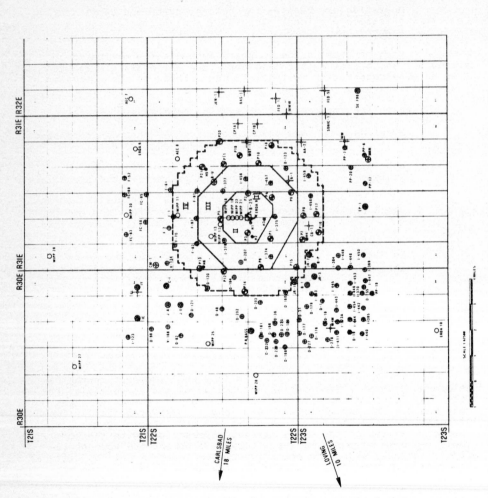

- 206 -

Subsequent sections will discuss the status of penetrations at and near the WIPP site and the assumptions that have been made in the safety assessments about plugging of these penetrations.

PENETRATIONS ASSOCIATED WITH WIPP

Siting studies for a nuclear waste repository must consider the location of existing drill holes within regions under evaluation. Sites for salt repositories such as WIPP are very likely to be located in regions that have been subject to exploratory drilling for hydrocarbons or minerals such as potash ore. The identification of those existing holes is essential in the characterization of a repository site.

In addition, site investigation will require the drilling of numerous holes to evaluate geologic and hydrologic characteristics or to further assess the potential for valuable minerals.

The WIPP site has been divided into four zones that will be controlled by the Department of Energy. Zone I consists of about 40 hectares (100 acres) and will contain most of the surface facilities. Zone II, an area of about 730 hectares (1800 acres), overlies the maximum extent of the underground development. Zone III, with a diameter of approximately 6.4 km (4 miles) and an area of 2500 hectares (6200 acres), is the area in which drilling and mining will be precluded unless further evaluation indicates it is acceptable. Zone IV, with a diameter of 10 km (6 miles) and an area of 4400 hectares (11,000 acres), would allow continuous or drill-and-blast mining under DOE restrictions, but no solution mining. Existing wells in Zone IV would also require sealing by DOE-prescribed methods, and new wells would be drilled and sealed in conformance with DOE standards.

Boreholes

The most credible threat to the WIPP repository imposed by boreholes is communication with deep aquifers. The criteria used in the identification of the current site included the avoidance of land within one mile of any boring into the Delaware Mountain Group or deeper formations.[4] Consequently, the current site does not include any deep holes within one mile of Zones I or II.

Site characterization activities for WIPP in southeastern New Mexico have included the drilling of 71 holes in the general vicinity of the site. Figure 4 shows the location of those exploratory drill holes and the existing industry holes that are within a 260 km^2 (100 mi^2) region surrounding the 4 zones that comprise the site.

The drill holes within the site, both existing industry and DOE exploration, can be generally categorized by such functions as: (1) stratigraphic, (2) hydrologic, (3) potash exploration, and (4) hydrocarbon exploration. A total of 66 holes are present within Zones I, II, III, and IV. Forty-five were drilled by DOE, while 21 were present when site investigation activities began. These drill holes are categorized by function and zone in Table II.

None of the holes in Zones I or II, shown in Table II, extend to the Delaware Mountain Group (DMG). The deepest (ERDA-9) is 880 m (2886 ft) deep, which just penetrates into the Castile Formation (861 m, 2825 ft), and stops 360 m (1200 ft) from the Castile/DMG contact. The holes listed in Table II are displayed in Figure 5 as a function of depth and distance from the center of the site. The only deep holes in Zone III are WIPP-12 and WIPP-13, stratigraphic holes drilled by DOE near the northern perimeter of Zone II and midway between the boundaries of Zones II and III, respectively. An oil exploration hole (BWU-1) that penetrates 4600 m

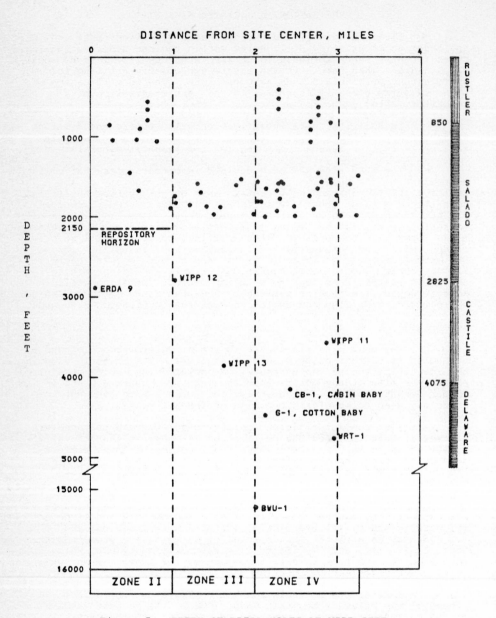

Figure 5.   DEPTH OF DRILL HOLES AT WIPP SITE

TABLE II
CATEGORIZATION OF DRILL HOLES WITHIN THE WIPP SITE

	STRATIGRAPHIC	HYDROLOGIC	POTASH	HYDROCARBON
ZONE I	1 (DOE)			
ZONE II	4 (DOE)	5 (DOE)	2 (DOE)	
ZONE III	2 (DOE)		5 (DOE) 7 (IND)	
ZONE IV	3 (DOE)	9 (DOE)	14 (DOE) 10 (IND)	4 (IND)

(15,000 ft) is the deepest hole on the site and is located near the inner perimeter of Zone IV approximately 1 mile from Zone II. In all 4 zones, only 8 holes penetrate as deep as the proposed repository horizon and none of the remaining 58 extend closer than several hundred feet of that level. The emphasis of the borehole plugging program at the WIPP site is thus concentrated on those holes that extend to or through the repository level, since no mechanism is envisioned by which fluids can reach the repository from holes that are terminated above the repository. Density stratification in a terminated hole will not allow continued dissolution and, moreover, hole closure in salt will ultimately seal the penetration.

Shafts

The design for the WIPP repository includes 4 shafts. Each extends to the repository level (665 m, 2150 ft) and is lined from the surface to the top of the salt (about 260 m, 850 ft) but unlined thereafter. These shafts are described as follows:[5]

Waste Shaft - 5.8 m (19 ft) in diameter, for transporting of waste

Ventilation Supply and Service Shaft - 4.9 m (16 ft) in diameter, for transporting personnel, materials, and equipment, and for primary air intake

Construction Exhaust and Salt Handling Shaft - 4.3 m (14 ft) in diameter, for salt removal and exhaust air duct for underground construction areas

Storage Exhaust Shaft - 3m (10 ft) in diameter, the exhaust air duct for underground waste storage areas

Site validation activities at the WIPP included plans for early development of two shafts for underground site validation and in situ experimentation.[7] These two shafts would ultimately be converted into the storage exhaust shaft and the waste shaft upon full construction of the repository. The latter shaft would be initially bored at 1.8 m (6 ft) diameter and subsequently expanded to full size. All shafts would be completely plugged upon cessation of operations. One of the shafts would also be coincident with the drill hole (ERDA-9) that currently penetrates through the Salado Formation, thus eliminating concern over two penetrations at that location.

CONSEQUENCE ASSESSMENTS

Determination of the potential hazard from disposal of nuclear waste is addressed by considering various scenarios in which the primary isolation mechanism--the media immediately surrounding the repository--is assumed to be breached. A series of these scenarios have been identified for WIPP.[8]

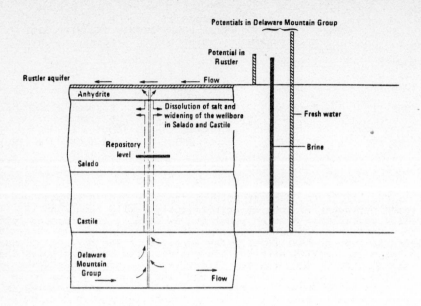

Figure 6. SCHEMATIC OF SCENARIO 1

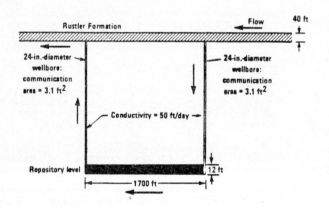

Figure 7. SCHEMATIC OF SCENARIO 2

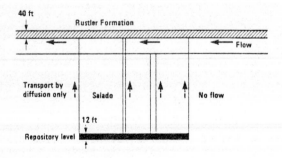

Figure 8. SCHEMATIC OF SCENARIO 3

Consequences of the radionuclide release in these scenarios have been determined and published.[1,5,9] Four of these scenarios are:

1. An open borehole that penetrated the repository allowing fluid transmission between lower and upper aquifers.

2. Two interconnected penetrations of overlying formations that allow fluids to enter the repository at one location and escape into the upper aquifer at another.

3. Diffusion of radionuclides from a flooded repository into overlying aquifers.

4. A direct access scenario in which a drilling operation intercepts a high concentration of radionuclides, thereby exposing the drill crew.

The first three of these scenarios are referred to as liquid breach and transport scenarios and can be examined to determine the role played by borehole plugging.

Consequence assessments are made in terms of the potential dose to an exposed individual in the nearby biosphere. Scenarios are analyzed by first specifying (a) the breaching event, (b) mechanisms for transport out of the repository, (c) the time of breach, and (d) the response of the buried medium to the releasing event. Secondly, the source term in terms of radionuclide inventory and physical and chemical condition of the waste at the time of breach must be determined. Thirdly, geosphere transport calculations are performed to determine the rate of radionuclide transport to the biosphere. In the WIPP analyses, the Rustler aquifers are assumed to be the link to the biosphere. Radionuclides are assumed to move from the repository through the Magenta and Culebra aquifers and to discharge into the Pecos River at Malaga Bend, a point 22 kilometers from the site. Finally, analyses are performed to determine the dose to invididuals who live near the assumed discharge point. These calculations translate the mass fraction of radionuclides in the aquifer water into whole body or specific organ doses considering ingestion, swimming, boating, food consumption, occupancy, and other pathways.

The three liquid breach and transport scenarios are depicted in Figures 6, 7, and 8. A brief description follows:

Scenario 1: A 9-inch uncased borehole allows up to 2.3 m^3/day (600 ft^3/day) of unsaturated brine (230,000 ppm) to enter the repository from the Bell Canyon, dissolve the salt and waste in the repository, and enter the overlying Rustler aquifer as a saturated brine (410,000 ppm). Under these assumptions, the hydraulic resistance of the wellbore as calculated using the Hagen-Poiseville law for laminar flow is negligible compared to the resistance of the aquifers. A pressure difference of 5.2 x 10^4 Pa (7.5 psia) between the Rustler and the Bell Canyon was assumed in the analysis.[4] The resulting initial velocity in the wellbore would correspond to that of a porous media with a permeability of 10^5 darcies.

Scenario 2: In this scenario, two 0.6 m (2-ft) diameter wellbores are assumed to fail such that their permeability is 50 times that of the Rustler aquifer, thus allowing fluid to enter the repository from the Rustler through one hole, dissolve the repository, and then reenter the Rustler through the other hole. Water entering the repository is assumed to contain 8000 ppm dissolved solids. After dissolution of the repository and the wastes, it reenters the Rustler saturated at 410,000 ppm. The conductivity assumed for the shafts and the borehole is equivalent to an effective permeability of approximately 20 darcies.

Scenario 3: In this scenario, penetrations into the repository are sufficient to allow the repository to flood, although no circuit is established to allow fluid flow. The only mechanism for waste transport is molecular diffusion in the liquid phase. An area of 2×10^5 m^2 (50 acres) is assumed to be open between the repository and the Rustler. This area is far larger than could credibly be achieved by any number of unplugged boreholes or shafts into the repository.

In all analyses for WIPP, no credit is given to the stability of the waste form beyond an assumed dissolution rate equivalent to that of the salt in the repository. Radionuclide transport in the aquifers is based on retardation parameters (sorption), which are derived from laboratory data and no credit it taken for sorption in the salt or associated minerals.[10] Transmissivity of the aquifers is varied, and the consequences considered are based on the maximum values measured for a given region. Doses are calculated for the maximally exposed individual.

Results

Results from the consequence analyses discussed above have been obtained for a repository based on the earlier DOE mission for WIPP, which contained both TRU waste and 1000 spent fuel assemblies[1] and for the current design with only TRU wastes.[5,9] Results of the latter case for the maximum scenarios are presented in Table III.

TABLE III

RESULTS OF CONSEQUENCE ASSESSMENTS FOR A TRU REPOSITORY

	Highest Annual Total Body Dose Rate (mrem/year)	Greatest Organ (Bone) Dose Rate (mrem/year)
Scenario 1	7.7×10^{-3}	1.3×10^{-2}
Scenario 2	1.7×10^{-3}	2.8×10^{-3}
Scenario 3	7.0×10^{-5}	1.2×10^{-4}

These results indicate that scenario 1 yielded the highest dose to individuals (7.7×10^{-3} mrem/yr whole body), which is about 0.005% of the dose from natural radiation in the United States.

Additional studies [3] have been performed to assess the sensitivity of the results in Table III to variations in radionuclide sorption in the fluid pathways. These studies concluded that, while sorption near the repositories plays an important role in reducing consequences, sites with favorable geologic and hydrologic settings such as WIPP can tolerate essentially no reliance on sorption and still achieve inconsequential predicted doses from releases from TRU wastes. Additional comparisons of the assumptions for waste form integrity, brine migration, and radionuclide sorption, and the consequence assessments performed here have also been developed.[11]

SUMMARY AND CONCLUSIONS

The Waste Isolation Pilot Plant has been designed for development at a site in southeastern New Mexico. The facility would allow for retrievable disposal of TRU waste and experimentation with defense high-level wastes. The proposed site has been identified by a process that included avoidance of any deep drill holes within one mile of the limit of underground workings. The 7700 hectare (19,000) acre area proposed for Department of Energy control does contain 21 holes drilled by the potash and petroleum exploration industry. In addition, 45 holes have been drilled by the DOE for geologic and hydrologic characterization of the site and for further estimates of potential potash reserves. Of these 66 holes, only 8 penetrate below the proposed repository level. Of these, the 4 that penetrate into underlying formations containing aquifers are greater than 1 mile from the underground workings. Development of the facility would include 4 shafts, the largest of which is 23 feet in diameter. All shafts and drill holes would be plugged in conformance with DOE standards before the repository is decommissioned.

Consequence assessments have been performed to evaluate the potential doses to man from hypothetical breaches of repository integrity. All of the scenarios evaluated have assumed large influxes of fluids from overlying or underlying formations. These influxes could be achieved only with completely unplugged or extremely permeable boreholes. Permeabilities greater than 10 darcies are necessary in all cases to achieve the conditions assumed. Even under these assumptions, the dose to a maximally exposed individual did not exceed 0.01% of natural background radiation dose.

The principal conclusion from these studies regarding borehole plugging is that criteria for sealing penetrations should be specific to a potential repository site and should be developed from comprehensive safety assessment studies that place requirements for borehole plugs into perspective with other factors that comprise the isolation system.

Although these bounding assessments indicate that a site with a favorable geologic and hydrologic setting such as WIPP can tolerate significant compromises in borehole or shaft seal integrity and in other potential man-made and natural barriers, an experimental program has been established to resolve the uncertainty associated with various technical issues. Specifically, a program has been initiated to develop and evaluate technology for sealing penetrations at the WIPP.[12,13] These studies will quantify the integrity of various sealing methods and determine the measure of additional safety assurance provided by competent borehole seals.

REFERENCES

1.   DOE/EIS-0026-D, Draft Environmental Impact Statement, Waste Isolation Pilot Plant, Vols. 1-2, U. S. Department of Energy, Washington, DC, 1979.

2.   Presidential Message to Congress, February 12, 1980, "Comprehensive Radioactive Waste Management Program," Weekly Compilation of Presidential Documents, Vol. 16, No. 7.

3.   Summary of Research and Development Activities in Support of Waste Acceptance Criteria for WIPP," SAND79-1305, Sandia National Laboratories, November 1979.

4.   Geological Characterization Report, Waste Isolation Pilot
       Plant (WIPP) Site, Southeastern New Mexico, Vols. 1-2,
       SAND78-1596, Sandia National Laboratories, December 1978.

5.   Waste Isolation Pilot Plant Safety Analysis Report, U.S.
       Department of Energy.

6.   Hunter, T. O., "Technical Issues of Nuclear Waste
       Isolation in the Waste Isolaton Pilot Plant (WIPP),"
       SAND79-1117C, Sandia National Laboratories. Presented at
       87th National Meeting American Institute of Chemical
       Engineers, Boston, MA, August 1979.

7.   Wowak, W. E. and A. R. Sattler, "Criteria and Preliminary
       Design for WIPP Exploratory Program Experiments,
       SAND79-2077, Sandia National Laboratories (to be printed).

8.   Bingham, F. W. and G. W. Barr, "Scenarios for Long-Term
       Release of Radionuclides from a Nuclear Waste Repository
       in the Los Medanos Region of New Mexico, SAND78-1730,
       Sandia National Laboratories, 1979.

9.   Tierney, M. S., et al., "Long-Term Safety Assessments of
       the WIPP: Final Consequence Analysis," SAND80-1116,
       Sandia National Laboratories (to be published).

10.   Dosch, R. and A. W. Lynch, "Interaction of Radionuclides
       with Geomedia Associated with the Waste Isolation Pilot
       Plant (WIPP) Site in New Mexico," SAND78-0297, Sandia
       National Laboratories, 1978.

11.   Hunter, T. O., "The Perspective of Waste Isolation
       Research Issues and Assessment of Consequences for
       Radionuclide Release, SAND79-1076C. Presented at the
       Materials Research Society International Symposium on the
       Scientific Basis for Nuclear Waste Management, Boston, MA,
       November 1979.

12.   Christensen, C. L. and T. O. Hunter "Waste Isolation Pilot
       Plant (WIPP), Borehole Plugging Program Description,
       January 1, 1979," SAND79-0640, Sandia National
       Laboratories, Albuquerque, NM, August 1979.

13.   Christensen, C. L., "Test Plan, Bell Canyon Test, WIPP
       Experimental Program, Borehole Plugging," SAND79-0739,
       Sandia National Laboratories, Albuquerque, NM, June 1979.

Discussion

R.D. ELLISON, United States

Has anyone estimated the changes that would occur to the consequence analysis if WIPP contained spent fuel or high-level waste insteed of only TRU wastes ?

T.O. HUNTER, United States

Yes, these studies were published in the draft environmental impact statement of the Spring of 1979. However, the current task for WIPP is limited to the assessment in the case of a TRU waste repository.

GENERAL DISCUSSION

R. PUSCH, Sweden

I have two questions of a general nature concerning salt formations. In Sweden we have been very much concerned with the movement of the earth crust itself. We know that the Russian Platform, or the Baltic Shield, as we prefer to call it, is a fairly stable area ; still minor displacements have taken place and we think that a buffer material with a swelling potential can withstand these displacements.

As far as I know, salt domes have been formed because the pressure of the overlying sediments on a salt bed has caused the salt to go up through a weak point in the overburden. Now my first question is : do you know with sufficient accuracy what the rate of movement is today, so that you won't find yourselves with a salt dome at the surface out of which 500 obelisks stick out ?

The other question is related to the creep properties of salt. Creep in salt, especially at high temperature, can be very rapid. This could produce the movement of individual canisters.

C.L. CHRISTENSEN, United States

Well, for the first question we need someone who's an expert on salt domes.

J. HAMSTRA, Netherlands

I'll try to answer it. It's true that the density difference between salt and overlying and surrounding sediments causes the upward movement of salt ; but to have additional upward flow a supply of salt from the mother bed is required ; it is questionable whether that supply is available today for many salt domes. Anyway for salt domes in the later stages of diapirism, the upward movement rate is in the order of millimetres per century and less ; therefore there is no need to worry about such extreme upward movements as you assume since they only occur in the full diapiric stage when you have the first breakthrough that initiates diapirism.

I think that's the answer to the first question. A recent German report about the theoretical stability of salt domes claims that a rate of millimetres per century can be considered to be geologically stable, since the earth crust isn't stable anyway. In the Netherlands we have a descending area of crust. Salt which was originally a few hundred metres from the surface is now situated at a depth of 3,500 m since the crust is continuously descending. So the earth's crust and the salt domes in it are not totally stable. As long as the salt is sufficiently covered by overlying sediments and maintains its plastic behaviour, it will easily adapt itself and maintain a certain geometric stability.

The second question is about the movement of the canisters. If there is movement it is of course by density difference. But the density difference isn't large. The density of salt is 2.6 and we expect the canisters to be 2.5.

R. PUSCH, Sweden

And the temperature would be ?

J. HAMSTRA, Netherlands

We aim at an average temperature for the area, but there is a difference between conceptual designs. We aim to keep temperatures below 100°C, about 80°C, to keep the whole area reasonably stable. If there is a movement, and there will be, it will be a very limited downward movement. There is no reason why it should be in all directions.

R. PUSCH, Sweden

Many of these salt domes may have inclined beds of anhydrite and other minerals.

J. HAMSTRA, Netherlands

We do not plan to have any boreholes going through that. We will avoid them by several hundred metres if we can.

The other question is that of criticality problems. This is only a problem which might occur for spent fuel and there is no intention at the moment in either of our concepts to dispose of spent fuel.

M.J. BARAINCA, United States

I'm with the Department of Energy. I'm not a technical expert, I'm a programme manager, but I may be able to shed some light on the problem. As part of the United States programme on salt domes, Louisiana State University has established a network of tiltmeters over some of the salt domes in Louisiana.

In addition, a precise levelling network was laid out by the Geological Survey to determine if there has been any rising in the domes. As you are aware, there are two kinds of domes in the United States. Those inland near the Gulf of Mexico, and those that are a little farther north. The domes which we are investigating are the more inland ones, and from our investigations there has been no perceptible rise. Of course, over geological time it's in the order of what Dr. Hamstra said, that is less than one mm per year. That is the answer to your first question. We have data which indicate that salt domes are pretty stable.

As to the second question - the U.S.G.S. has raised this with regard to what might happen if you put a heated canister into a salt dome. Will the heat of that canister cause a China syndrome in the repository and cause it to melt down ? They considered it to be a valid question and it was published in the U.S.G.S. Circular by Dr. Stewart.

As a result of this question and other questions related to brine migration, the Department of Energy has initiated a series of in situ tests in the Avery Island Salt Mine. Periodical surveys have been made of canisters which have been heated for one year to see if there is any perceptible movement, and none has been detected. The information on the first question is available in the Quarterly Report, Louisiana State University (LSU), and the second piece of information, concerning canister movement, is available in an Interim Status Report on Avery Island published by Respec.

J. HAMSTRA, Netherlands

Do you know the heat load of these canisters at Avery Island ?

M.J. BARAINCA, United States

I'm not positive, but I think in the order of 5 kW.

J. HAMSTRA, Netherlands

Is 250°C still the maximum temperature ?

M.J. BARAINCA, United States

I believe so.

F. GERA, NEA

On this question of the movement of salt domes, I think I will have to say something, since a few years'ago I made a study of exactly this problem. I can confirm that the geologic evidence shows that when domes are undergoing uplift, the rates of uplift may be up to 1 to 2 mm per year. I think that is the figure you mentioned Dr. Hamstra.

J. HAMSTRA, Netherlands

In the older literature there are figures stated of millimetres per year in the diapiric stage. A recent German report by Dr. Jaritz, claims that for the younger salt domes of the German part of the Zechstein basin, even in the diapiric stage, the maximum upward movement was no more than 0.2 to 0.4 mm per year. Data on movements of 2 mm that are stated do in fact include the formation of caprock from anhydrite, and that causes additional rates which haven't anything to do with the salt movement itself. So the 2 mm that is incorporated in most of the hazard analyses up until now as a possible maximum of upward movement is a high figure.

F. GERA, NEA

The question is if it is 2 mm or if it is 0.2 mm, so we are dealing with an order of magnitude of possible difference. These 1 to 2 mm per year values that are found in the older literature are said to be an upper limit because when you look at the geologic evidence and the rate of movement derived by the shape and by the thinning of the flanking sediments, you arrive at much lower uplift values. It is normally possible by looking at the stratigraphic relationship between the domes and the surrounding sediments to find out when the fastest period of growth ended. For example, for the internal salt domes of the Gulf Coast region, it appears that growth finished a long time ago. For the younger ones closer to the coast, including Avery Island and all the five island domes, the geologic relationships indicate very recent and possibly current growth, but it might be difficult to pick such movement up using the kind of measurements that can be carried out in very short time periods. But even if they are growing, it is possible to establish upper limits and, in a separate assessment, it is possible to come up with scenarios and I doubt that we will have the "obelisks" that Professor Pusch suggested. What might happen could be that the salt containing the canisters could migrate to the caprock where salt dissolution had taken place. Then you will end up with a situation similar to a granite environment where you have the possibility of ground water reaching the waste. But you can calculate how many thousands of years, or how many hundreds of thousands of years, this would take, based upon the best available estimate of dome uplift.

R.D. ELLISON, United States

I have a question for Prof. Pusch. We do a lot of work
with bentonite slurry walls as cut-offs for dams and chemical waste
disposal areas, and we find that the behaviour of bentonite is
extremely sensitive to the water quality and sometimes we find the
bentonite wall doesn't work at all, its a complete failure. We have
to look at the source of the water and also the potential leachates
from the system that we are working with in the event that over a
long period of time you'd get a dispersion effect, and I just wonder
if maybe in Sweden the prediction of water quality is fairly simple
because your geology is relatively uniform. Would you change your
views as to it being a super material if you were in a geology where
the water quality changed with the horizon, or maybe even more
importantly, if the water quality were to change with a change in
climatic conditions such that the raising and lowering of water
levels would cause a disruption of the water quality.

R. PUSCH, Sweden

I would say that when you use bentonite for cleaning
purposes and in excavations for instance, or in dams, you use much
lower densities. What kind of density do you have ?

R.D. ELLISON, United States

It's much lower.

R. PUSCH, Sweden

Yes, I thought so. And then the changed properties really
mean something. If we have a bentonite suspension with very low
solid content, and add to that a salt to a salt concentration of
1 %, then the whole thing flocculates and settles and you get clean
water. We also know from a long series of tests that if we add water
with a high organic content, it also changes its properties very,
very much. Sewage water is not suitable for saturating bentonite.
Under the best circumstances, we expect that the variation in water
quality will produce very small changes or variations in the physical
properties, but that might be due to the fact that we are using a
high density.

R.D. ELLISON, United States

Even if you were in a different geology, where you have
five different rock types, would you still consider using bentonite ?

R. PUSCH, Sweden

I think you've certainly got to see what the condition of
the water is ; salinity, acidity, pH and other factors, and I would
say that the idea might not be applicable everywhere.

R.D. ELLISON, United States

You're advancing bentonite as being a super sealant and
you have strong justifications for that, but if someone else were
to develop an alternative material, would you consider using
bentonite alternated with that to get an even better material ?

R. PUSCH, Sweden

That's in fact what we do in the gaps and holes where we have the bentonite. When we want to seal off the whole thing we use bulkheads and in between we have the other material.

I'm concerned about the problem of determining the quality of our borehole plugging, especially in long boreholes. How do you know once you have filled the whole borehole with the sealant that there isn't the odd metre left here and there, especially where you have had cracks or something in the rocks which form natural cavities. I think that is one of the main points and I just want to take the opportunity to say that in the method we have suggested it would be possible, since the perforated pipe is electrically conducting, to apply some electrodes which are separated from the pipe, so that we can measure the electrical conductivity. This is in fact a very good measure of both the water content and the degree of homogeneity. We know exactly the geometry of the system, when we put the pipes down and that every single pipe is filled with bentonite.

C.L. CHRISTENSEN, United States

I think that idea sounds excellent.

I think we also may find from some of the work that we have done that it's not clear that you have to fill the whole bore-hole. If you have void spaces then that may not be such a bad problem, but we haven't reached the end of this work yet. Some other people may have some comments.

M. GYENGE, Canada

I would like to ask Prof. Pusch how he would ensure that the perforated tube is concentric.

R. PUSCH, Sweden

It sure is not concentric, but that doesn't matter. It doesn't matter whether the pipe is situated right in the centre or has moved to the side a little.

D.M. ROY, United States

Well, I'm not sure I really want to get back to chemistry, but the normal degradation process with granite would be illite, hydromuscovite, etc., then kaolinite. How long would it take to exchange this sodium bentonite for a non-swelling phase ?

R. PUSCH, Sweden

That would take enormous periods of time. It is all a matter of thermodynamics. If we find ourselves in that particular area in the pH diagram where we have moderate pressure, where the pH is between say 7 and 10.5, where the chemical environment is moderate, then we will have that montmorillonite forever. It's funny to say that I wouldn't rely on that stuff myself if I didn't know that in every rock excavation we have done in Sweden in granite for the last ten years, we have found these zones which contain montmorillonite, where the conditions are exactly what they would be at 500 metres depth and they've been there for millions of years. Well, that reinforces experimental evidence and is of primary importance.

SESSION 3

Chairman - Président

R.D. ELLISON
(United States)

SEANCE 3

SANDIA BOREHOLE PLUGGING PROGRAM FOR THE WASTE ISOLATION PILOT PLANT (WIPP)

C. L. Christensen
Experimental Programs Division 4512
Sandia National Laboratories
Albuquerque, New Mexico   87185
United States of America

ABSTRACT

This paper presents the current Borehole Plugging Program (BHP) at Sandia National Laboratories and addresses the four major functional tasks considered necessary for a successful program: (1) project management; (2) identification of candidate materials; (3) development of test instrumentation; and, (4) field test program. The scope of the current program on wellbores, cementitious grouts, and available technology is compared to planned extension to shafts and drifts, alternate materials, and improved instrumentation. The relationship of this program with assessments of long-term suitability performed in the geochemical program is presented. A brief presentation of current and planned field tests is given along with projections for future activities and results.

BACKGROUND

The current Borehole Plugging Program (BHP) has been in progress since October 1978. Prior efforts within the Sandia Waste Isolation Pilot Plant (WIPP) were devoted to establishing a baseline materials development program using cementitious materials that could be candidates when a systematic program for wellbore plugging was established. During the original phase of the WIPP plugging program, the ERDA-10 wellbore in southeastern New Mexico (SENM) was plugged in October 1977. Four stages of plugging material were used to demonstrate the capability of emplacing a segmented-by-mix plug under actual field conditions.[1] The current program at Sandia is sponsored by the Office of Nuclear Waste Isolation (ONWI) as the bedded salt element in their overall generic plugging program. Appendix A shows a generalized stratigraphy for the WIPP site. While the immediate application of results is directed at WIPP-related issues, the overall concept of the program is designed to support generic salt formation repositories. Certain field test techniques under development will be applicable to plugging efforts in all media.

PROGRAM BASIS

The current Sandia BHP is motivated by results obtained from consequence assessment scenarios previously described.[2,3] Thus for repositories sited in soluble media, intrusion of ground waters leading to salt dissolution and subsequent entrainment and transport of the radionuclides to the biosphere pose the threat scenario. While the WIPP site hydrology is such that the dose to a maximally exposed individual is a small fraction of natural background according to consequence assessment evaluations, nevertheless it seems prudent to control these fluid intrusions into the repository as an added safety feature. Additionally, factors which provide the basis and motivation for a comprehensive BHP program include: state laws requiring the plugging of abandoned wells, the general public's apprehension about nuclear waste repositories, application to other sites without the favorable hydrologic setting, and a general lack of data on the effectiveness of plugging efforts. An outline of the program is contained in Appendix B. The presentation in this paper is shown on B-1, while B-2 summarizes the objectives of the program.[4]

TASKS

The issues identified to date within the BHP lead to the definition of the four major tasks described below that are required to attain the planned effectiveness. The function of these tasks is to identify, assess, resolve, and document the relative impact of these issues as applied to a repository site in bedded salt strata found in SENM. The documentation of this program will provide the basis for addressing the generic issues as they become known. Some of the current issues under consideration in the program are listed on B-3. A brief discussion of each task follows.

Management and Design

This task (B-4) provides the central coordinating function for the BHP tasks and includes budget allocations and priorities, required quality assurance activities, and program documentation. Contact with other WIPP tasks and outside agencies is provided within this task.

Program needs, guidance, and direction are defined in this task as is overall responsibility for support and implementation. Primary program emphasis at this time is an understanding of wellbore plugging requirements and an evaluation of current efforts to meet the requirements. In the conceptual stage at this time are multimedia plugs providing for the separate functions of sealing, radio-nuclide sorption, and filling, along with modular emplacements of these multimedia plugs for protection of the seal against earth motions.

Field and laboratory activities conducted within the program are implemented under a Quality Assurance Program to provide confidence in the results obtained during testing and to establish the basis for industrial plugging efforts at repository sites.[5]

Expansion of the progam to include shaft sealing and drift backfilling to provide compartmentalized isolation within the repository is now in the planning stage and will be emphasized in 1982.

Materials

This task (B-5) includes the development, testing, and selection of appropriate materials to support BHP program needs. In the current stage of development, cementitious materials are the primary plugging candidates. These include both fresh water and brine mix grouts and various additives. The identification of cementitious materials as acceptable plugging materials would simplify the BHP considerably since this technology now exists. While cementitious plugs are considered adequate for sealing in the short term, their long-term suitability requires further evaluation.[6,7]

Responsibility for determining the long-term suitability of plugging candidates, however, is also included within the materials task. To support this effort, a joint cooperative program is being conducted in geochemical analysis and testing by Sandia, ONWI, the Pennsylvania State University, and the U. S. Army Corp of Engineers Waterways Experiment Station (WES) at Vicksburg, MI. This effort will initially examine the proposed grout plug candidates to evalute long-term (300 years plus) suitability; future expansion of this program to include alternate materials is possible depending on the initial evaluation of the grout work.[8,9,10]

At this time, only cementitious materials are under development at Sandia as plug (sealing) candidates; however, expansion to include other materials into the program either as seal, getter, or filler candidates is inherent within the materials task and will be assessed as the program continues and when the appropriate technology can be developed.

Instrumentation

This task (B-6) includes assessing instrumentation requirements, identifying instrumentation sources, achieving required accuracies, developing new systems if necessary, and providing the expertise to obtain the data required. While specifically directed at field test requirements, any and all needs for instrumentation in the laboratory testing program are included in the task responsibility. It is important to note that the current BHP program is designed to evaluate candidate plug performance, and the instrumentation task supports this effort. Once the adequacy of a plugging technology is assured, subsequent testing and monitoring of the repository plugs may not be necessary, thus lessening the instrumentation task role.[11,12]

Current task emphasis is on existing technology, primarily oil field techniques with modifications to enhance their application to the BHP. When inadequacies or nonexistent techniques are encountered, the instrumentation task will initiate the appropriate development. Current efforts include improved logging techniques for nondestructive assessment of plug performance and downhole television (DHTV) use for assessing wellbore conditions in preparation for plugging.[13]

Field Test

This task (B-7) provides in situ evaluations of plug candidate materials and emplacement techniques. It provides for comparison of results in the field environment with those observed in the laboratory and allows the development of the reasons for observed differences. The need for field testing is evident considering the many in situ variables that cannot be modeled adequately in the laboratory and the higher acceptability of field results by scientific and lay observers.

The Bell Canyon Test (BCT) (B-8), in which a six-foot plug is emplaced in a thick anhydrite bed and the natural hydrostatic pressure of 1800 psi is applied across the plug to determine the in situ effectiveness of the plug formation complex, is underway.[14] The plug will be evaluated at its original length and subsequently lengthened to a total of 18 feet to assess predictions for scaling length to performance.[15,16]

Planned tests in FY80 and FY81 include the Potash Core Test (PCT)[17] (B-9) and the Surface Wellbore Test Bank (SWTB)[18] (B-10). In the PCT, an existing plug in a potash borehole will be selectively overcored to determine the quality of the plug/formation performance. Laboratory analysis will provide required data for evaluating the suitability of the current materials program. The SWTB consists of two phases: Phase one will emplace a bank of candidate mix plugs for in situ curing and subsequent overcoring at multiyear intervals for performance evaluation; Phase two is designed to provide insight to techniques and effectiveness of plugging aquifers. Future tests in the conceptual phase at this time include the Diagnostic Test Hole (DTH)[4] (B-11), the Media Quantification Experiment (MQE)[4] (B-12) and additional undefined plugging exercises in selected wellbores near the WIPP site. The DTH will provide an open center plug for in situ nondestructive testing; the MQE will provide for in situ formation testing at a mining horizon.

PROGRAM SCOPE

The foregoing discussions indicate the current efforts underway within the BHP; the basic approach is to use available technology, i.e., oil field cementing, instrumentation, and testing procedures, modified as necessary to attain the program objectives for bedded salt strata. Current emphasis is on laboratory and field wellbore testing for short-term (approximately 50-300 years) validity. Long-term analysis and identification will be determined in a companion geochemical program.

Planned expansion of the program includes shaft sealing and drift backfilling and sealing, and inclusion of alternate sealing material candidates as the need is identified and the technology becomes available. B-13 displays the concepts described here.

PROGNOSIS

Current information suggests that cement technology now in use will be adequate in the short term; long-term stability is expected. Planned plug configuraton involves multimedia components for sealing, adsorption and void filling; these in turn can be "stacked" to provide wellbore segment isolation.

There are no fundamental issues identified to date which would preclude cement technology. Long-term monitoring of the wellbores is not planned and will probably not be required. Final plug design will lead to a quality-assured plugging technology using industrial practices (B-14).

REFERENCES

1.   Gulick, C. W., Jr., Borehole Plugging Program, Plugging of ERDA-10 Drill Hole, June 1979, SAND79-0789, Sandia National Laboratories, Albuquerque, NM 87185, June 1979.

2.   Hunter, T. O., Role of Borehole Plugging in the Evaluation of the Waste Isolation Pilot Plant, SAND80-0502C, Sandia National Laboratories, Albuquerque, NM, prepared for the Workshop on Borehole and Shaft Plugging, May 1980, Columbus, OH.

3.   Draft Environmental Impact Statement (DEIS), Waste Isolation Pilot Plant, DOE/EIS-0026-D, Volume 1, Chapter 9, U. S. Department of Energy, April 1979.

4.   Christensen, C. L. and Hunter, T. O. Waste Isolation Pilot Plant (WIPP), Borehole Plugging Program Description, January 1, 1979. SAND79-0640, Sandia National Laboratories, Albuquerque, NM 87185, August 1979.

5.   Waste Isolation Pilot Plant (WIPP) Quality Assurance Program Plan (QAPP), Sandia National Laboratories, Albuquerque, NM, April 1979 (controlled distribution).

6.   Gulick, C. W., Jr., Borehole Plugging Materials Development Program, Report No. 2, SAND79-1514, Sandia National Laboratories, Albuquerque, NM 87185, February 1980.

7.   Gulick, C. W., Jr., Bell Canyon Test (BCT) Cement Development Report, SAND80-0358A, Sandia National Laboratories, Albuquerque, NM 87185, prepared for the Workshop on Borehole and Shaft Plugging, May 1980, Columbus, OH.

8.   Buck, Alan D., and Burkes, J. P., Examination of Grout and Rock from Duval Mine, New Mexico; Miscellaneous Paper SL-79-16, Structures Laboratory, U.S. Army Engineer Waterways Experiment Station, Vicksburg, MI, July 1979, Final Report.

9.   Scheetz, B. E., et al., Characterization of Samples of a Cement-Borehole Plug in Bedded Evaporites from Southeastern New Mexico, ONWI-70, July 1979, Materials Research laboratory, The Pennsylvania State University, University Park, PA.

10. Lambert, S. J., A strategy for Evaluating the Long-Term Stability of Hole-Plugging Materials in Their Geological Environments, SAND80-0359C, Sandia National Laboratories, Albuquerque, NM 87185, prepared for the Workshop on Borehole and Shaft Plugging, May 1980, Columbus, OH.

11. Cook, C. Wayne, Instrumentation Development Report for the Waste Isolation Pilot Plant (WIPP) Borehole Plugging Program (BHP), SAND79-1902, Sandia National Laboratories, Albuquerque, NM 87185, November 1979.

12. Cook, C. W. and Lagus, P. L., Bell Canyon Test (BCT) Instrumentaton Development, SAND80-0408C, Sandia National Laboratories, Albuquerque, NM 87185, prepared for the Workshop on Borehole and Shaft Plugging, May 1980, Columbus, OH.

13.   Christensen, C. L., Statler, R. D. and Peterson, E. W.,
      Downhole Television (DHTV) Applications in Borehole
      Plugging, SAND80-0459C, Sandia National Laboratories,
      Albuquerque, NM 87185, prepared for the Workshop on
      Borehole and Shaft Plugging, May 1980, Columbus, OH.

14.   Christensen, C. L., Test Plan, Bell Canyon Test, WIPP
      Experimental Program, Borehole Plugging, SAND79-0739,
      Sandia National Laboratories, Albuquerque, NM 87185, June
      1979.

15.   Statler, R. D., Bell Canyon Test--Field Preparation and
      Operations, SAND80-0458C, Sandia National Laboratories,
      Albuquerque, NM 87185, prepared for the Workshop on
      Borehole and Shaft Plugging, May 1980, Columbus, OH.

16.   Peterson, E. W., Analysis of Bell Canyon Test Results,
      SAND80-7044C, prepared for the International Borehole
      Plugging Symposium, Columbus, Ohio, May1980.

17.   Christensen, C. L., Test Plan, Potash Core Test, WIPP
      Experimental Program, Borehole Plugging. July 1979,
      SAND79-1306, Sandia National Laboratories, Albuquerque,
      NM 87185.

18.   Christensen, C. L. and Goodwin, R. H., Test Plan, Surface
      Wellbore Test Bank, WIPP Experimental Program, Borehole
      Plugging. July 1979, SAND79-2211, Sandia National
      Laboratories, Albuquerque, NM 87185 (to be published).

Discussion

W. FISCHLE, Federal Republic of Germany

What is the length of the packers used in the Bell-Canyon test ?

C.L. CHRISTENSEN, United States

Standard Lynes 66 inches packers were used. Actuel wall contact length of rubber element is approximately 53 inches.

BELL CANYON TEST - FIELD PREPARATION AND OPERATION

R. D. Statler
Engineering Projects Division 1133
Sandia National Laboratories
Albuquerque, New Mexico 87185 (United States)

ABSTRACT

This paper describes the field preparations and operations conducted during
the Bell Canyon Test. An abandoned exploratory hole was reentered, recon-
ditioned, and converted into a test bed for evaluating the in situ performance of a cement grout plug while a pressure differential of 1800 psi was
applied. A brief history of the abandoned well and the work involved in
reconditioning and deepening the well to attain an acceptable test condition
is discussed. The "as-built" test configuration is presented along with a
review of the initial test design and the field changes required to accommodate the formation conditions determined during wellbore characterization
tests. Finally, some problems encountered in adapting typical oil field
services and practices to a "laboratory" experiment will be addressed.

1. INTRODUCTION

The Bell Canyon Test (BCT) was conducted to determine
the in situ performance of a cement grout plug while subjected to
a naturally occurring, high pressure fluid source. The test concept provided for a plug of known standard constituents to be
placed with existing oilfield techniques and evaluated over an
extended series of instrumented operations.

The test plan provided for rehabilitating an existing
well, deepening to develop an acceptable test environment, and
then setting up the experiment [1].

The experimental setup included six tracer gas capsules
with timed release mechanisms to be emplaced below a minimum
length grout plug, isolating a high pressure fluid source. An
instrumentation system was developed to measure the transit time
for fluids and gas pulses to travel from below the grout plug to a
test zone immediately above [2, 3].

The field operations included the reentry, rehabilitation, and deepening of an existing exploratory borehole, the testing of the hole to establish and characterize the test bed, the
emplacement of the grout plug, and subsequent measurements to
determine the plug performance.

2. WELLBORE REENTRY, REHABILITATION AND DEEPENING

An abandoned exploratory hole AEC-7, originally drilled
in April 1974 by the Oak Ridge National Laboratory, was selected
for the BCT. In the ORNL program, a surface conductor was placed
to 49 ft.. 8-5/8 in. casing placed and cemented into the Salado
Salt at 1016 ft. and continuous core then taken to a total depth
(TD) of 3918 ft. The 7-13/16 in. open hole was capped at the
surface and abandoned after topping off with salt saturated brine
May 23, 1974 [4].

Sandia National Laboratories in support of the Department of Energy (DOE) undertook a workover of the wellbore in June
1975. Attempts were made to test selected zones in the Salado
formation but these attempts failed because of bridging at about
1649 ft. Hydrologic tests were then made of the Culebra formation
after setting a bridge plug in the casing at 1002 ft. and perfora-
ting between 880 to 930 ft. Tests showed fluid heads at 800 ft.
with about 300 psi (about 20 atmospheres) shut-in pressure. Following the tests, the hole was capped and abandoned once again
until February 1979. Objects known to be in the wellbore at this
time were two short lengths of tubing, a 6.87-ft.-long packer, and
the bridge plug.

The Sandia National Laboratories project group responsible for the BCT arranged for support of a DOE contractor, Fenix
& Scisson, Inc., to contract for and administer the required
drilling. Fenix & Scisson (F&S) is a drilling engineering firm
under prime contract to the DOE at the Nevada Test Site and is
skilled in the many facets of the drilling technology.

The Verna Drilling Company was contracted by F&S to
mobilize their Rig No. 14 at the AEC 7 pad on March 19, 1979. The
rig had a 94 ft. mast, 12 ft. substructure, 5000 ft. of 4-1/2 in.
drill pipe, 18 each 6-1/4 in. drill collars and related equipment.
After one day of rigging up, the drilling string was picked up and
operations begun. Figure 1 shows the drill-pad layout.

The first phase of borehole rehabilitation consisted of
a fishing operation to remove objects left from the previous

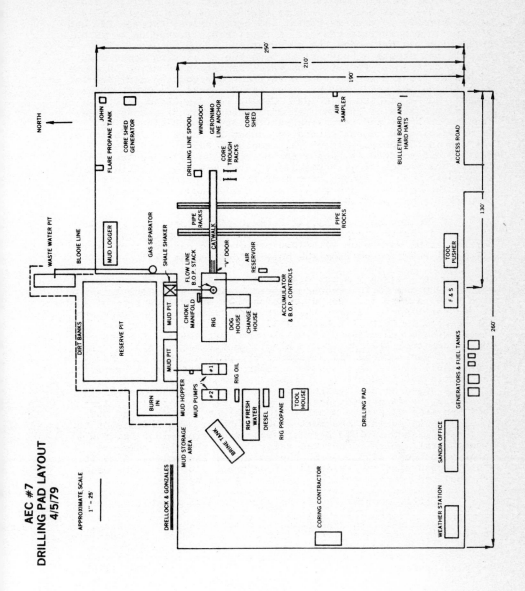

AEC #7
DRILLING PAD LAYOUT
4/5/79

APPROXIMATE SCALE

1" = 25'

NORTH

FIGURE 1.

- 235 -

operation. Typical fishing tools were used which included impression blocks, an overshot with jars and bumper subs, milling tools, magnets, and junk baskets. After two days of fishing, the hole was made ready to squeeze cement the perforations between 880 and 930 ft. in the surface casing. A retrievable packer on 4-1/2 in. drill pipe was set above the perforation. Pumping water at 3 barrels per minute at 350 psi through the drill pipe broke down the perforations. Perforations were then squeezed using 50 sacks of DOWELL Class H cement with a 2% friction-reducing additive and a 2% fluid-loss additive, followed by 100 sacks of DOWELL Class H cement and 2% calcium chloride. Cement was displaced in stages to a final pressure of 350 psi (about 23 atmospheres) before releasing the pressure and reversing out the drill pipe with brine. After waiting on cement curing for about 20 hours, a 7-7/8 in. bit was run in the hole. Cement was tagged at 849 ft. and then drilled out to 1002 ft. where the bridge plug had been left in 1975. The bridge-plug retrieval required milling, drilling, and subsequent pushing to the bottom of the hole before recovery was made. The hole was then worked and cleaned and conditioned to prepare for gas permeability measurements.

Following the initial permeability tests, the hole was again conditioned and made ready for the coring operation. Several trips were made with junk baskets and magnets to be sure all metal was removed from the hole before coring was begun.

A 7-13/16 x 4 in. diamond core bit with a 50-ft. barrel was made up and coring began on April 5, 1979, with Core 1 cut from 3926 to 3932 ft. Coring continued at an average rate of 43 ft/day until April 22 when the Castile/Bell Canyon interface was intercepted in Core 14 between 4532 and 4583 ft. Coring was interruped to run a standard Drill Stem Test (DST). A Lynes, Inc., test tool was run with a single packer set at 4531 ft. and DST No. 1 run against the bottom of the hole at 4583 ft. No appreciable fluids or gases were observed so the coring operation was continued through Core 17 taken between 4664 and 4714 ft. Since it appeared the desired high-pressure zone was present in the Ramsey Sands between 4635 and 4681 ft., a Lynes, Inc., test tool was made up with a single packer set at 4609 ft., and DST No. 2 was run against the bottom of the hole. The results of the test established a fluid pressure source of about 1850 psi ($\sim$ 127 atmospheres). A third DST was later run to test the complete interval between 4495 ft. and the bottom of the hole at 4714 ft. Fluid pressure at the selected cement grout plug test zone was extrapolated to be about 1811 psi ($\sim$ 124 atmospheres).

After completion of the DSTs, a 7-7/8 in. bit was picked up and the hole was drilled from 4714 ft. to a drillers TD of 4734 ft. The hole was then conditioned and made ready for logging. The reentry, rehab and deepening program was essentially completed on May 1, 1979, and the drill pipe was laid down. Preparations were made for continuing with wellbore testing, plug emplacement and plug performance evaluations. The lithology of AEC-7 is shown in Figure 2.

The core was logged as it was taken by geologists from USGS and F&S. It was photographed, sleeved, and boxed in the field and placed in file in the WIPP core library in Carlsbad, New Mexico. Selected samples were shipped to the Waterways Experiment Station, the Pennsylvania State University, ORNL, and the Dowell Corporation for tests of compatibility with plug mixes.

3.   WELLBORE TESTING AND CHARACTERIZING

After it was established that the AEC-7 hole qualified as an acceptable test bed for the BCT, the field operations

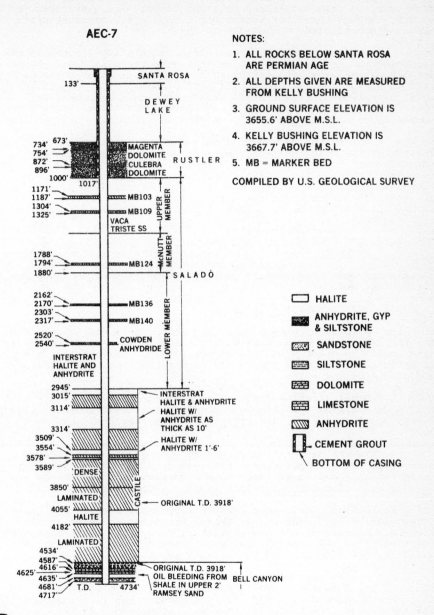

AEC-7

NOTES:

1. ALL ROCKS BELOW SANTA ROSA ARE PERMIAN AGE
2. ALL DEPTHS GIVEN ARE MEASURED FROM KELLY BUSHING
3. GROUND SURFACE ELEVATION IS 3655.6' ABOVE M.S.L.
4. KELLY BUSHING ELEVATION IS 3667.7' ABOVE M.S.L.
5. MB = MARKER BED

COMPILED BY U.S. GEOLOGICAL SURVEY

SANTA ROSA

133'

D E W E Y
L A K E

734' 673'
754'
872'
896'
1000' 1017'

MAGENTA
DOLOMITE
CULEBRA
DOLOMITE   R U S T L E R

1171'
1187'   MB103
1304'   MB109   UPPER MEMBER
1325'
VACA TRISTE SS

McNUTT MEMBER

1788'
1794'   MB124
1880'   S A L A D O

2162'
2170'   MB136
2303'
2317'   MB140   LOWER MEMBER

2520'   COWDEN
2540'   ANHYDRIDE

INTERSTRAT HALITE AND ANHYDRITE

2945'   INTERSTRAT HALITE & ANHYDRITE
3015'
3114'   HALITE W/ ANHYDRITE AS THICK AS 10'

3314'   HALITE W/ ANHYDRITE 1'-6'
3509'
3554'
3578'
3589'   DENSE

3850'   CASTILE
LAMINATED   ORIGINAL T.D. 3918'
4055'
HALITE
4182'

LAMINATED

4534'
4587'
4616'
4625'
4635'   ORIGINAL T.D. 3918'
4681'   T.D.   4734'   OIL BLEEDING FROM BELL CANYON
4717'   SHALE IN UPPER 2'
                          RAMSEY SAND

☐ HALITE

▨ ANHYDRITE, GYP & SILTSTONE

▨ SANDSTONE

▨ SILTSTONE

▨ DOLOMITE

▨ LIMESTONE

▨ ANHYDRITE

▯ CEMENT GROUT

＼ BOTTOM OF CASING

🔲 Sandia National Laboratories

FIGURE 2.

changed from primarily a "drilling operation" to more of a "well completion and production operation." The Verna rig was rigged down and demobilized. A Hopper Hoist double-drum double derrick work-over rig from Mack Chase Drilling Company was contracted for by F&S. Around-the-clock operations ceased and subsequent operations were conducted only on day shifts.

The testing and characterizing of the AEC-7 borehole was conducted in various ways with several objectives in mind. A primary objective was to establish baseline data which could be used in evaluating the grout plug performance. Other objectives were to examine different measurement techniques and consider their usefulness in experiments such as the BCT. Some tests were conducted to satisfy other WIPP-related activities.

While some testing did take place with the Verna drilling rig on the hole, most of the test operations were done with the work-over rig using 2-3/8 in. tubing. Testing techniques included both standard oilfield services and special test operations not commonly used in the oilfield. The standard testing services, conducted while the drilling rig was on the hole, are listed below with a brief description of their purpose in the BCT:

3.1 Drill Stem Test

Three DSTs were run to establish the formation pressures in the zones of interest and to obtain fluid samples. It was these tests that were used to qualify the AEC-7 hole as an appropriate test bed.

3.2 Geophysical Logs

A suite of geophysical logs were run to aid in determining the lithologic boundaries within the hole as well as to obtain data useful to other WIPP-related activities. Caliper logs and a directional survey were run to aid in defining the geometrical conditions of the hole.

Other wireline logging was conducted at various times to aid in determining specific hole conditions. These included fluid level, fluid densities, and tubing collar/marker positions downhole.

In the preparation period for the grout plug emplacement, some special operations were conducted to aid in characterizing the hole and to evaluate hole conditions. Listed below are descriptions of those operations and their purposes:

Permeability Test--Gas permeability measurements were conducted in several zones in the Salado formation in which steady-state gas acceptance and pressurization decay rates were monitored. Low-level air (~ 7 atmospheres) was injected into the formation and decay rates determined to evaluate formation permeability.

There were initial plans for air permeability measurements of the formation immediately above the BCT plug zone at 4489 ft., but, because of the difficulties encountered in evacuating all the fluids from the hole, they were abandoned. It was observed during evacuation operations that the production of fluids seemed to increase as did the intensity of the H_2S gas that had been logged during the coring operation. The concern over the H_2S production and the continual bleeding of fluids into the hole above the test zone made it necessary to redesign the testing program and prepare to conduct the experiment with fluids in the hole.

Television--The efforts to understand the production of fluids and gases while evacuating the hole led to the hiring of a downhole television survey from a local service company. The closed circuit video system was operated with a wireline hoisting unit and was capable of operating in air or clear fluids to a depth of 3000 ft. [5].

The survey showed some startling cross sections of the borehole and proved to be invaluable in selecting optimum packer seats for setting inflatable packers used in formation tests.

Hydrologic Testing--The possible presence of fluids above the planned BCT test zone, however minute, was of sufficient concern that the test operations were interrupted to conduct a series of hydrologic tests. These tests were made to determine the quantities of fluids being produced and the specific origin of the fluids. Gamma, gamma neutron (noncompensated), and caliper logs were run to select the probable producing zones for testing. A Lynes retrievable bridge plug was placed at 4300 ft. to guarantee isolation from the Bell Canyon. A Lynes testing and treating tool was made up on 2-3/8 in. tubing with nominally 100 ft. spacings between hydraulically inflatable packers. The instrumentation and test procedure were similar to those used in standard DSTs. Nine tests were run at the selected intervals. Possible producing zones were identified with permeabilities estimated in the microdarcy range. The total production of fluids across the interval between 1017 ft. and 4300 ft. was estimated to be less than 0.5 gal/min. The integrity of the casing seat at 1017 ft. and the bridge plug setting at 4300 ft. were verified. The fluids "produced" were considered to be mostly drilling or testing fluids that may have been pressured into the more permeable zones during evacuation operations.

Plug Zone Integrity Tests--The objective of these tests was to evaluate the candidate zones selected for plug emplacement after examining the core specimens and the geophysical logs. The tests also provided opportunity to proof-test the integrated test hardware systems which are discussed in more detail in the following sections and references 2 and 3.

The test set-up consisted of positioning a Lynes retrievable hydraulically inflatable packer at the designated plug zones to simulate the cement grout plug. A proto-type instrumentation package, containing tracer-gas and packer-deflate systems, was attached to the bottom of the packer. Both tracer gas and deflate systems were controlled by valves actuated by small explosive charges timed to detonate after downhole emplacement.

The test sequence began with the actuation of the gas release system which released a tracer gas from the Freon family into the wellbore below the packer. A differential pressure load of several hundred psi was applied across the packer by lowering the hydrostatic head in the tubing above the packer. During a waiting period of about 48 hours, the fluid column rise in the tubing was monitored and the fluid immediately above the packer was sampled for presence of tracer gas. The packer-deflate system then actuated permitting the fluids in the packer to be released into a reservoir in the instrumentation package and the packer unseated. Fluid samples were again taken.

The actuation of the valves for the tracer gas and the packer-deflate system was detected with geophones lowered into the hole. The successful opration of the systems was confirmed since fluid samples swabbed from the hole before and after the packer deflated showed respectively the absence and presence of the tracer gas. Gases from the Freon family were used so the wellbore would not be contaminated for the use of the Sulfur Hexaflouride (SF_6) tracer intended for use in the final grout plug instrumentation.

While more accurate permeability values may have been obtained by other tests or by running these tests longer, given the many other unknowns at the time, the measurements were considered adequate to qualify the plug zone at 4490 to 4495 as acceptable for the BCT.

4.   TEST BED PREPARATIONS

The initial BCT design concept included the plan to evacuate all fluids from the hole after plug emplacement. Monitoring instrumentation systems were designed to operate in a dry, open hole above the plug and monitor fluid and gas arrival from below the plug. Plans were developed to place a 5-ft. cement grout plug of known constituents in the dry, relatively impermeable anhydrite bed at the base of the Castile formation, above the Bell Canyon. A releasable bottom packer was used to position the bottom of the plug. A system for the automatic deflation and release of the packer was developed so that it would automatically deflate, after an appropriate time for the cement to cure. The packer was expected to drop away after deflating, thereby exposing the plug to Bell Canyon formation pressure.

To prevent the cement from encapsulating the top of the packer and hold it from dropping out of the way, a 5-ft. sand plug was included as a buffer between the grout plug and the bottom packer.

The instrumentation package attached to the bottom of the packer was developed by Systems, Science and Software (S^3) of San Diego, California, under contract to Sandia Laboratories. This package provided a reservoir into which the fluids from the inflatable bottom packer could flow when a preset timer actuated a small, explosive-driven valve. Also included in this bottom instrumentation package were six pressurized capsules of SF_6 tracer gas. The capsule controls were individually actuated by pre-set timers to release tracer gas into the wellbore below the grout plug in a planned sequence. An upper instrumentation system was developed by Sandia National Laboratories and S^3 to be lowered into the well on 2-3/8 in. tubing and record the arrival of the tracer-gas pulses and fluid buildup on top of the plug (Figure 3).

The initial design concept was reevaluated after it was determined a dry hole above the plug could not be maintained. It was decided that, except for the "dry-hole" upper instrumentation system, the BCT could be conducted in a fluid-filled hole.

The final plug emplacement began September 24, 1979. A Lynes 6-5/8 x 66 in. inflatable bridge plug was made up with the S^3 bottom packer instrumentation package attached below. A long tailpipe was attached below the S^3 package to prevent the complete assembly from dropping more than a few feet from the bottom of the grout plug when the release mechanism deflated the bridge plug.

This complete bottom hole assembly, measuring over 220 ft. long, was lowered into the hole on 2-3/8 in. tubing until the upper shoulder of the rubber element of the bridge plug was 4502 ft. below RKB[1].

----

[1]RKB stands for Rig Kelly Bushing. It is traditional to reference depths in drill holes below this point on a rig. RKB on AEC-7 was 12 ft. above ground surface.

Test Bed Configuration

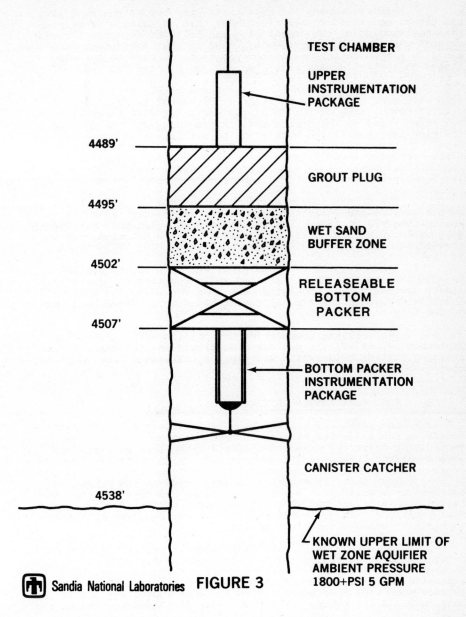

TEST CHAMBER

UPPER INSTRUMENTATION PACKAGE

4489'

GROUT PLUG

4495'

WET SAND BUFFER ZONE

4502'

RELEASEABLE BOTTOM PACKER

4507'

BOTTOM PACKER INSTRUMENTATION PACKAGE

CANISTER CATCHER

4538'

KNOWN UPPER LIMIT OF WET ZONE AQUIFIER AMBIENT PRESSURE 1800+PSI 5 GPM

Sandia National Laboratories   FIGURE 3

A dump bailer, made up of three each 5 in. x 10 ft. sections, was used to lower about 2-1/2 ft^3 of clean sand and release it on top of the bridge plug. The cement for the grout plug was being mixed at the wellhead while the sand was being dumped.

Before pumping the cement mix, the top of the sand plug was tagged and found to measure higher than expected. Since precise positioning of the cement plug was believed important, preparations were made to remove enough sand so that the cement plug could be placed in the designated position. A sand bailer was rigged on a wireline and about two gallons of sand removed.

Before the cement was placed downhole, several practice runs were made with the dump bailer. Four shallow holes were drilled on the edge of the drill pad and lined with 8-in.-diameter PVC pipe capped on the bottom end. A sand plug was poured into the pipe and then filled with brine to simulate a downhole environment. The dump bailer was filled with volumes expected to make a 5-ft.-high plug in an 8-in.-diameter hole and then operated in each of the holes. This exercise proved useful and resulted in the final placement of 3 ft^3 downhole.

The cement slurry was pumped from the mixer into buckets on the surface and while the dump bailer was held at an angle against the rig platform, was poured into the dump bailer. The dump bailer was then carefully picked up with a hoisting unit and lowered into the hole to tag the sand plug.

After tagging the sand plug, the dump bailer was pulled up slightly and the cement released while continuing to pull slowly upwards. The dump bailer was then worked up and down in short jerks, to be certain all cement had been released, and then pulled from the hole.

The test bed preparation was considered complete as of September 26, 1979. The pad was demobilized and the hole was left standing full of brine while waiting for the cement grout plug to cure.

5. PLUG PERFORMANCE TESTING

The work-over rig was usually mobilized a day in advance to set up for scheduled plug performance testing. Three types of tests using primarily standard oilfield techniques and services were conceived.

Test operations were naturally keyed to the scheduled releases of tracer gas. A release schedule had been selected and timers set in an attempt to avoid holiday periods and other potential conflicts. This, in general, gave about 5-1/2 weeks between releases to conduct each test series.

While there were some variations in the test setups and sequences, they generally followed the same pattern. That is, Geophones were lowered downhole to monitor the audio signal of the explosive valve releasing a pulse of tracer gas to confirm that release occurred on schedule. A retrievable packer was run in the hole on 2-3/8 in. tubing and set 10 to 12 ft. above the top of the grout plug. This served to isolate the fluid in the annulus between the tubing and the hole above the packer. With the tubing opened to the test zone below the packer, the tubing was swabbed to lower the hydrostatic head on the grout plug to a few hundred psi. This placed a significant differential pressure load across the plug and stimulated any flow from the Bell Canyon around and through the plug. Fluid buildup on top of the plug was measured by monitoring the level of the fluid column in the tubing and by standard oilfield pressure gages installed in the hardware above and below the packer (Figure 4).

After an appropriate time to allow the fluids and gases to flow, the packer was released and retrieved from the hole. Tubing was run in the hole open only on the bottom and swabbed. This served to take fluid samples from immediately above the plug. These samples were then analyzed to determine presence of tracer gas.

The fluid buildup tests gave a gross measurement of the integrity of the grout plug but a better measurement came from shut-in tests which usually followed the gas sampling and fluid buildup cycles.

The shut-in tests were conducted in very much the same way as the fluid buildup tests except that, after swabbing the tubing to reduce the head on the plug, valves were actuated to shut-in the lowered pressure above the plug. Pressure gages were used to record annulus pressures, tubing pressures, and test-zone pressures. Test-zone pressures were also monitored at the surface in real time by using recently marketed Lynes CWL gages with surface readouts. The real time surface readouts were valuable because they would very quickly indicate whether the test setup was tight or leaking. Pressure rise curves were plotted and used to indicate the relative permeability of the grout plug.

6.   CONCLUSIONS

The field operations of the BCT produced some important information for use in future borehole plugging experiments. The BCT did meet its objectives. It verified that it is possible to plug a borehole penetrating a unit of high hydrostatic head using a minimum length cement grout plug of known standard constituents.

An important byproduct of the BCT is the knowledge and experience obtained which allows development of techniques for subsequent testing.

Some difficulty was experienced in translating laboratory standards with their tradition of accuracy into practical field operation plans which could be handled by the "oil patch" services. Valuable lessons were learned in how to "calibrate" the enthusiastic responses of "oil-patch" service personnel. One cannot place enough importance on the need to recognize the difference between units of measurements among laboratory and field personnel when designing such an experiment.

The Bell Canyon Test will serve as an important model for future borehole plugging experiments.

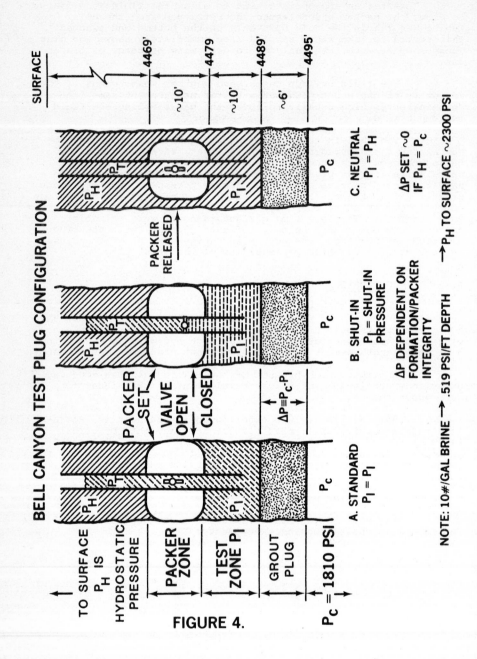

FIGURE 4.

BELL CANYON TEST PLUG CONFIGURATION

NOTE: 10#/GAL BRINE → .519 PSI/FT DEPTH

P_H TO SURFACE ~2300 PSI

Acknowledgments

The overall Bell Canyon experiment began in mid-March 1979 and continued for a little over a year. During that time over 75 suppliers and contractors in the Permian Basin provided services and materials, all of which contributed to the success of the BCT. Their enthusiastic and rapid responses deserve much appreciation. Special recognition is extended to the fine support of the Fenix & Scisson staff and particularly the outstanding work of D. L. Bradley and Matt Wilson of F&S. Dale Fain, Field Engineer of Lynes, Inc., also contributed significantly to the testing program.

References

1   Christensen, C.L., Test Plan, Bell Canyon Test, WIPP Experimental Program, Borehole Plugging, SAND79-0739, Sandia National Laboratories, Albuquerque, New Mexico, June 1979.

2   Cook, C.W. and Lagus, P.L., Bell Canyon Test (BCT) Instrumentation Development, SAND80-0408C, Sandia National Laboratories, Albuquerque, New Mexico, prepared for the International Borehole Plugging Symposium, Columbus, Ohio, May 1980.

3   Cook, C.W., Instrumentation Development Report for the Waste Isolation Pilot Plant (WIPP) Borehole Plugging Program (BHP), SAND79-1902, Sandia National Laboratories, Albuquerque, New Mexico, November 1979.

4   Stubbs, P.J., Geologist Report, prepared for Union Carbide Corp. for AEC-7, June 1974.

5   Christensen, C.L., Statler, R.D., and Peterson, E.W., Downhole Television (DHTV) Applications in Borehole Plugging, SAND80-0459C, Sandia National Laboratories, Albuquerque, New Mexico, prepared for the International Borehole Plugging Symposium, Columbus, Ohio, May 1980.

ANALYSIS OF BELL CANYON TEST RESULTS

E.W. Peterson
Systems, Science and Software
La Jolla, California 92038

C.L. Christensen
Sandia National Laboratories
Albuquerque, New Mexico 87185

ABSTRACT

Performance of the Bell Canyon Test (BCT) borehole plug has been monitored for a seven-month period. Data indicate the 2.1-meter-long plug-formation system limited the Bell Canyon flow to 5×10^{-3} STB/D. Complete test results, analytical/numerical data analysis techniques, and associated plug performance evaluation assessments are included herein. Extension of conventional oil-field technology to formations having the microdarcy-type permeabilities of those examined in the BCT emphasize the measurement limitations inherent within the current technology. Problems encountered in performing reliable, reproducible, microdary range permeability measurements are therefore identified, data discrepancies are discussed, and the resulting impact on plug performance assessment is defined.

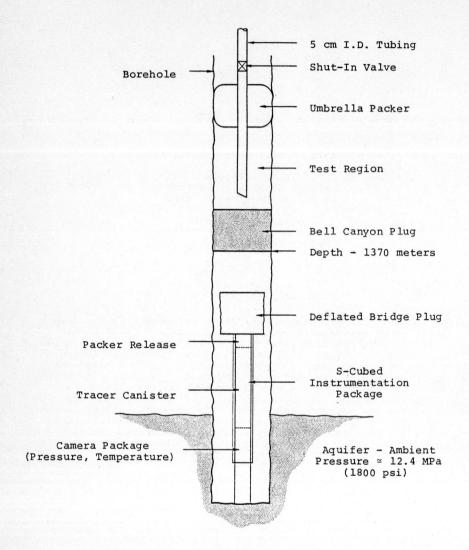

Figure 1

Schematic Showing Configuration Used for Evaluation
of Bell Canyon Plug Performance

INTRODUCTION

A major task of the Sandia [1] borehole plugging (BHP) program has been to evaluate a potential plug design in a true in situ environment. In September of 1979, the Bell Canyon test was initiated. As part of this program, a 1.8 meters in length by 20 centimeters in diamter grout plug was installed in in an anhydrite layer intersecting the AEC-7 borehole near the proposed Waste Isolation Pilot Plant (WIPP) site in southeast New Mexico. In February of 1980, the plug length was extended to 5.5 meters. This plug served to isolate the upper borehole regions from the high pressure (12.4 MPa) Bell Canyon aquifer located below the 1370 meters emplacement depth. Plug performance has now been monitored over a seven month period.

The primary objectives of the field test program were to assess the plug/formation system's effectiveness in blocking flow, and to characterize (i.e., size, permeability, porosity, fracture extent, etc.) any observed flow region. The additional plug length was added in order to more fully assess emplacement techniques, to verify performance repeatability, and to evaluate scaling effects. Evaluations of the capability of standard oil field technology to satisfy data aquisition requirements common to waste isolation applications were also made.

Measurements were made of both the fluid volumetric flow rate and flow velocity through the plug/formation system. Volumetric flow rates were determined from results of both pressure build-up and fluid build-up tests. These data may be roughly interpreted in terms of an effective system permeability. Flow velocities were determined by measuring tracer arrival times at the plug top surface, following six discrete tracer releases pre-programmed to occur below the plug at forty-day intervals. These later data not only provide positive identification of the fluid source, but also give a measure of the largest flow path (i.e., maximum fracture width). The combined flow volume and velocity data characterize system uniformity. In addition, knowledge of the flow path size and associated flow velocity can be important for modeling radionuclide migration.

This paper presents, in detail, results of the first Bell Canyon Test (BCT). The plug/formation system test methods and measurement techniques are described. Analytical/numerical data analysis methods are presented, and specific test results are given. Finally, conclusions describing the isolation effectiveness of the system in this true in situ environment are presented. Performance of the extended 5.5 meters long plug will not be reported here, as testing of this system is not complete.

TEST DESCRIPTION

The test configuration used for evaluation of the Bell Canyon plug performance is shown in Figure 1. A review of the initial test design, and the formation conditions leading to selection of this final configuration is given by Statler [3].

Briefly, the plug was installed in the AEC-7 borehole to isolate the upper regions of the borehole from the Bell Canyon aquifer, which has a 12.4 MPa shut-in pressure and a 3.7×10^7 cm^4/day (240 STB/D) production capability. The source temperature and pressure are continuously monitored by a remote (there are no cable penetrations through the plug) instrumentation package [4], scheduled for recovery upon test completion. This package also contains six canisters pre-programmed to release tracer gas at forty-day intervals. Measurements were to be made of both the volumetric rate and the velocity of fluid flowing from the aquifer through, or around, the plug into the upper wellbore. Because the upper portions of the borehole produce fluid, an umbrella packer was installed (as shown in Figure 1) in order to provide an isolated test region for measurement purposes. The pressure in the wellbore annulus above this packer was

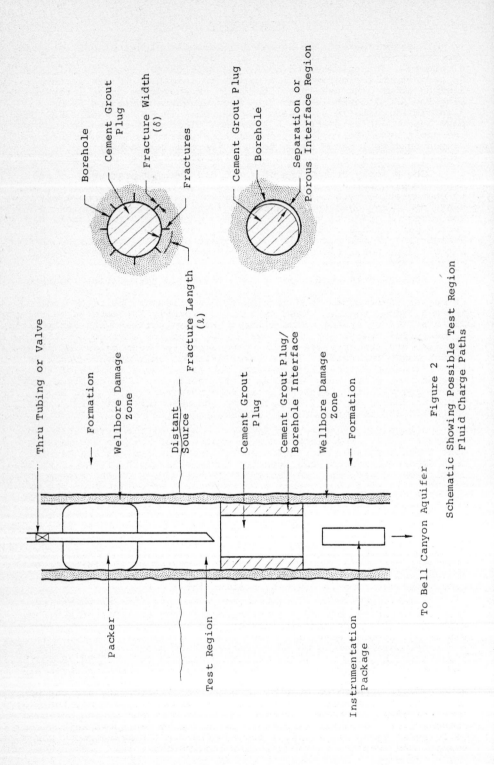

Figure 2

Schematic Showing Possible Test Region
Fluid Charge Paths

approximately 15.8 MPa.

Since plug performance is evaluated from measurements of fluid intrusion into the test chamber, it is important to identify possible flow paths into the region. As shown in Figure 2, there exist four distinct flow paths from the aquifer through, or around, the plug into the test region. These include: the cement plug; the plug/borehole interface region in which some porous microstructure may exist, either as a result of bonding imperfections or chemical reactions; a possible damage region along the borehole wall resulting from drilling and coring operations; and the undisturbed formation surrounding the plug. It is, in fact, just the flow occurring along these paths, defined as the plug/formation system, which must be measured.

Unfortunately, there exist additional possible paths along which fluid can enter or leave the test region. These include flows from above the umbrella packer through the surrounding formation or wellbore damage region, and those occurring along some formation discontinuity leading to a distant source. Clearly, the experimenter has little control of these mechanisms. Finally, test measurements may be influenced by leakage from the umbrella packer and attached tube string assemblies. This may include leakage from the packer, past valves, or into the test region through joints in the 2-inch tubing string leading to the surface. Flow occurring along these paths must be differentiated from that occurring through the plug/formation system.

Three specific types of tests were carried out to evaluate the plug/formation system performance. These include the fluid build-up, pressure build-up, and tracer flow tests discussed in detail in the following paragraphs. Because standard oil field technology was used for data acquisition, the measurement sensitivity and its associated impact on plug performance assessment is also discussed.

Fluid Build-up Test

The fluid build-up test measures total volumetric fluid flow into the test region. For these tests, the packer-valve-tube assembly was configured with the packer seated and the valve open, as shown on Figure 3a. Fluid was removed from the tube to a depth such that the pressure on the top surface of the cement plug was less than .2 MPa. A sensitive pressure-gauge [3] was placed in this column, and the fluid level rise was determined from the measured pressure increase.

With this configuration, a pressure differential of approximately 12.4 MPa was maintained across the plug. The test, therefore, measures the largest possible flow rates obtainable with this system. Even so, the test may be considered quasi-static, as measured volumetric flow rates are extremely small compared to the Bell Canyon production capability. Similarly, measured changes in the fluid column pressure above the plug are negligible, compared to the driving pressure.

Unfortunately, the fluid build-up test is also subject to the largest interpretation error. In this test, the 2-inch tubing is unloaded above the umbrella packer, yet the annulus is completely brine-filled. To obtain good test results, leakage from the annulus into the two-inch tubing, through the approximately 150 connecting joints, must be small compared to the measured plug/formation system flows of less than one liter/day. Fluid build-up test data has been subject to large scatter.

Pressure Build-up Test

The pressure build-up (shut-in) test also measured volumetric flow into the test region. In this test, fluid is first removed from the tubing until the pressure on the top surface of the cement grout plug is less than .2 MPa. The shut-in valve (shown on Figure 3b.) is

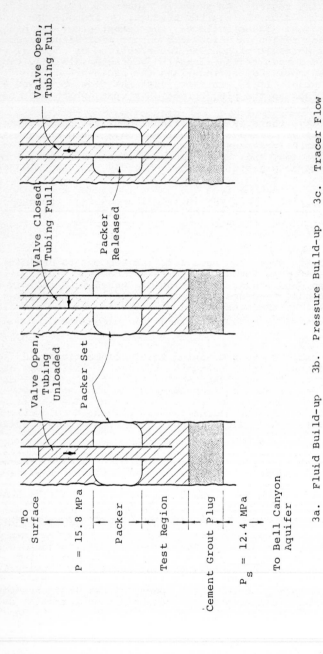

Figure 3

Schematic Showing the Packer-Valve-Tube Assembly
Configurations Used for the Fluid Build-up,
Pressure Build-up, and Tracer Flow Tests

3a. Fluid Build-up   3b. Pressure Build-up   3c. Tracer Flow
          Test   Test   Test

then closed. Any subsequent inflow into the test region is accompanied by a build-up in pressure. The test is continued until the test region pressure approaches that of the Bell Canyon aquifer. Volumetric flow rates are then interpreted from the resulting pressure history data.

The pressure build-up test also provides a means for identification of the flow source. If there exists a single source, the final test region pressure will approach that of the source. Since the 12.4 MPa aquifer pressure obtains below the plug, while the annulus pressure above the packer is 15.8 MPa, these two potential sources and their relative flow contributions can be identified, provided there exist no formation thief zones in communication with the test region.

During the fluid build-up test, the shut-in valve is closed, and both the tubing and annulus are filled with brine. The shut-in valve has, therefore, replaced the tubing as a potential leak source. Although shut-in valve performance has been somewhat inconsistent, pressure build-up tests have proven more repeatable than the fluid build-up tests.

Tracer Flow Tests

Positive evidence of flow through the plug/formation system was provided by the tracer tests. In addition, these tests were used to determine maximum flow velocities and flow channel dimensions.

Six discrete tracer releases were pre-programmed to occur below the plug at forty-day intervals during the test period. During each test, a 700 cubic centimeter (at 12.4 MPa) bubble of N_2, containing a 3% concentration of SF_6 as the tracer, was released below the plug [4]. Subsequent detection of this tracer above the plug was taken as positive confirmation of flow through the plug/formation system.

Maximum fluid flow velocities were determined from measurements of the time between tracer release and its first detection above the plug. This time was taken as roughly (within a factor of two) representative of that for fluid flow, since all major flow paths may be assumed saturated prior to tracer release, and fluid must first be pushed through these paths before significant tracer penetration can occur.

Tracer flow and fluid build-up tests were generally performed simultaneously, since the fluid build-up test configuration yields maximum flow rates and, therefore, minimizes test times. Tracer arrival at the top of the plug is determined by analyzing fluid samples for evidence of SF_6. These samples are taken at specified times, either by swabbing or by re-circulating fluid using the configuration shown in Figure 3c. Clearly, the fluid build-up and tracer flow operations must be interrupted for this procedure.

The initial test design, as reported by Statler [3], was based on having a dry borehole above the plug. Under these conditions, gas in the test region would be continually sampled to determine both tracer arrival times and concentration build-up rates. However, SF_6 and N_2 have a water solubility of .005 cc/cc and .025 cc/cc, respectively, at standard conditions. Thus, tracer arrival times may be determined by analysis of gas dissolved in the fluid samples. It should be noted [4] that the S-Cubed field tracer detection instrumentation is sensitive to approximately one part SF_6 in 10^{12} parts nitrogen or air.

Some portion of the tracer gas may go into solution below the plug. Analyses have shown that even if the fluid below the plug is fully saturated with the SF_6/N_2 gas, the resulting multi-phase flow will produce first arrivals of tracer within thirty percent of that time calculated, assuming the gas must simply displace water within the porous system prior to gas breakthrough. This was considered to be within the accuracy of the test method.

Measurement Sensitivity

A 12.4 MPa differential pressure across a one microdarcy permeability plug/formation system of 1.8 meters in length by 20 centimeters in diameter, produces a flow of about 10 cm^3/day (10^{-4} STB/D). Longer plugs would have correspondingly lower flow rates. Understandably, conventional oil field technology has not addressed problems inherent in performing reliable, reproducible, quantitative measurements in this regime.

Quartz crystal gauges [3] (which are the most pressure-sensitive devices available in the well-logging industry) are adequate for pressure build-up test monitoring. In addition, these gauges satisfactorily measured the changes in fluid column heighth (i.e., through observation of pressure changes) occurring during most of the fluid build-up tests reported here. However, they do not have the accuracy required to monitor a fluid build-up test for a one microdarcy system, unless the test period extends numbers of months.

Typical oil field service flow test tools are not designed to operate under the low flow conditions obtaining in the borehole plugging environment. Results are, therefore, often compromised by leakage through tube joints and/or system shut-in and control valves. Unfortunately, the test tools lack those features which allow either control or measurement of this leakage. Because of this lack of system flow control, injectivity tests cannot be performed at the required low flow rates.

Oil field service test tools, themselves, have not been examined to determine system-compliance, packer-deformation under changing vertical loads, packer motion occurring during setting or testing procedures, or test region over-pressure induced when setting straddle packer systems. These effects may influence data interpretation or, if severe, completely negate the test results.

The most severe limitation of the oil field service test tools is their inability to distinguish between predominantly horizontal flow into the formation, and vertical flow as may result from wellbore damage. These tools are incapable of providing the precise resolution necessary for wellbore characterization prior to plug installation. If these studies are not performed, subsequent tests cannot distinguish between flow contributions resulting from the plug, formation, and/or wellbore damage regions. Wellbore characterization tests could be performed, using guarded straddle packer systems, with associated tracer and flow control capabilities (with modification for fluid-filled holes), analagous to that used by Peterson and Lagus [5] in measuring in situ permeabilities of rock salt.

In terms of the present test series, use of existing technology has influenced the results, as follows. Pressure build-up test data was satisfactory. In those tests, flow from the high pressure upper annulus through the packer-valve-tubing assembly could be identified. In cases where this leakage dominated flow through the plug/formation system, the test would be terminated, the system re-worked, and a second test performed. Leakage from the annulus cannot be identified when performing fluid build-up tests. In that case, all flow was attributed to the plug/formation system. Results of these tests were inconsistent. The actual plug/formation flow rate should, therefore, be taken as less than, or equal to, the lowest measured value. Wellbore characterization tests simply compound those problems apparent in fluid build-up testing, since packer assemblies are located at both ends of the test region.

DATA ANALYSIS

The intent of the data analysis is to characterize (i.e., determine the permeability, porosity, cross-section area, fracture extent, etc.) those flow paths originating below the plug which penetrate into the test region (shown on Figure 2). Available data include the experimentally-measured test region pressures, water accumulation rates and tracer arrival times, and the annulus and Bell Canyon pressures. A classic unfolding problem, therefore, exists in which a number of constants, representing the formation characteristics and flow path geometries, need to be evaluated, given a specified set of boundary conditions. A rigorous derivation of the equations defining the flow through this plug/formation system is given in the following paragraphs.

Equation Derivation

Consider flow from the Bell Canyon aquifer (shown in Figure 2) through and/or around the plug into the test region. When the pore space is completely saturated and flow velocities are slow, flow is dominated by viscous effects, and the simple diffusion or darcy flow models described by Bear [6] and Bird [7] are valid. Under these conditions, the area-averaged fluid velocity $\bar{q}$ is proportional to the pressure gradient, giving

$$\bar{q} = \frac{\kappa}{\mu} \nabla P \quad , \tag{1}$$

where κ and μ represent the formation permeability and the fluid viscosity, respectively. The corresponding fluid particle velocity (i.e., tracer particle average velocity) is just

$$\bar{v} = \bar{q}/\phi \quad , \tag{2}$$

where ϕ is the connected porosity. Conservation of mass within the system requires

$$\phi \frac{\partial \rho}{\partial t} + \nabla \cdot (\rho \bar{q}) = w(t) \quad , \tag{3}$$

where ρ and $w(t)$ represent fluid density and a source of flow rate, respectively. If the dynamic response is to be evaluated, an equation of state is required to close the system of equations. For a fluid, this may be approximated as

$$P = P_\infty + C \left(\frac{\rho}{\rho_\infty} - 1 \right) \quad , \tag{4}$$

where C is the compressibility, and the subscript ∞ represents standard conditions. Equations (1) through (3) may be combined to give

$$\phi \frac{\partial P}{\partial t} = \nabla \left\{ \frac{\kappa}{\mu} \left[(P - P_\infty) + C \right] \nabla P \right\} + \frac{C}{\rho_\infty} w(t) \quad , \tag{5}$$

Equation (5) may be used to model flow within the system (shown in Figures 1 and 2). This flow may occur through a porous formation, as previously described, or along narrow fractures (such as those illustrated in Figure 2). For problems of interest here, the fracture widths are small, as are flow Reynolds numbers, there-

- 255 -

fore, viscous effects again dominate. Under these conditions, the average fluid velocity, and hence the average tracer velocity (the tracer leading edge moves with twice this velocity), along a fracture of width (δ) [8] is

$$\overline{v} = - \frac{\delta^2}{12\mu} \nabla P \quad .$$ (6)

It follows, in a very crude sense [6], from Equations (1) (2), and (6), that

$$\kappa = \frac{\phi \delta^2}{12} \quad .$$ (7)

Fracture characteristics may, therefore, be expressed in terms of equivalent permeabilities and porosities, and the flows may be described using Equation (5).

The preceding formulation may also be modified to model non-darcy flows, where viscous-inertial, or possibly turbulent, losses are important. In such applications, the permeability appearing in Equation (5) is expressed in terms of its value measured under viscous flow conditions and the local fluid velocity and density [9, 10].

Solutions to Equation (5) are obtained using a two-dimensional (e.g., either axisymmetric or cartesian coordinates), finite element numerical code, developed by Systems, Science and Software (S-Cubed). The numerical description is general in the sense that κ, ϕ, and μ may be taken as functions of position, pressure, temperature, or time. Hence, multi-dimensional effects, anisotropic permeabilities, wellbore damage, non-darcy flow (as may exist in larger fractures), thermal effects, etc., may be considered.

One-Dimensional Models

Unfortunately, a knowledge of the Bell Canyon and annulus pressures, the test region pressure history or input flow rate, and the tracer flow time is insufficient to completely characterize the complex flow system shown in Figures 1 and 2. It is, therefore, informative to examine the plug/formation system response in terms of one-dimensional quasi-steady models.

Pre-test sensitivity studies by Peterson and Lie [11] show the time required to attain a steady-state pressure field in this fluid-saturated system, following some pressure excursion, is short compared to test times of interest. In addition, those studies also indicated that if the predominant flow zones are within the near wellbore vicinity (i.e., a wellbore radius), the fluid and volumetric flow rates may be reasonably-well approximated one-dimensionally. The one-dimensional quasi-steady models used for analysis of the fluid build-up, pressure build-up, and tracer flow tests are given in the following paragraphs.

Assume, in the fluid build-up test, that the Bell Canyon fluid below the plug is the source of all flow entering the test region. It follows from Equation (1), that the one-dimensional, quasi-steady, volumetric flow rate (Q) is

$$Q = \frac{\kappa (P_s - P) A}{\mu L}$$ (8)

where P, P_s, A, and L represent the test region pressure, Bell Canyon pressure, flow region cross-section area, and plug length, respectively. Under these conditions, the fluid build-up data provide a measure of the area-permeability product κA, since all re-

maining terms in Equation (8) are known. It can also be seen from
Equation (6), that for fracture flow these data can be related to
$\delta^3 \ell N$, where ℓ and N represent the fracture length and total number,
respectively (see Figure 2).

The time (T) between tracer release and its first detection
above the plug is roughly the ratio of plug length to particle velocity (it is assumed that the path is not tortuous). Equations (1)
and (2) show that this time is a measure of the flow zone permeability to porosity ratio

$$\frac{\kappa}{\phi} = \frac{\mu L^2}{T\,(P_s - P)} \quad . \tag{9}$$

If the flow occurs along fractures then, according to Equation (6),
this time provides a rough measure of fracture width

$$\delta^2 = \frac{12\mu L^2}{T\,(P_s - P)} \quad . \tag{10}$$

Pressure build-up test results may also be examined in terms
of a simple one-dimensional model. For quasi-steady flow into the
enclosed fluid-filled test region, the resulting pressure history is
given by

$$P = P_o + \frac{C}{V} \int_o^t Q_T \, dt \quad , \tag{11}$$

where V and P_o represent the test region volume and initial presure,
respectively. If all flow into the test region comes from the Bell
Canyon aquifer and annulus regions, then the total flow may be written as

$$Q_T = \eta\,(P_s - P) + \eta_a\,(P_a - P) \quad . \tag{12}$$

where $\eta = \kappa A/\mu L$, and where P_a and η_a represent the annulus pressure
and corresponding flow coefficient, respectively. The far right-
hand term, representing the annulus contribution, again assumes a
volumetric flow rate proporational to the pressure gradient. Solving
Equations (11) and (12) yields

$$\frac{P_F - P}{P_F - P_o} = \text{EXP}\,[-(\eta + \eta_a)\,t] \quad , \tag{13}$$

where the final test region pressure (P_F) is given by

$$P_F = \frac{\eta P_s + \eta_a P_a}{\eta + \eta_a} \quad . \tag{14}$$

The final pressure is seen to depend on that of both the annulus and
Bell Canyon regions. The pressure build-up test can, therefore, be
used for source identification. The flow coefficient $\eta = \kappa A/\mu L$
(which, again, defines the permeability-area product) can be determined from Equations (13) and (14). In fact, the volumetric flow
contributions attributable to each source region can be determined.

The one-dimensional formulations serve to clearly identify
those system characteristics which can readily be identified from

Table I

Summary of System Integrity and Plug
Zone Characterization Data

Initiation Date	Measured Test Region Inflow Rate * (cm³/day × 10³)	Flow Path Permeability-Area (κA × 10⁻¹⁸ cm⁴)	Maximum Flow Channel Width ** (δ cm × 10⁻⁵)	Remarks
05/03/79 (Depth 1388 m)	—	—	—	Camera package check only. Bell Canyon fluid pressure and temperature were measured for a 60-day period, and found to remain constant at 12.7 MPa and 31°C, respectively.
08/07/79 (Depth 1370 m)	—	—	—	Mechanical failure in tubing string, package dropped to bottom of hole.
08/31/79 (Depth 1370 m)	—	—	—	Tracer release system operation satisfactory. Bridge plug release valve operation satisfactory. Packer system leaked (premature deflation). Freon C318 not detected in air above fluid column. Freon C318 detected in fluid at ≈ 1370 meters depth.
09/10/79 (Depth 1370 m)	4.1	11	< 2.1	Freon 12 did not flow past bridge plug during the 56-hour test period.
09/20/79 (Depth 1320 m)	1.8	5	—	Tracer not used.

* This represents the inflow rate obtained, given a 12.4 MPa pressure differential across the length of the plug zone.

** The Freons 13B1, C318, and F12 were used as tracers on the 8/7, 8/31, and 9/10 tests, respectively.

the measured data. These properties include the product of the flow
zone permeability times area (κA), the ratio of flow zone permeability to connected porosity (κ/ϕ), the largest flow channel width
(δ), and the product of the total fracture aperture times the fracture width squared ($\delta^3 \ell N$). The number of maximum-width fractures
required to provide the measured flow rate can be approximated from
this last expression. As a result, flow through fracture systems
typical of those shown in the Figure 2 inserts can be distinguished
from flows through regions where the micro-structure approximates a
more classical porous media.

TEST RESULTS

The test program may be divided into two portions for discussion purposes. These include the system integrity and wellbore characterization tests performed prior to plug installation, and tests
performed to evaluate isolation characteristics of the 1.8 meters in
length plug. Performance of the 5.5 meters in length plug will not
be reported here, as system testing has not been completed.

System Integrity and Wellbore Characterization Tests

Prior to installation of the 1.8 meters in length by 20 centimeters in diameter cement grout plug, system integrity and plug zone
characterization tests were performed. The system configuration
generally appeared as shown in Figure 1, with the exception that the
inflated bridge plug was positioned at the cement grout plug location. Tests results are summarized in Table I.

The intent of the system integrity tests was to check operation
of the instrumentation canister [4], which contained the packer release, tracer release, and camera package for measuring source pressure and temperature. Down-hole camera package operation was successfully demonstrated for a sixty-seven-day period, during which
time the Bell Canyon source pressure and temperature were found to
remain constant at 12.7 MPa and 31°C, respectively [5]. No data
was obtained from the August 7 test, as a mechanical link in the tube
string parted during installation, allowing the package to settle to
the bottom of the hole. Successful operation of the tracer and
bridge plug release valves and timing mechanisms were demonstrated
during the August 31 test.

The packer system (i.e., probably the seating valve) leaked
during the August 31 test, allowing premature packer deflation. Plug
zone formation characteristics could, therefore, not be measured.
Tracer gas sampling was, however, performed during this test. Air
samples were collected at the top of the fluid column, located at
300 meters in depth. These samples, when analyzed, contained no evidence of the Freon C318 tracer gas. The released gas had dissolved
in the approximately 1070-meter column of drilling fluid above the
release point. However, fluid samples collected at the depth of the
bridge plug were found to contain substantial quantities of Freon
C318. All subsequent tracer detection was done through fluid sampling.

Plug zone formation characteristics were measured during the
September 10 test using Freon 12 in N_2 as a tracer. Fluid build-up
rates measured during this test are shown on Figure 4. If all flow
entering the test region passed around the bridge plug through the
wellbore damage region shown on Figure 2, then $\kappa A = 11 \times 10^{-10}$ cm^4,
for this flow path (i.e, $\kappa = 335$ microdarcies, if A equals the wellbore cross-section area). There was no tracer detected in the region above the packer during the fifty-six hour test period. Subsequent sampling after packer deflation indicated that the fluid below
the packer did contain Freon 12. Failure of the tracer to penetrate
into the test region during this time period indicates that κ/ϕ is
less than 3.6×10^{-3}, and that the width of the largest flow channel
(δ) is less than 2.1×10^{-5} cm.

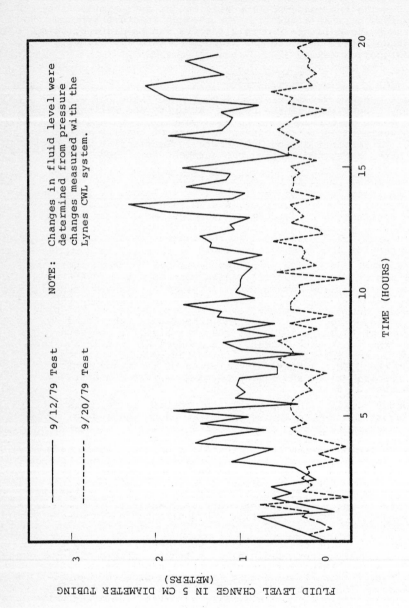

Figure 4

Fluid Build-up Test Data Obtained During
Plug Zone Characterization Studies

A fluid build-up test was also performed at a depth of 1230 meters, in order to evaluate formation properties at a position other than that intended for plug installation. This test yielded $\kappa A = 5 \times 10^{-5}$ cm^4 (see Figure 4).

The fluid build-up tests performed during the plug zone characterization studies are subject to the uncertainties previously discussed. In particular, data interpretation must be based on the total volumetric flow rate into the test region. Contributions to this flow may result from system leakage or formation flow around the upper and/or lower packers. There is no way to distinguish the relative contributions. The measured values, therefore, provide only an upper bound on formation permeabilities in the plug region.

Initially, the plug zone verification studies were to have been performed using a guarded straddle packer system [5], which can provide the necessary flow control. This system was configured to operate in the AEC-7 borehole under dry conditions. The production of fluids in AEC-7 precluded these measurements at the plug location [3].

Plug Performance Test Results

Performance characteristics of the 1.8 meters in length cement grout plug were monitored over a four-month period, beginning October 9, 1979. Results of the fluid build-up, pressure build-up, and tracer flow tests are summarized in Tables II, III, and IV, respectively. Data were interpreted using fluid properties of $\rho = 1.2$ gm/cc, $\mu = 1.57 \times 10^{-2}$ poise, and $C = 3.26 \times 10^3$ MPa, as determined by Sandia Laboratories [12].

Five fluid build-up tests were performed. Data obtained during four of these tests are shown on Figure 5. The October 19 test was continued for a total of 144 hours, with the fluid rise continuing at, roughly, the rate shown on Figure 5. The fluid level was tagged at 1330 meters and 1296 meters, taken on December 21 and January 11, respectively. These tags represent the data for the last test shown on Table II.

Results obtained during these fluid build-up tests are consistent with those obtained during the plug zone characterization tests. However, the October 9 and 10 tests definitely indicate lower test chamber inflow rates, and are probably more representative of the plug/formation system performance. These data have been interpreted to obtain the approximate values shown on Table II. The pressure fluctuations observed during the October 9 and 10 tests do not represent variations in fluid level, but reflect noise in the CWL monitoring system [3]. Since these fluctuations are large compared to the pressure increase resulting from the rise in fluid level over the twenty-hour test period, the data are difficult to interpret.

Data obtained during the three pressure build-up tests are shown on Figure 6. These data can be well-represented by the simple one-dimensional model defined by Equation (13), using those values of κA shown on Table III. Results of the pressure build-up tests were very consistent, and indicate that the plug/formation system responds as if there existed an approximately fifty microdarcy flow path of cross-section area equal to that of the wellbore.

During the December 9 test, a significant time lapse occurred between shut-in and the onset of the rapid pressure build-up phase. Also, at late times the pressure continued to increase above the Bell Canyon value. The delayed initial response is thought to be the result of an air bubble trapped in the tubing which extends into the test region below the umbrella packer. A slight flow from the annulus region can account for the higher late time pressure. Both of these effects can be modeled (as illustrated in Figure 6) using the simple one-dimensional formulation. Subsequently, the packer-valve-tube assembly was modified to eliminate the possibility of air entrapment, and the January 20 test was performed for data verification

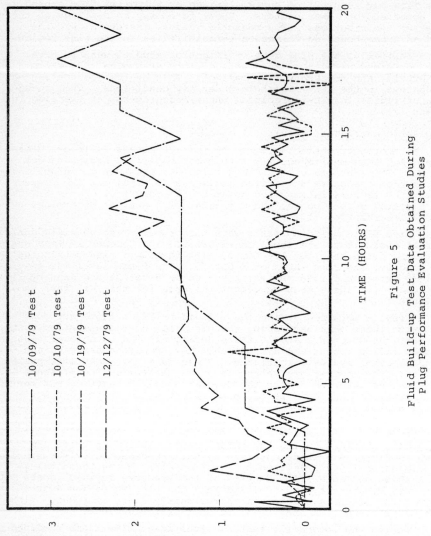

Figure 5

Fluid Build-up Test Data Obtained During
Plug Performance Evaluation Studies

Table II

Summary of Fluid Build-up Test Results

Test Initiation Date	Test Duration (Hours)	Measured Test Region Inflow Rate * (cm³/day x 10³)	Flow Path Permeability-Area (κA x 10⁻¹⁸ cm⁴)
10/09/79	18	≤ 0.6	1.6 (51 μdarcy) **
10/10/79	18	≤ 0.6	1.6 (51 μdarcy)
10/19/79	144	4.6	12.0 (385 μdarcy)
12/12/79	16	7.1	20.0 (607 μdarcy)
12/21/79	504	3.3	8.9 (275 μdarcy)

* This represents the inflow rate obtained, given a 12.4 MPa pressure differential across the 1.8 meter long cement grout plug.

** Numbers in parentheses indicate the permeability for a flow path whose cross-section area is taken equal to that of the wellbore.

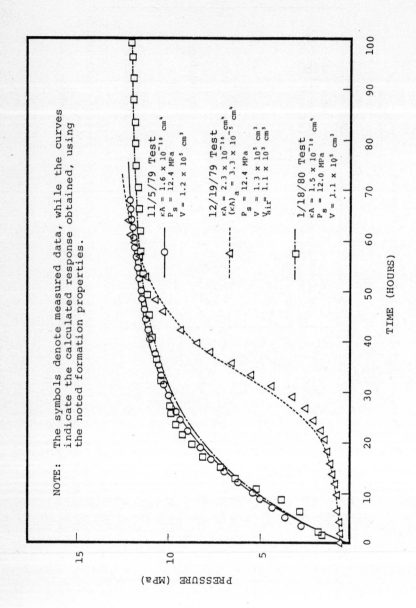

NOTE: The symbols denote measured data, while the curves indicate the calculated response obtained, using the noted formation properties.

11/5/79 Test
$\kappa A = 1.6 \times 10^{-10}$ cm⁴
$P_s = 12.4$ MPa
$V = 1.2 \times 10^5$ cm³
○

12/19/79 Test
$\kappa A = 2.3 \times 10^{-10}$ cm⁴
$(\kappa A)_a = 3.3 \times 10^{-5}$ cm⁴
$P_s = 12.4$ MPa
$V = 1.3 \times 10^5$ cm³
$V_{air} = 1.1 \times 10^3$ cm³
△

1/18/80 Test
$\kappa A = 1.5 \times 10^{-10}$ cm⁴
$P_s = 12.0$ MPa
$V = 1.1 \times 10^5$ cm³
□

Figure 6

Pressure Build-up Test Data Obtained During
Plug Performance Evaluation Studies

Table III

Summary of Pressure Build-up Test Results

Test Initiation Date	Test Duration (Hours)	Measured Test Region Inflow Rate * (cm³/day x 10³)	Flow Path Permeability-Area (κA x 10⁻¹⁸ cm⁴)
11/02/79	67	.61	1.6 (51 μdarcy) **
12/18/79	62	.83	2.3 (70 μdarcy)
01/18/80	100	.55	1.5 (46 μdarcy)

* This represents the inflow rate obtained, given a 12.4 MPa pressure differential across the 1.8 meter long cement grout plug.

** Numbers in parentheses indicate the permeability for a flow path whose cross-section area is taken equal to that of the wellbore.

Table IV

Summary of Tracer Flow Test Results

Tracer Release	Time Until First Arrival * (Hours)	Permeability To Porosity Ratio $(\kappa/\phi \times 10^{11}\ cm^2)$	Maximum Flow Channel Width $(\delta\ cm \times 10^5)$	Maximum Channel Flow Velocity (m/day)
10/09/79	< 68	> 2.2	> 1.5	> .6
10/30/79	≈ 36	3.3	2.0	1.2
12/09/79	≈ 36	3.3	2.0	1.2
01/18/80	No Sampling			

* This represents the time until first arrival, given a continuous 12.4 MPa pressure differential acting across the 1.8 meter long cement grout plug.

purposes.

It is important to note that there is no evidence of flow from the high pressure annulus into the test region during the November 2 and January 18 pressure build-up tests. This strongly suggests that, at least in the area where the umbrella packer is positioned, well-bore damage is slight. If one assumes a damage area equal to the wellbore area, these pressure build-up tests are consistent with an equivalent permeability of less than five microdarcies. Note also that the flow observed from the annulus region during the December 18 test probably occurs through the packer-valve-tube assembly. Leakage through this assembly is probably responsible for the higher flow rates measured during the fluid build-up tests.

A history of the tracer sampling tests is presented in Table V. The approximately thirty-six hour time interval between gas release and detection at the top surface of the plug was well-established after the December 9 series of tests. Further sampling was, therefore, discontinued.

Tracer flow data indicate (see Table IV) a maximum flow channel width and velocity of approximately 2×10^{-5} cm and 1.2 meters/day, respectively. If all flow is assumed to occur through fractures whose length (measured in the horizontal plane) equals that of the wellbore circumference, then approximately 4000 such fractures are required to provide the measured volumetric flow rate. This suggests that flow through the plug/formation system does not occur through a small number of fractures, but through a region where the microstructure approximates, to a reasonable extent, a porous medium.

When interpreted in terms of flow through a classical porous medium, the tracer data indicate a permeability to porosity ratio of $\kappa/\phi = 2.7 \times 10^{-10}$ cm^4. If the cross-section area (A) of the flow path through the plug/formation system (see Figure 2) were known, the associated permeabilities and porosities could be determined. There is evidence that plug permeabilities are small, and that flow primarily occurs through a permeable microstructure at the plug bore-hole interface [13]. Required permeabilities and porosities of such an interface zone are shown in Figure 7 as a function of flow zone cross-section area.

It is informative to calculate the test region pressure history, assuming the primary inflow occurs through an interface zone (see Figure 2) with secondary flow occurring through a wellbore damage zone. The cross-section areas of these zones are taken as equal to 10% and 100%, respectively, of that of the wellbore. Results of these calculations are intended to verify the adequacy of the one-dimensional analytical models, and to demonstrate that measured test results are, indeed, consistent with the assumption of small flow zones.

The results of two calculations are shown on Figure 8. Both use the flow geometry and formation properties shown on Figure 9. Interface zone permeability and porosity values were selected from Figure 7, while damage zone properties were arbitrarily chosen for demonstration purposes.

The solid curve shown on Figure 8 compares favorably with both the measured data and previous one-dimensional analysis (see Figure 6). In this calculation, communication between the wellbore damage region and high pressure annulus region was prevented by assuming an impermeable horizontal plane along the top of the grid. If flow from the annulus region is allowed, the pressure build-up curve is noticeably changed (see dotted line on Figure 8), and the final pressure exceeds that of the Bell Canyon aquifer.

Again, these solutions are not unique. Any reasonable combination of flow zone properties consistent with the κA and κ/ϕ values defined in Tables III and IV will provide test region pressure histories consistent with the measured data. It should again be noted

Table V
History of Tracer Sampling Tests

Tracer Release Date	Sampling Date	Tracer Detected	Remarks
10/09/79	10/08/79	No	Background check.
	10/12/79	Yes	Arrival had occurred in < 68 hours. *
	10/18/79	Yes	
	10/27/79	Yes	Wellbore fluid replaced.
10/30/79	10/29/79	Trace	Background check.
	10/31/79	Trace	
	11/01/79	Yes	Arrival had occurred in = 36 hours. *
12/09/79	12/06/79	Trace	Background check.
	12/11/79	No	
	12/12/79	No	
	12/13/79	Yes	Arrival had occurred in = 36 hours. *
01/18/80			No further tracer samples were taken.

* Arrival times represent the tracer transit time, given a continuous 12.4 MPa pressure differential across the 1.8 meter cement grout plug. The actual wellbore pressure history is used to determine this value.

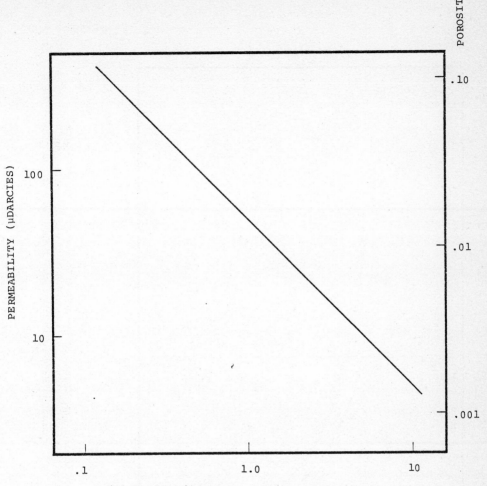

Figure 7

Relationship Between Flow Zone Permeability,
Porosity, and Cross-section Area, Based on
One-Dimensional Analysis of Measured Data

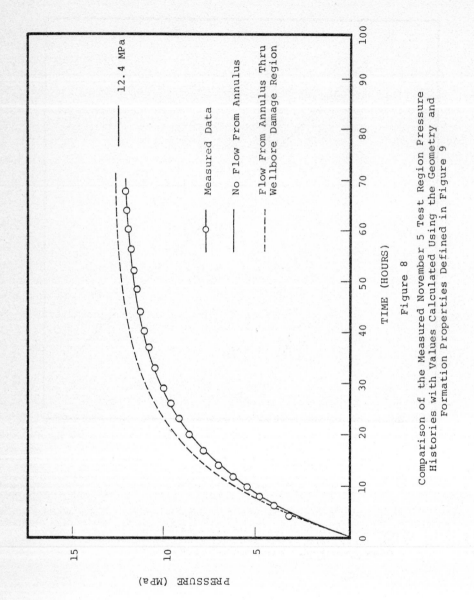

Figure 8

Comparison of the Measured November 5 Test Region Pressure
Histories with Values Calculated Using the Geometry and
Formation Properties Defined in Figure 9

that combinations of wellbore damage and formation regions having κA greater than approximately 1.5×10^{-10} cm^4 will result in test region pressures exceeding those observed during the November 5 and January 18 tests. In that sense, the pressure build-up tests have provided a measure of the maximum wellbore damage which may exist in the umbrella packer region.

CONCLUSIONS

The 1.8 meters in length by 20 centimeters in diamter cement grout plug, installed in anhydrite at a depth of 1370 meters in the AEC-7 borehole, limits flow from the high pressure (12.4 MPa) Bell Canyon aquifer to 0.6 liters/day. Tracer flow studies are consistent with a flow channel possessing a maximum width not greater than roughly 2×10^{-5} centimeters in which the maximum flow velocity is approximately 1.2 meters/day. This suggests that predominant flow is not through a finite number of fractures, but through a region whose microstructure approximates a porous medium. If the flow path cross-section area is taken as equal to that of the wellbore, the corresponding permeability and connected porosity would be fifty microdarcies and 0.018, respectively. Plug performance remained consistent throughout the relatively short four-month test period.

Measured data indicate flow through the formation surrounding the umbrella packer (positioned approximately four meters above the cement grout plug) is at least one order of magnitude less than that through the plug formation system. If the medium surrounding the cement grout plug is similar, flow must occur through the plug or plug/wellbore interface. It is of interest to note that if the cross-section area of the wellbore damage region surrounding the umbrella packer is equal to that of the wellbore, it must have a permeability of less than five microdarcies.

Standard oil field service testing tools and practices were found to lack the precision required to perform wellbore characterization tests. This technology can be used for pressure build-up testing to determine plug/formation system performance under conditions obtained in the tests reported here. However, significant test tool modifications are required if one microdarcy plug/formation systems are to be examined.

REFERENCES

1.   Christensen, C.L. and Hunter, T.O., "Waste Isolation Pilot Plant (WIPP), Borehole Plugging Program Description", January 1, 1979. SAND79-0640, Sandia National Laboratories, Albuquerque, NM 87185, August 1979.

2.   Christensen, C.L., "Test Plan, Bell Canyon Test, WIPP Experimental Program, Borehole Plugging", June 1979, SAND79-0739, Sandia National Laboratories, Albuquerque, NM 87185.

3.   Statler, R.D., "Bell Canyon Test - Field Preparation and Operations", presented at the Workshop on Borehole and Shaft Plugging, Columbus, Ohio, May 7, 1980.

4.   Cook, C.W., Lagus, P.L., and Broce, R.C., "Bell Canyon Test (BCT) Instrumentation Development", presented at the Workshop on Borehole and Shaft Plugging, Columbus, Ohio, May 7, 1980.

5.   Peterson, E.W., and Lagus, P.L., "In Situ Permeability Testing of Rock Salt", presented at the Third Invitational Well Testing Symposium, Lawrence Berkeley Laboratory, Marcy 26 - 28, 1980.

6.   Bear, J., "Dynamics of Fluids in Porous Media", American Elsevier Publishing Co., New York, 1972.

7.   Bird, R.B., et al., "Transport Phenomena", John Wiley and Sons, New York, 1960.

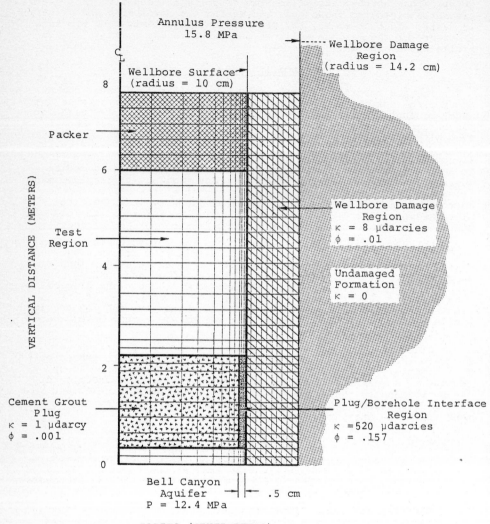

Figure 9

Calculational Grid for Modeling Plug/Formation
System Showing Interface Flow Zone and
Wellbore Damage Region

8.   Schlichting, H., "Boundary Layer Theory", McGraw Hill, New York, 1960.

9.   Wong, S.W., "Effect of Liquid Saturation on Turbulence Factors for Gas-Liquid Systems", Journal of Canadian Petroleum, October-December, 1970.

10. Holditch, S.A., and Morse, R.A., "The Effects of Non-Darcy Flow on the Behavior of Hydraulically Fractured Gas Wells", Journal of Petroleum Technology, October 1976.

11. Peterson, E.W., and Lie, K., "Analysis of Bell Canyon Test Plug Seepage", Systems, Science and Software Report (SSS-4-79-3952), February 1979.

12. Cook, C.W., private communication, April 1980.

13. Gulick, C.W., Boa, J.A., and Mather, K., "Bell Canyon Test (BCT) Cement Development Report", presented at the Workshop on Borehole and Shaft Plugging, Columbus, Ohio, May 7, 1980.

Discussion

W. FISCHLE, Federal Republic of Germany

      How long did the tests (fluid build up, pressure build up, tracer flow) last ?

E.W. PETERSON, United States

      With the 1.8-meter-plug fluid build up tests lasted between 1 and 21 days, pressure build up tests lasted between 60 and 100 hours. Tracer arrival was found to occur within ~36 hours, however, intermittent sampling was continued over a number of days to confirm these measurements.

W. FISCHLE, Federal Republic of Germany

      What was the composition of the mixture of the cement grout plug ?

E.W. PETERSON, United States

      The details of the BCT-IFF grout mixture are given in the report SAND 80-0358. Briefly the composition was :

	% by weight
Class H cement	53.1
Fly ash	18.1
Expansive additive	7.1
Dispersant (D65)	0.1
Defoamer (D47)	0.02
Water	21.6

G.M. IDORN, Denmark

      Is the special character of the "interface region" of the cement plug towards the wall of the borehole due to chemical interaction between the hardened cement mixture and the constituents of the host rock ?

E.W. PETERSON, United States

      This problem will be addressed in the following presentation by C.W. Gulick.

A.M.L. BOULANGER, France

      Les propriétés pétrophysiques du massif aux abords du puits ont été perturbées par les opérations de forage et les opérations de mise en place du bouchon. Quelles techniques utilisez-vous pour éliminer ces perturbations et comment vérifiez-vous que les valeurs que vous mesurez sont représentatives du massif vierge ?

E.W. PETERSON, United States

      Engineering problems involved in borehole drilling and plug installation are recognized. These operations were accomplished using standard oil field technology. The intent of the test was to measure the flow characteristics of the plug/formation system. This system includes the plug, the plug/borehole interface region, the borehole damage region and the surrounding virgin formation.

The flow through the entire system was measured. Contributions arriving through the virgin formation were negligible. If these contributions had been significant, the final shut-in pressure, observed during the fluid build up test, would have approached that of the annulus (i.e. 15.8 MPa).

R.D. ELLISON, United States

Is it possible that instead of 10^{-5}-cm-wide continuous cracks in the disturbed zone, there could be larger but disconnected cracks ? This may give similar flow velocities, but would be easier to seal if necessary.

E.W. PETERSON, United States

The cracks could be larger but their paths must be more tortuous to be consistent with the measured data. Given a 10^{-4}-cm-wide crack the associated path length must be 10 times that of the plug. The flow velocity in this larger crack also increases by a factor of 10. This larger crack with a faster fluid flow may be conceptually easier to plug, however, because of the increased tortuosity the plugging efficiency may not increase by a factor of 10. The crack size cannot be arbitrarily increased. At about the 10^{-4} cm value, one crack will account for the total measured volumetric flow. Clearly, easy to plug crack geometries which are semi-disconnected can be hypothesized. The efficiency of plugging such semi-disconnected cracks is an open question.

T.O. HUNTER, United States

Would you state the resolution of the guarded packer permeability apparatus ?

E.W. PETERSON, United States

The guarded straddle-packer system used to measure the in situ salt bed permeabilities in the AEC-7 borehole, and which was described in report SSS-R-79-4084, had a resolution of ~.1 μdarcy.

BELL CANYON TEST (BCT) - CEMENT DEVELOPMENT REPORT

C. W. Gulick
Sandia National Laboratories
Albuquerque, New Mexico 87185 (United States)

J. A. Boa, Jr., and A. D. Buck
Structures Laboratory
U.S. Army Engineer Waterways Experiment Station
Vicksburg, Mississippi 39180 (United States)

ABSTRACT

The Borehole Plugging (BHP) materials development
program which has been underway at WES under
Sandia sponsorship for about five years is
reviewed. Development testing data for candidate
grout mixtures for the BCT plug are presented.
Field batching, mixing, and placement operations
are discussed. Data from field samples molded
during the two plug placements include strength,
expansion, compressional wave velocity, dynamic
modulus, density, and porosity. Microstructure
and composition are compared for grout samples at
ages of a few weeks and one year.

RESUME

1.   BACKGROUND AND INTRODUCTION

        Cementing materials development studies for the Borehole
Plugging Program (BHP) have been underway at the U.S. Army Engineer
Waterways Experiment Station (WES) Structure Laboratory (formerly
Concrete Laboratory) since 1975. The work is sponsored by the
Waste Management Technology Department of Sandia National Labora-
tories which has technical responsibility for the studies support-
ing development of the Waste Isolation Pilot Plant (WIPP) in
southeastern New Mexico.

        The purposes of the BHP are [1, 2]

        1. Development of techniques for plugging boreholes
through and adjacent to underground waste disposal facilities

        2. Sealing boreholes with plugs which will prevent
movement of fluids and gases toward or through the salt beds

        3. Providing plugs which will maintain their integrity
for time periods comparable to the life of the rock formations in
which they are used.

        Several review meetings have concluded that the only
material and technique currently available for plugging boreholes
is high quality cementing [1, 3]. These conclusions were the
impetus and are the basis for the long-range development and testing program at WES. The two continuing tasks of the Materials
Development Program have been to

        1. Combine an extensive laboratory program with field
supervision and quality control of grouting operations

        2. Continue study of representative field samples pre-
pared at the surface and, when possible, cores of downhole plugs.

        Table I lists some of the desirable grout properties
that candidate plugging grouts should possess. The effect of
water/cement ratio on the quality of cement products has long been
recognized. Pumpable grouts generally require far more water than
is needed for complete hydration of the cement. The reduction in
water/cement ratio has been studied and achieved by various combinations of the following items: coarseness of the grind of the
cement, water reducers including superplasticizers and turbulence-
inducing compound, retarder, and temperature of the grout slurry.
The lowest possible water/cement ratio has a maximum effect on
improving the first four properties: low permeability, low porosity, high density, and high strength (Table I).

TABLE I

Desirable Grout Properties

1. Low permeability

2. Low porosity

3. High density

4. High unconfined compressive strength

5. Expansive potential

6. Isotropy and homogeneity

7. Pumpability and adequate working time

8. Stability and durability

Expansive cement systems have been included to provide a positive expanding force against the rock surface of the borehole after hardening when shrinking during cooling takes place. This expansive force should reduce the microfractures in the destressed annulus of rock around the plug. It should help achieve a "tighter" interface between the plug and the rock to improve bonding resistance to plug movement and the impermeability of the grout plug, particularly at the interface. Two expansive systems have been used during the program - Type K expansive cement and the DOWELL* Self Stress Cement.

Both fly ash and natural pozzolans have been included in the studies to determine their effects on permeability and durability. Fly ash has been included in all mixtures (30% by volume replacement for cement) to reduce early age temperature rise and to contribute to later age strength gain and durability by combining with calcium hydroxide to form the more stable and durable calcium silicate hydrates.

Salt (sodium chloride) has been included in some of study grout mixtures because of the evaporite rock sections at the WIPP site. The salt brine mixing water is necessary to prevent dissolution of the evaporite rock at the interface while free water is present in the grout.

Cements low in tricalcium aluminate (C_3A) which are generally used in the Southwest have been the basis for the grout mixtures. Low C_3A content is a major factor in the resistance of cement mixtures to sulfate attack. Fly ash in the grout mixture is also a factor in improving resistance to sulfate attack. The lowest possible water/cement ratio and its effect on the gel-pore structure of the grout also has a significant effect on the stability and durability of the grout plug.

Results of the studies are available in References 4 and 5. A third report (number 3) in the series is in preparation and will include data through three years of exposure. More comprehensive and detailed studies of cement hydration phases and interactions with wall rocks of boreholes are being addressed in the Geochemical Program by Steven Lambert at Sandia Laboratories, by Katherine Mather at WES and Dr. Della M. Roy at Penn State University.

2.   BCT GROUT MIXTURE DEVELOPMENT

The first review and planning meeting for the Bell Canyon Test [6] was held in January 1979. Based on previous laboratory development work at WES, Class H cement and a proprietary additive of Dowell Division of the Dow Chemical Company were proposed for the grout mixture. During the next month twenty-four grout mixtures were formulated and tested for time of efflux, workability, density, and strength (under accelerated curing). From these studies a grout mixture designated as DHT-1(9) was selected as having an optimum combination of properties. The mixture proportions in percent by weight are shown in the first column of Table II. Class H cement was chosen because of its coarseness (and low water demand) and appropriate amount of C_3A for expansivity. The mixture also included 10.5% expansive additive, 30% fly ash (by volume), salt (30% by weight of mixing water), and a superplasticizer. Salt brine mixing was used in the development mixtures because of unknown conditions of the surface of the wall rock at the plug location. Previous testing had indicated the desirability of including salt in the mixing water for plugs at locations where salt was the host rock and brine drilling mud had been used.

─ ──────
*Trademark of the Dow Chemical Company

TABLE II

Salt Grout Mixtures

Proportions (Wt %)	DHT-1(9)	BCT-1F
Class H Cement	50.6	50.1
Expansive Additive	5.3	6.7
Fly Ash	17.8	16.9
Salt (NaCl)	5.8	6.5
Superplasticizer	1.3	-
Dispersant		0.2
Defoamer	0.02	0.02
Water	19.2	19.5
Properties		
Water/Cementitious Ratio	0.26	0.26
Fluid Density, g/cm^3	2.06	2.04
Fluid Density, lb/gal	17.2	17.0

Using the mixture proportions developed by WES, Dowell adjusted the mixture by substituting a powder dispersant instead of the liquid superplasticizer. Mixture proportions for this mixture designated BCT-1F are shown in the second column of Table II. The expansive additive was also raised to 13% (by weight of cement). The water/cementitious ratio (including cement expansive additive and fly ash) was similar for both mixtures, and the fluid densities were considered acceptable. WES continued testing physical properties of the DHT-1(9) mixture through the month of April. Unconfined compressive strength of the 5 cm (2 in.) cubes cured at 53°C (128°F) was 57 MPa (8375 psi) at 14 days' age. Expansion prisms conforming to ASTMC 806-75 were cast and cured in different environments. Linear expansion of the bars was generally in the range of 0.06 to 0.10% for all bars inundated in fresh or brine water at 56 days' age.

During late April the plug location in the hole was selected to be in the anhydrite sequence. Boxes of anhydrite core were shipped from Carlsbad to WES and also to Penn State University (PSU), Oak Ridge National Laboratory (ORNL), and Dowell. At this time the BCT-1F mixture was selected as the plugging grout mixture. Quantities of the materials in the mixtures had been previously provided to the PSU and ORNL Laboratories so that all four laboratories would be using the same materials and mixture.

The temperature at the plug location was expected to be 53°C (138°F) which had been the reported temperature in another drill hole near the WIPP site at a comparable depth. The compressive strengths at 14 days' age were above 26 MPa (3800 psi) and as high as 34 MPa (5000 psi). Push-out bond strengths for grout-filled holes in the anhydrite were above 2.8 MPa (400 psi) and as high as 6.9 MPa (1000 psi), depending upon laboratory curing conditions and size and age of specimens. All of the expansion data, push-out shear bond strengths, and compressive strength data showed the BCT-1F grout mixture to be adequate and were in general agreement with previous development data.

During June and July 1979 the results of permeability tests on anhydrite cores with grout-filled holes showed a wide variability in results: These range from 10^{-3} to 10^{-6} darcy. Samples of BCT-1F grout consistently had water permeabilities less than 1×10^{-6} darcy. Anhydrite cores with grout-filled holes showed leakage at the interface and evidence of white powder (halite crystals) at the interface on some samples, particularly for samples cured at lab ambient temperatures of about $22^{\circ}C$ ($72^{\circ}F$). The push-out bond strength for these samples was still in excess of 2.5 MPa (360 psi). This showed that push-out bond strength was not an adequate measure of the quality of the grout/rock interface. Recent temperature measurements in the AEC-7 hole at the plug location depth indicated an ambient temperature of about $32^{\circ}C$ ($90^{\circ}F$), unexpectedly low for the depth of about 1370 m (4500 ft).

The continuing laboratory grout development program had included comparable freshwater grouts while the major development effort for BCT was the evaluation of the BCT-1F salt grout. A grout mixture with fresh mixing water was formulated as an alternate, subject to further testing. The freshwater grout mixture was designated as BCT-1FF. Relative proportions of its dry materials (except salt) were about the same as the BCT-1F grout mixture. Mixture proportions and properties are shown in Table III. The water/cementitious ratio was increased to provide an acceptable viscosity and pumpability of the grout mixture which reduced fluid density to 1.98 g/cm^3 (16.5 lb/gal).

TABLE III

Salt and Freshwater Grout Mixtures

Proportions (Wt %)	BCT-1F	BCT-1FF
Class H Cement	50.1	52.2
Expansive Additive	6.7	7.0
Fly Ash	16.9	17.6
Salt (NaCl)	6.5	–
Dispersant	0.2	0.2
Defoamer	0.02	0.02
Water	19.5	23.0
Properties		
Water/Cementitious Ratio	0.26	0.30
Fluid Density, g/cm^3	2.04	1.98
Fluid Density, lb/gal	17.0	16.5

Studies at the three laboratories (WES, PSU, and ORNL) [7] showed that the freshwater grout had a higher strength and more expansions than the salt grout. Push-out bond strengths in anhydrite cores were equal to or greater than for the salt grout. When split along the length of the specimen, the freshwater grout samples showed a tighter adherence of grout to rock, a reduction in permeability at the interface, and no discernible leakage paths. On the basis of these studies the BCT-1FF mixtures were chosen for the BCT plug.

Samples were prepared late in August for permeability testing of the grout and interface in anhydrite rock cores. The type of specimen previously used for push-out bond strength tests was used for water permeability tests - 5.7 cm (2.25 in.) diameter grout-filled hole in 10.2 cm (4 in.) diameter anhydrite cores. The core was restrained in a 15.2 cm (6 in.) diameter steel ring with the same grout filling the annulus between anhydrite rock and steel ring. Three curing conditions were used through about 10 days' age and are described at the bottom of Table IV. In all cases the samples were surrounded by brine water while the grout was still fluid for the 10 days of curing.

TABLE IV

Water Permeability Tests
BCT-1F Grout Lab Samples

Age, Days	Water Permeability, 10^{-6} darcy		
	Curing Condition		
	1	2	3
25	2.26	13.1	0.88
50	1 37	3.3	0.73
82	1.49	2.09	0.56
110	2.25	4.12	0.73
210	2.84	4.76	0.58

Curing Conditions

1. Curing temperature curve to 66°C (150°F) at 12 hours, reduced to 38°C (100°F) at 24 hours and remains constant at 38°C for 10 days. Pressure of 10 MPa (1500 psi) applied to fluid grout in the anhydrite core and confining ring through hardening and to 10 days' age.

2. Constant temperature of 38°C (100°F) and pressure of 10 MPa (1500 psi) applied to fluid through hardening and continuing to 8 days' age.

3. Constant temperature of 100°F and ambient pressure to 10 days' age.

The samples were then placed in the permeability apparatus and subjected to freshwater at 1.4 MPa (200 psi) applied to the grout-filled central hole and to about a 1.3 cm (0.5 in.) annulus of the anhydrite core. WES equipment for permeability is described in Reference 8.

Permeability values are listed in Table IV. The sample under the third curing condition of constant 38°C (100°F) temperature and no additional pressure has maintained the lowest permeability values - less than 1.0×10^{-6} darcy through 210 days. The other two curing conditions were intended to bracket the downhole curing temperature and curing conditions for the

field plug. The small sample size and confining ring may not provide as much restraint (when removed from the curing chamber and
prepared for permeability testing) as the in situ field plug.
Values of less than 5×10^{-6} darcy after 48 days and through 210
days' age for all three samples show evidence of a tight contact
between the grout and the anhydrite. Water permeability values of
grout samples have typically been less than 0.1×10^{-6} darcy.
These samples will continue to be tested at later ages.

3.   FIELD OPERATIONS DURING PLUGGING

        Field mixing of cement grout for the two BCT plugs took
place on September 26, 1979, and February 14, 1980. Although the
actual volume for each short plug length was less than 0.3 m^3
(10 ft^3), batching and mixing operations were standardized for a
volume of 200 sacks (cement plus fly ash), yielding about 5.4 m^3
(190 ft^3) of grout slurry. These mixing operations were used as
trials in the development of quality control procedures for future
and larger borehole plugs.

        Batching and blending of the dry materials took place at
the DOWELL plant in Artesia, New Mexico, during the day previous
to mixing at the BCT site. Operation of the batch plant was
inspected by Sandia and WES personnel, and the calibration dates
of scales and equipment were noted. A sample of each of the
materials used in the grout mixture was obtained, and the sources
and dates of shipment were recorded. Dry materials were weighed,
blended, and loaded into a tier of a bulk truck for transport to
the site. Each weighing operation was closely monitored by WES
personnel.

        On-site mixing was done in one tank of a twin-tank,
paddle mixing system with a 6.3 m^3 (225 ft^3) capacity. Carlsbad city water was pumped into the tank and used as the mixing
water. The dry material was added, and the final amount of water
was added during mixing for proper viscosity of the mixture. For
both mixing operations the final fluid density of the grout slurry
was 2.11 g/cm^3 (17.6 lb/gal). Measurements of density grout
temperature, air content, and time of flow through a flow cone
were made periodically for 3-1/2 hours. Grout slurry properties
suitable for pumping (as measured by the flow cone) were maintained for 3-1/2 hours.

        Weight percentages of the grout mixtures for the two
field operations and the lab development mixture are listed in
Table V. Planning for field mixing has always involved using the
minimum amount of water to achieve maximum density and suitable
flow characteristics of the grout slurry for adequate pumpability.
Resulting water/cementitious ratios were 0.28 and 0.27 for the
field mixtures compared to 0.30 for the lab mixture. Batching of
about 9000 kg (20,000 lb) of dry materials resulted in normal
slight differences in the amount of each component as noted in the
table. This slight variation is within acceptable quality control
limits. Uniform fluid densities for the two field mixing operations attest to the successful control of batching of dry materials.

        A dump bailer was used to place 0.085 m^3 (3 ft^3) of grout
slurry at the plug location for the September 1979 placement [9].
Preparation of the hole was limited to circulation and replacing
all of the brine fluid in the hole with clean brine water before
placing the packer assembly and sand. The dump bailer was filled
with grout slurry and lowered to the sand surface. It was raised
about 0.3 m (1 ft), and the frangible glass nose window was broken
to allow the grout to flow out and displace the brine water upward.
The volume of grout should have formed about a 2 m (6.8 ft) plug.
Measurements have indicated that the hardened plug length is about
1.87 m (6 ft).

TABLE V

BCT-1FF Grout Mixtures

Proportions (Wt %)	Lab	Field 9/26/79	Field 2/14/80
Class H Cement	52.2	53.1	52.7
LITEPOZ* 3 Cement Extender (Fly Ash)	17.6	18.1	18.2
Expansive Additive	7.0	7.1	7.5
Dispersant (D65)	0.2	0.1	0.2
Defoamer (D47)	0.02	0.02	0.02
Water	23.0	21.6	21.4
Properties			
Water/Cementitious Ratio	0.30	0.28	0.27
Fluid Density, g/cm^3	1.98	2.11	2.11
Fluid Density, lb/gal	16.5	17.6	17.6

---

*Trademark of the Dow Chemical Company

A more normal cementing operation was selected for the plug addition in February 1980. Hole preparation consisted of circulating the hole through tubing with clean brine water to replace all the fluid in the hole while the grout was being mixed. The bottom of the tubing was located about 0.3 m (1 ft) above the top of the first plug. After circulating the hole, 3.2 m^3 (112 ft^3) each of a chemical wash and spacer designed by Dowell were pumped down the tubing immediately followed by about 1.9 m^3 (67 ft^3) of grout slurry at a slow rate of about 0.3 m^3/min (11 ft^3/min). The grout slurry was displaced with 2.7 m^3 (96 ft^3) of brine water to equalize the level of grout in the tubing and annulus. After raising the tubing 6 m (20 ft), the excess grout above that depth was circulated out by pumping brine water down the tubing and forcing the excess slurry up the annulus. Subsequent field measurements have indicated that the additional hardened plug length is about 3.6 m (12 ft) instead of the intended 5.5 m (18 ft).

WES personnel cast a large number of expansion prisms, cylinder, and cube specimens for strength and other physical properties. Specimens were cured in a tank of brine water obtained during circulating the hole. Lack of a temperature control for the heater during September 1979 allowed fluctuations of as much as 10°C (18°F) below the intended temperature of 38°C (100°F). During the first 24 hours the temperature in the tank reached 71°C (160°F) because of the exotherm of the tightly packed samples. Water temperature in the tank in February 1980 was more closely controlled to within + 2°C (4°F). A larger tank, removal of insulation, and cooler starting temperature of the water reduced the peak temperature during the first 24 hours to about 49°C (120°F). Expansion prisms were measured daily. Some compressive strength testing at early ages was done at the site on a portable testing machine in February. Samples were shipped by truck to the Vicksburg lab at 7 days' age for the first plug and 21 days' age for the second.

4.   LAB EVALUATION OF FIELD SAMPLES

WES personnel performed some testing of the samples at the WIPP site before shipment to Vicksburg. Expansion bars were measured daily through 14 days, weekly through 28 days, and then periodically for the long term durability studies. A portable compression machine was available for the second pour so that early age strength gain could be determined. After arrival at the Vicksburg lab, tests for compressional wave velocity, dynamic modulus of elasticity, compressive strength, density, and porosity have been performed.

Unconfined compressive strength of the grout is comparable for the two field mixing and plugging operations as shown in Table VI. Strengths at 14 days' age are almost identical.

TABLE VI

Physical Properties, BCT-1FF

Unconfined Compressive Strength, MPa (psi)

Age/Days	9/26/79 Plug	2/14/80 Plug
1	–	33 (4,840)
3	–	58 (8,410)
7	–	73 (10,610)
14	75 (10,870)	74 (10,670)
21	–	78 (11,350)
28	–	79 (11,445)
56	93 (13,440)	
90	88 (12,750)	

Compressional Wave Velocity, m/s (ft/sec)

Age/Days	9/26/79 Plug	2/14/80 Plug
19	3,945 (12,945)	–
28	–	4,197 (13,770
62	4,005 (13,140)	
99	4,149 (13,612)	

Dynamic Modulus, E 10^3 MPa (E x 10^6 psi)

Age/Days	9/26/79 Plug	2/14/80 Plug
19	16.3 (2.36)	–
28	–	26.3 (3.81)
62	16.5 (2.40)	
99	16.8 (2.44)	

Density and Porosity at 105 Days' Age for 9/26/79 Plug

	Dry density (g/cm^3)	Grain Density (g/cm^3)	Porosity %
3x13 Cylinder, Top	1.878	2.715	30.83
3x13 Cylinder, Bottom	1.724	2.668	35.38
Grout in Permeability Specimen:			
T1	1.908	2.709	29.57
T2	1.911	2.860	33.18
B2	1.910	2.803	31.86
B1	1.900	2.773	31.48

Compressional wave velocity data show a slightly higher value for the second plug at 28 days' age. The second plug dynamic modulus determined by the resonant frequency method is about 50% higher than any previous field or lab data at a comparable age of 28 days. Testing is planned to determine the static modulus of elasticity of unconfined specimens during compression testing for comparison. Density and porosity data are consistent with the previous laboratory studies of borehole plugging development mixtures.

TABLE VII

Expansion Data
BCT-1FF Field Pours

Bar Size and Restraint, Age	Linear Expansion, %		Volumetric Expansion, %	
	9/26/79	2/14/80	9/26/79	2/14/80
2.5 cm (1 in.) Unrestrained				
3 days	+.010	+.271	− .04	+ .10
7 days	+.026	+.372	− .25	+ .22
14 days	+.059	+.416	+ .82	+ .54
28 days	+.063	+.455	+2.01	+ .29
56 days	+.086		+2.37	
90 days	+.098		+2.74	
5.1 cm (2 in.) Restrained				
3 days	+.013	+.049	+ .10	+ .48
7 days	+.006	+.086	+ .42	+ .51
14 days	+.016	+.090	+ .40	+ .71
28 days	+.010	+.091	+1.20	+ .38
56 days	+.015		+1.25	
90 days	+.024		+1.62	
5.1 cm (2 in.) Unrestrained				
3 days	+.001	+.325	+ .18	+ .60
7 days	+.005	+.435	+ .68	+ .68
14 days	+.022	+.446	+ .72	+ .97
28 days	+.014	+.457	+1.56	+ .61
56 days	+.018		+1.40	
90 days	+.018		+1.71	
7.6 cm (3 in.) Restrained				
3 days	−.001	+.209	+ .08	+ .66
7 days	−.011	+.267	+ .15	+ .75
14 days	+.014	+.276	+ .16	+ .96
28 days	+.001	+.268	+ .70	+ .52
56 days	+.010		+1.32	
90 days	+.010		+ .78	
7.6 cm (3 in.) Unrestrained				
3 days	−.004	+.283	− .06	+ .70
7 days	+.004	+.386	− .02	+ .86
14 days	+.017	+.388	+ .09	+1.03
28 days	−.006	+.383	+ .56	+ .66
56 days	−.001		+1.04	
90 days	+.001		+ .86	

Expansion data in Table VII are average values for two or more prisms. Initial datum for each bar was established at about 21 to 25 hours' age when the prism was demolded. A length comparator conforming to ASTM C-490 was used to measure length changes of the restraining rod or small metal bolts embedded in the ends of the unrestrained prisms. Volume changes were measured by comparing the weight of bars immersed in water to the weight in air. The 5.1 cm (2 in.) restrained prisms conform to ASTM C 806-75 and have the most restraint by a 0.64 cm (0.25 in.) diameter threaded rod. The relative restraint factor is about 4 to 1 for the smaller prism compared to the larger prism.

The 7.6 cm (3 in.) prisms from the first plug show the least expansion; however, the volumetric expansion is comparable to the second plug samples at 28 days' age (Table VII). All prisms show positive expansions after 7 days' age. One factor that might account for generally lower expansions of the first plug samples was a peak temperature in the curing tank during the first 24 hours before demolding of at least 71°C (160°F). More closely controlled curing temperatures for the second plug samples should have resulted in more representative expansion data for the grout mixture. Data show good positive expansion through at least 90 days' age for the first plug. These values for inundated prisms are upper limits of expansivity and will be monitored particularly for long term durability under exposure to brine water. Prisms have been cured by two other methods to try to prevent moisture addition and moisture loss. Protection by two layers of a "paint-on" membrane and double layers of plastic bags have not prevented moisture loss for samples placed just above the water level or some water addition for submerged protected samples. These data will be included and evaluated in a future report.

5.   LONG TERM DURABILITY STUDIES

Five grout mixtures were selected for the beginning of long term durability studies. Samples were molded between August 1976 and February 1977, and have been inundated in a tank of brine water that simulated the underground water at the WIPP site. During plugging operations of an exploratory drill hole near the WIPP site, ERDA No. 10, samples of the three grout mixtures were molded and shipped to WES for testing and inclusion in the long term studies. These samples have been cured in tanks of brine water at 53°C (129°F) except for the freshwater grout samples from plug 1 which have been cured at lab ambient temperature. Similarly, samples from the two plugs of the BCT have been included in the long term durability studies.

The physical properties of the samples as measured by compressional wave velocity, dynamic modulus, density, porosity, strength, and expansion continue to show good durability and no evidence of deterioration. The earlier data from these studies are included in References 3 and 4. A third report in the series is in preparation.

The Petrography and X-Ray Branch of WES has been performing studies in support of grout development and durability studies and as a prelude to the more comprehensive Geochemical program. Figures 1 and 2 show partial X-ray diffraction patterns of freshwater and saltwater grouts from the plugging of ERDA 10. In Figure 1 the composition of salt grout from plug 1 at about 16 days and 1 year is compared. For this nonexpansive portland cement grout there is a little ettringite at 16 days, but by 1 year it is gone, and the amount of tetra-calcium aluminate dichloride-10-hydrate (chloroaluminate) has increased. Chloroaluminate is formed at the expense of the ettringite. The amount of calcium hydroxide (CH) has decreased with age, but it is still present. AF stands for residual calcium aluminoferrite from the cement. HY is a small amount of what may be hydrogarnet (5.4A).

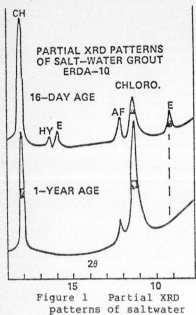

Figure 1   Partial XRD
patterns of saltwater
grout samples from
ERDA 10 plug 1.

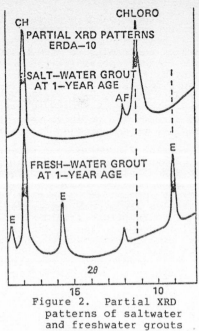

Figure 2.   Partial XRD
patterns of saltwater
and freshwater grouts
from ERDA 10 plugs
1 and 4.

In Figure 2 the freshwater grout from plug 4 and the
saltwater grout from plug 1 are compared at 1 year. The 1-year-
old saltwater grout is the same in both figures. The significant
difference is the lack of ettringite in the saltwater grout and
the lack of chloroaluminate in the freshwater grout.

An additional study was made to investigate the composition and integrity of a plug taken from the Duval potash mine at a
depth of 305 m (1000 ft) [10]. While no XRD pattern of the
17-year-old sample of a saltwater grout is shown, its composition
would be similar to the 1-year-old ERDA 10 saltwater grout. The
Duval grout also contained some halite (NaCl) and sylvite (KCl).

The next five figures (3 through 7) are SEM micrographs
of ERDA 10 grouts at different ages and, for comparison, one
normal consistency portland cement paste at an early age. The
magnification of the five figures (except Figure 4) is about 2000X.
Figure 4 is at about 4800X. These were selected to show

1. typical porous microstructure of this saltwater
    grout at an early age (Figures 3 and 4)

2. tabular crystals believed to be chloroaluminate in
    open space in Figure 4. This is the hydration
    product of portland cement when salt (NaCl) is
    present.

3. typical dense microstructure of what are probably
    crystals of chloroaluminate.

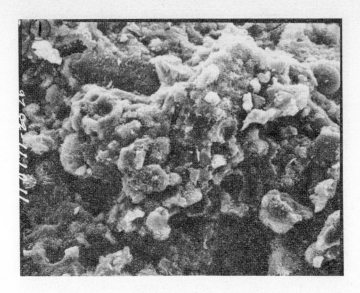

Figure 3.   ERDA 10 Salt Grout Core, Plug 1,
28 Days' Age (X1700)

        The grout in Figure 3 is a 28-day-old core sample from
the salt grout core from plug 1 which had 30% salt by weight of
mixing water (X1700).   The open structure is evident.   CSH coating
on residual cement grains is shown and probably CH is present at
right center.

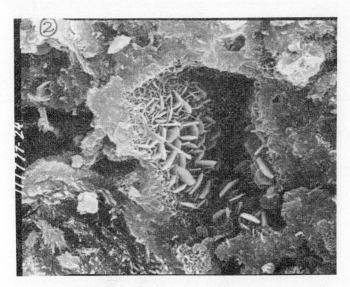

Figure 4.   ERDA 10 Salt Grout Sample, Plug 3,
28 Days' Age (X4750)

        The grout in Figure 4 is a 28-day-old sample molded from
the saltwater grout of plug 3 which was fully saturated--36% by
weight of mixing water (X4750).   The open structure is again
evident.

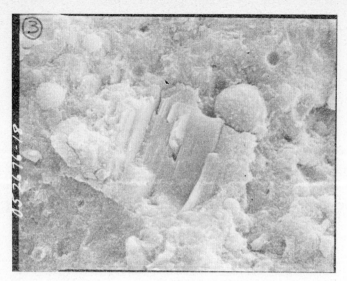

Figure 5.   Normal Consistency Portland
Cement Paste, 28 Days' Age (X2000)

The grout in Figure 5 is a 28-day-old sample of
hydrated, normal consistency portland cement paste with 30% fly
ash (X2000). This is the same volume replacement as the ERDA 10
grout mixtures. The dense structure is evident. Massive CH is
present in the center. Fly ash spheres and CSH can also be seen.

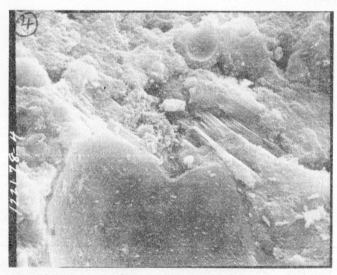

Figure 6.   ERDA 10 Salt Grout Core, Plug 1,
Age - One Year (X2100)

The grout core sample in Figure 6 is the same as in
Figure 3 but at 1 year (X2100). After curing in brine water at
53°C (128°F) for 1 year the structure is now more dense. CSH
and CH can be seen, and a relatively large cavity is at bottom
center. No chloroaluminate can be seen, but XRD shows it to be
present.

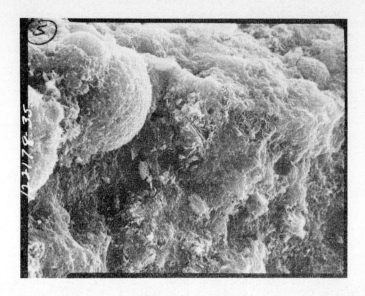

Figure 7.   ERDA 10 Freshwater Grout, Plug 4,
Age - One Year (X2300)

The sample in Figure 7 is freshwater grout from plug 1
in ERDA 10, cured in freshwater at lab ambient temperature for 1
year (X2300). The structure is also dense when compared to
Figures 5 and 6. CSH residual fly ash spheres and massive calcium
hydroxide can be seen.

Figure 8.   Simulated Borehole Specimen (SBS)
for BCT

During the BCT development, samples with BCT-1FF grout were cast in anhydrite cores to simulate and study the grout/rock interface and are called simulated borehole specimens (SBS). Figure 8 shows the appearance of such a sample. It is necessary to remove the top surface to see the actual contact. The next four figures (9 through 12) illustrate some of the problems in trying to study this contact surface. As a result of this work it is believed that such specimens are satisfactory simulations and that the way to study the contact between grout and rock is to

1. Keep the metal restraining ring in place to prevent stress relief cracking (Figure 9)

2. Examine a fracture surface (Figure 10)

3. Examine the contact with a steromicrosope (Figure 11)

4. Keep the specimens wet to avoid drying cracking (Figure 12)

5. Do not examine with SEM unless precautions can be taken to prevent the cracking that occurs with drying.

This type of specimen (SBS) is used in the laboratory to determine the permeability along the contact of grout to rock. Examination of specimens before and after this testing can serve to characterize the contact before testing and determine if it changed during testing.

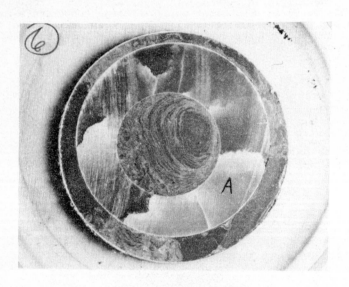

Figure 9. Sawed Surface of SBS Specimen
with Restraining Ring Removed (X0.7)

A sawed surface is shown in Figure 9. When the metal restraining ring was removed, the anhydrite rock cracked (radially) in several places within a few hours because of stress relief. The grout is not cracked. Its "marbled" appearance is caused by pouring the grout through brine water standing in the center hole and annulus. The white lines in the grout are believed to be regions of higher water content or salt concentrate.

A fracture surface of the same specimen with the metal restraining ring still in place is shown in Figure 10. The anhydrite rock did not crack.

A portion of the contact of grout and anhydrite and a sawed surface of another SBS is shown in Figure 11 (X6). The metal restraining ring is still in place, and no cracking is seen. There is evidence from XRD that the white material at the contact contains some gypsum (hydrated calcium sulfate). If so, it indicates some dissolving of the anhydrite rock and later precipitation of gypsum in the contact surface.

The same sample as in Figure 11 is shown in Figure 12 after overnight vacuum drying at 60°C (16°F). A prominent crack at the contact can be seen. Several radial cracks are present in the grout but may not be evident in the figure.

XRD work has shown that penetration of chloride into the freshwater grout during storage of the SBS specimens under brine is limited to the upper 1/4 in. of the specimen in the samples which have been examined.

Figure 10. Fractured Surface of SBS Specimen with Restraining Ring in Place (X0.7)

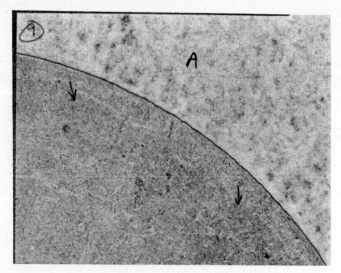

Figure 11.   Portion of Figure 10
             Before Drying (X0.6)

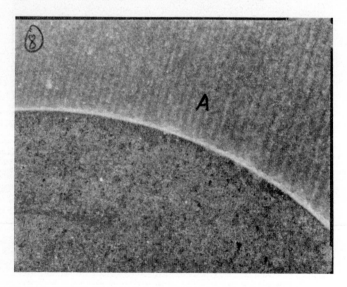

Figure 12.   Same as Figure 10 after Over-
             night Vacuum Drying (X0.6)

6. CONCLUSIONS

The continuing laboratory grout development program provided mixture data to rapidly respond to the BCT plugging requirements. Both saltwater and freshwater grouts with acceptable physical properties in the fluid and hardened states were developed. The freshwater BCT-1FF grout was chosen because of permeability tests showed a higher leakage for the saltwater grout at the interface, and because it had higher strength and expansivity than the saltwater grout. Field batching, mixing, and placement of the grout at the plug location were satisfactory. The freshwater grout mixture maintained adequate flow characteristics for pumpability for 3-1/2 hours during the two field operations. Physical property and expansivity data for the field samples are in general agreement with the lab development data [11 12]. The high density, low water/cementitious ratio expansive grout (BCT-1FF) is considered an excellent candidate for plugging boreholes at most locations (except through halite sections).

A large number of samples were obtained for inclusion in the long term durability studies and the geochemical programs.

References

1   Gulick, C. W., Jr., "Borehole Plugging - Materials Development Program," SAND78-0715 (Albuquerque, NM: Sandia National Laboratories, June 1978).

2   Christensen, C. L., and Hunter, T. O., "Waste Isolation Pilot Plant (WIPP), Borehole Plugging Program Description, January 1, 1979," SAND79-0640 (Albuquerque, NM: Sandia National Laboratories, April 1979; controlled distribution).

3   D'Appolonia Consulting Engineers, Inc., "Development of Plan and Approach for Borehole Plugging Field Testing," Report No. ONWI-3 (October 1978).

4   Boa, J. A. Jr., Borehole Plugging Program (Waste Disposal), Report 1 "Initial Investigations and Preliminary Data," U.S. Army Engineer Waterways Experiment Station, Miscellaneous Paper 6-78-1, January 1978 (Sandia Report SAND77-7005).

5   Gulick, C. W., Jr., et al, "Borehole Plugging - Materials Development Program," Report 2, SAND79-1514 (Albuquerque, NM: Sandia National Laboratories).

6   Christensen, C. L., "Bell Canyon Test, WIPP Experimental Program, Borehole Plugging," SAND79-0739 (Albuquerque, NM: Sandia National Laboratories, June 1979).

7   Moore, J. G., et al, "Cement Technology for Plugging Boreholes in Radioactive Waste Repository Sites: Progress Report for the Period October 1, 1978, to September 30, 1979," ORNL-5610 (Oak Ridge, TN: Oak Ridge National Laboratory).

8   "Method of Test for Water Permeability of Concrete," CRD-C 48-78, Handbook for Concrete and Cement, U.S. Army Engineer Waterways Experiment Station.

9   Statler, R. D., "Bell Canyon Test - Field Preparation and
    Operations," SAND80-0458C, prepared for the International
    Borehole Plugging Symposium, Columbus, OH, May 1980 (Albu-
    querque, NM: Sandia National Laboratories).

10 Buck, A. D., and Burkes, J. P., "Examination of Grout and
    Rock from Duval Mine, New Mexico," Miscellaneous Paper
    SL-79-16, U.S. Army Engineer Waterways Experiment Station
    (July 1979).

11 Christensen, C. L., "Sandia Borehole Plugging Program for
    the Waste Isolation Pilot Plant (WIPP)," SAND80-0390C, pre-
    pared for the International Borehole Plugging Symposium,
    Columbus, OH, May 1980 (Albuquerque, NM: Sandia National
    Laboratories).

12 Peterson, E. W., "Analysis of Bell Canyon Test Results,"
    SAND80-7044C, prepared for the International Borehole Plug-
    ging Symposium, Columbus, OH, May 1980 (Albuquerque, NM:
    Sandia National Laboratories).

Discussion

W. FISCHLE, Federal Republic of Germany

How long are plugging materials tested in brine ?

What sort of brine is this ?

C.W. GULICK, United States

Material samples are stored from the time of casting into the foreseeable future. Some samples have been stored and tested for more than 3½ years. Samples are continually being added.

For some samples brine water is simulated ground water, that is water having the same composition as the major aquifers above and below the WIPP site.

Field samples from the BGT are stored in brine formed by circulating water through the ground near the WIPP site to dissolve and absorb the chemicals present (and used for filling drill holes to maintain stability).

M. GYENGE, Canada

In the formulation of the grout mix, is the mechanical strength of the grout an essential requirement ?

C.W. GULICK, United States

By achieving low permeability the grout will have a low water/cement ratio which will automatically give adequate strength, greater than 3500 psi.

BELL CANYON TEST (BCT) INSTRUMENTATION DEVELOPMENT

C. W. Cook and C. B. Kinabrew
Sandia National Laboratories
Albuquerque, New Mexico 87185 (United States)

P. L. Lagus and R. D. Broce
Systems, Science and Software
La Jolla, California 92039

ABSTRACT

This report discusses the instrumentation used to assess the performance of the BCT plug and addresses the future direction of
instrumentation development. The BCT was initiated with a timed-
release packer system below the plug; six time-released tracer gas
bottles were included to provide a method of measuring permeability through the plugged region. A geophone package installed
above the plug verified the initial operation of the below-the-
plug system. In addition to tracer gas arrival times, fluid
buildup and shut-in pressure measurements were made with current
oil field systems--DMR (Digital Memory Recorder) and CWL (Conductor Wire Line)--to evaluate plug performance. A probe for measuring discrete fluid levels, fluid conductivity pressure, and temperature above a borehole plug has been developed. The conductivity measurement provides a rough measure of dissolved salts that
can be useful in identifying the source of the water. Development
has started on two probes that can be grouted in place to monitor
the leakage of brine plast a plug. A wireline closed circuit TV
capable of operating underwater in 5000-foot-deep holes is being
set up to ascertain the wellbore conditions.

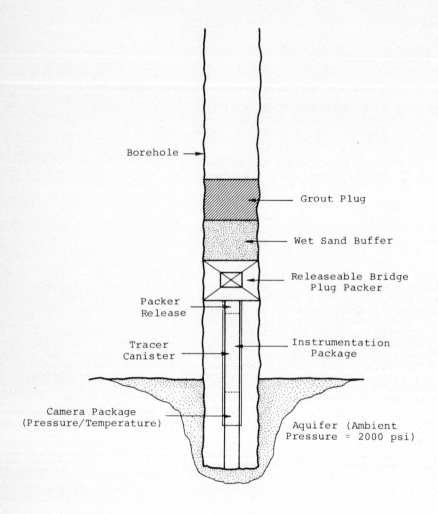

Borehole ——→

Grout Plug

Wet Sand Buffer

Releaseable Bridge
Plug Packer

Packer
Release

Tracer
Canister

Instrumentation
Package

Camera Package
(Pressure/Temperature)

Aquifer (Ambient
Pressure ≃ 2000 psi)

Figure 1. Schematic Showing Instrumentation
Canister Package

1.   INTRODUCTION

    The objective of the Bell Canyon Test (BCT) was to verify
that a borehole penetrating a zone with a large hydrostatic head in
the Bell Canyon (Ramsey Sands) could be successfully plugged [1].
A minimum-length cement grout plug was installed to isolate the
production zone from the remainder of the borehole. To assure
that the plug was being loaded by the production zone, a packer
which released after the cement had cured was installed below the
plug. A schematic illustrating the installation is shown in
Figure 1.

    A remote instrumentation package and time-released
electronegative tracer gases were included in the below-the-plug
package. The instrumentation package was designed to continuously
monitor fluid pressure and temperature below the base of the grout
plug. The tracer gases provided a means for measuring the permeability through the plugged region.

    Instrumentation was required to monitor the migration of
fluids (tracer gas and brine) through the plugged region so that
the performance of the plug could be evaluated. The initial test
plan for the BCT called for removing all of the fluids above the
plug so that the plug would have to withstand the full head of the
production zone. However, the production of fluids in the wellbore made this approach impossible, and the instrumentation developed for the initial test plan was not used.

    This report discusses the instrumentation package
installed below the plug, the monitoring of the system below the
plug, the current oil field systems used to monitor fluid migration, the instrumentation developed for the initial test plan, and
probes under development for future applications.

2.   INSTRUMENTATION BELOW THE BCT PLUG

    2.1 General description

    The instrumentation canister was designed by Science,
Systems and Software (S^3) to provide measurements over a 1-year
period, while submerged in a saturated salt brine where the hydrostatic pressure was about 2000 psi. The overall system, as shown
schematically in Figure 2, is 27 feet long and is divided into
three modules: temperature/pressure measurement module, tracer
gas release module, and packer fluid release module. The canister
housing was made from sections of ARMCO 5.5 in. diameter by 15.5
pounds per linear foot Seal-Lok casing. Individual modules were
separated from each other by suitably ported blind couplings.
Special adapter couplings were attached to the top and bottom of
the instrumentation canister. A jack leg made from 5.5 in. casing
was connected to the bottom, while a length of 2-3/8 in. production tubing was used to connect the canister to the bridge plug
packer. The detailed designs of the three modules are described
in the following sections.

    2.2 Temperature pressure measurement

    Two redundant temperature and pressure monitoring
devices are included in this module. The basic device consists of
an Environmental Devices Corporation Type 109 Recording Bathythermograph, which was modified to record temperature and pressure
values measured by dial gages. The Recording Bathythermograph
consists of a special, shutterless, fixed-frame, 16mm movie
camera, a four-digit LED frame counter, and a timer module which
can select a picture-taking interval. This unit was modified at
S^3 to allow filming one frame every 8 hours. Since the camera

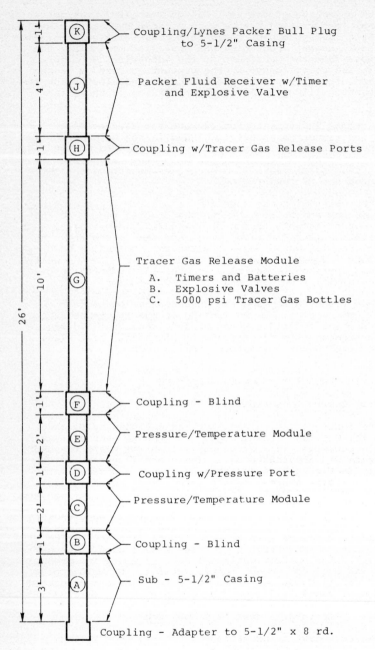

Figure 2. Instrument Package -
Below Bridge Plug

is designed to be maintained in a light-proof housing, there is no shutter. In operation, the film advances about 7cm at each time interval. During this advance, the LED frame counter advances by one count. Halfway through the time interval, four incandescent lamps are momentarily illuminated, and the LED interval counter is energized. When this happens, the dial gage and LED display readings are recorded on a single film frame. During the next interval, the film is again advanced about 7cm, and the exposure sequence repeated. The film used is a 16mm, magazine-loaded Tri-X emulsion on a mylar base, furnished by Environmental Devices Corporation. The mylar base allows slow film-feed speeds, and also permits a higher operating temperature (130°F). At a framing rate of one exposure every 8 hours, one standard film magazine was sufficient for 426 days of data recording.

Each camera unit was powered by four 3-V lithium D cells, which provide sufficient power for 7 years. The dial thermometer was a Marshall Model 20-240-F thermometer, with a reading accuracy of + 1^{O}F and a full scale of 240°F. Pressure was monitored using a Marshall Model 78C 3000 psi dial pressure gage, with a reading accuracy of $\pm$ 1% of full scale.

Two independent pressure and temperature monitoring systems provided the redundancy necessary to ensure that usable data would result during the 1-year measurement interval. The two systems were mounted on opposite ends of a bull plug. This plug was ported to allow direct pressure gage communication with the formation fluid. The temperature gages were permanently affixed to the bull plug by using thermally conductive epoxy. A drawing of the redundant pressure/temperature monitoring module is shown in Figure 3.

This instrumentation canister will be retrieved from the borehole upon completion of a test sequence. Film magazines will then be removed, and the film developed, to provide complete pressure, temperature, and timing data. A typical single exposure showing the LED frame counter, the dial temperature gage, and the dial pressure gage is presented in Figure 4.

2.3 Tracer gas release module

The tracer gas release module contains six compressed gas bottles. Gas is released from these bottles by means of timer-controlled Conax explosive valves. In practice, each bottle is loaded to a pressure of 5000 psi with an electronegative tracer gas (sulfur hexafluoride, SF_6) diluted in nitrogen. As initially designed, the tracer gas release module was to provide three sets of redundant (two-bottle) tracer gas releases. However, for reasons of experimental expediency coupled with the strong desire to extend the time period over which data on the permeability characteristics of the emplaced borehole plug could be taken, it was decided to use each of the six compressed gas bottles individually. Thus, six single releases of tracer gas were planned. Each release used a single Conax 5000 psi pressure cylinder, a Conax explosive valve, a timer board, a battery pack consisting of two 3-V lithium cells, a check valve, and attendant tubing to allow tracer gas from a given bottle to exit from the top of the release module. The individual tracer gas release lines are connected to a single, common gas release line, which in turn connects to gas release orifices in a ported blind coupling (labeled as "H" in Figure 2), affixed to the top of the module. A needle valve in the common release line was adjusted to allow gas to bleed out over a period of a minute following detonation of the Conax valve. This slow release was designed to prevent a sudden shock to the formation. All gas release lines above the check valves were filled with oil, and the ports in the bull plug were filled with silicone grease.

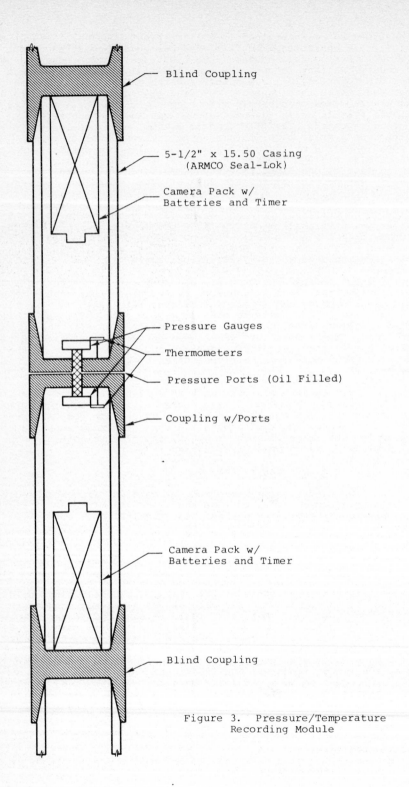

Blind Coupling

5-1/2" x 15.50 Casing
(ARMCO Seal-Lok)

Camera Pack w/
Batteries and Timer

Pressure Gauges

Thermometers

Pressure Ports (Oil Filled)

Coupling w/Ports

Camera Pack w/
Batteries and Timer

Blind Coupling

Figure 3. Pressure/Temperature
Recording Module

Figure 4. Photograph from Pressure/
Temperature Probe

Figure 5. Tracer Gas Release Module

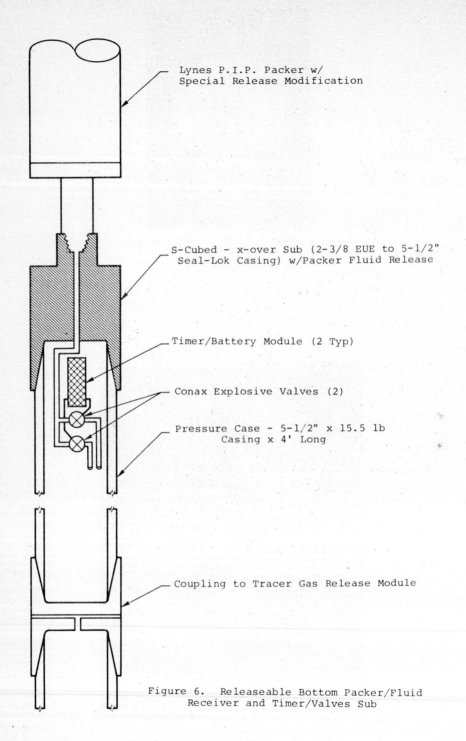

Lynes P.I.P. Packer w/
Special Release Modification

S-Cubed - x-over Sub (2-3/8 EUE to 5-1/2"
Seal-Lok Casing) w/Packer Fluid Release

Timer/Battery Module (2 Typ)

Conax Explosive Valves (2)

Pressure Case - 5-1/2" x 15.5 lb
Casing x 4' Long

Coupling to Tracer Gas Release Module

Figure 6.   Releaseable Bottom Packer/Fluid
Receiver and Timer/Valves Sub

Detonation of each Conax valve is controlled by an individual GNAT Computers, Inc., Selectable Timer. This timer is a custom-designed crystal time base possessing a switch-selectable time interval from 0 to 999 days in 1-day increments. Timing accuracy is 0.1% of elapsed time. The timer is powered by two lithium D cells, which provide sufficient energy to power the timer for 2 years. The cells also provide the electrical signal to fire the Conax valve. Timer boards were encapsulated in Dow Corning Silicone Conformal Coating Compound to reduce the possibility of moisture damage.

The timers and gas bottles were mounted on a 3/16-in.-thick by 5-in.-wide and 10-ft.-long steel mounting plate. Semicircular standoff plates welded perpendicular to the mounting plate ensured that, when the timer and bottle-mounting plate was inserted into a section of ARMCO casing, the plate could not flex within the casing.

Figure 5 shows a photograph of the timers and gas bottles affixed to the mounting plate. Adjacent to this assembly is the section of Seal-Lok casing which houses the tracer gas release module.

2.4 Packer fluid release module

A fluid-set bridge plug packer was used to provide isolation from the high aquifer pressure for the time period required to emplace and cure the grout plug. The packer fluid release module was designed to release this bridge plug packer after a predetermined number of days. Release was accomplished by allowing the packer fluid to drain into a captured reservoir contained in the instrumentation canister. To accomplish this a steel mounting plate, similar to the one used in the tracer gas release module, was bolted to the bottom of a crossover sub. Two timers and a battery pack, identical to those used in the tracer gas release module, were bolted to this mounting plate. A fluid release line extending to the fluid-containing reservoir was plumbed to include two Conax explosive valves in parallel. The other end of this line terminated in a drain hole drilled in a specially made crossover sub. This crossover sub connects the bridge plug packer to the instrumentation canister by means of 2-3/8-in. tubing. The packer setting fluid was drained through this tubing.

In practice, the packer fluid is pressurized against the Conax valves within the release module when the packer is set in the borehole. After a preselected time period, the individual timers provide a detonation signal to each Conax explosive valve. Actuation of either Conax valve opens the fluid release line, and allows drainage of the packer fluid into the captured reservoir. A drawing of the packer fluid release module is presented in Figure 6.

2.5 Experimental experience

Preliminary tests were performed to check various features of the bottom-hole instrumentation canister and to ensure that the modules would perform as anticipated. Four preliminary tests were performed as documented in Table I. An example of the pressure and temperature data obtained with the canister is provided in Figure 7. The fluctuations recorded between days 38 and 48 were caused by hydrologic testing being performed above the bridge plug. All modules performed their tasks without major equipment problems.

Table I

BOTTOM-HOLE INSTRUMENTATION CANISTER TESTS

Test #	Date Initiated	Duration	Bridge Plug Release	Pressure/Temperature Measurement	Tracer Gas	Comments
1	05/01/79	67 days	No	Yes - 2 cameras	Nitrogen	Test pres/temp measurement system and gas bottle release module.
2	08/07/79	-	-	Yes - 1 camera	Freon 13B1	Dropped package.
3	08/31/79	8 days	Planned	Yes - 1 camera	Freon C-318	Bridge plug packer leaked. No Freon C-318 detected in air above fluid column. Freon C-318 detected in fluid near bridge plug.
4	09/08/79	4 days	Yes	Yes - 1 camera	Freon 12	Freon detected after packer released.

2.6 Conclusions

The initial testing demonstrated that the basic system concept was sound. On September 24, 1979, the entire bottom-hole canister was placed into AEC-7, with the bridge plug set at a depth of 4495 feet. A grout plug was subsequently emplaced above the bridge plug. As of early April 1980, the evidence indicates the fluid release module performed satisfactorily, allowing the bridge plug packer to deflate. In addition, five tracer-tagged compressed gas bottles have vented. SF_6 has subsequently been detected in the formation fluid immediately above the grout plug, thereby providing positive evidence of flow of fluid through the plug/borehole system. Arrival times associated with tracer detection also provide valuable information on the plug permeability. These arrival times also provide qualitative evidence on the extent of fracture in the grout plug/formation system.

At present, the first phase of the WIPP Borehole Plugging Program is approaching completion. Subsequently, the bottom-hole canister will be retrieved, and the downhole pressure and temperature data recovered and analyzed.

3. GEOPHONE MONITORING OF BELOW-THE-PLUG FUNCTIONS

Initiation of the BCT required that the packer, below the plug, release automatically after the cement grout had cured. It was desired to monitor this operation remotely. A Geo Space Corporation Model 20-D geophone with a downhole amplifier designed by Sandia Labs was selected to monitor the below-the-plug functions. The geophone and amplifier are shown in Figure 8. Figure 9 is the circuit used with the geophone. A blue-quad cable (tough cable with four conductors and an overall shield) was used for the electrical connections. After the geophone system was attached to the blue-quad, the entire assembly was potted in the package shown in Figure 10. The holder for mounting the geophone to the tubing is also shown.

Because of the production of fluids in the upper portions of the wellbore, all tests of the BCT plug required that a packer be set above the plug. The tubing used to install the packer was also used for installing the geophone. The geophone system was attached to the tubing just above the packer. This arrangement successfully monitored both the packer and initial tracer-gas releases. Two subsequent tracer-gas releases were not monitored because of brine leaking into the blue-quad splice. After the blue-quad splice failures, gas releases 4 and 5 were successfully monitored with a geophone installed near the top of the wellbore. Blue-quad will not be used on future tests where the wellbore has a hydrostatic level of more than 100 psi. In future testing involving large hydrostatic heads, the geophone will be adapted to a wireline tool using oil field systems.

4. FLUID MIGRATION PAST PLUG

No attempt will be made in this report to describe the operation of the oil field systems used to measure fluid migration past the plug, but the application of these systems will be described briefly. Basically two approaches were used to measure flow past the BCT plug. The first was to set a packer above the plug and monitor the hydrostatic head increase in the tubing used to install the packer. After installation the tubing is open to the region above the plug which is isolated by the packer. Fluid leakage is then monitored with a conductor wireline (CWL) connecting a pressure transducer to a surface readout. A problem with this system is that it is impossible to distinguish slight leaks in the tubing joints from the desired plug leakage measurements.

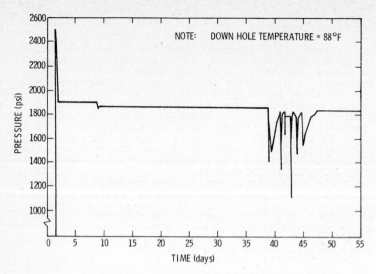

FIGURE 7. DOWN HOLE PRESSURE MEASURED WITH REMOTE INSTRUMENT PACKAGE

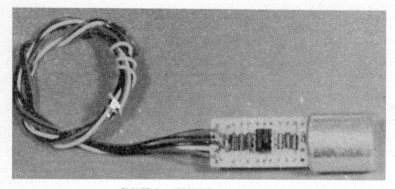

FIGURE 8. GEOPHONE AND AMPLIFIER

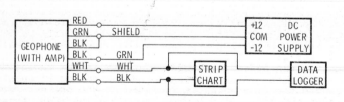

FIGURE 9. GEOPHONE CIRCUIT

FIGURE 10. GEOPHONE PACKAGE AND HOLDER

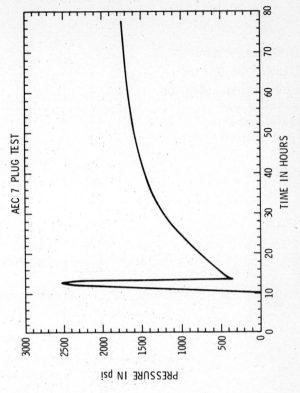

FIGURE 11. SHUT-IN TEST DATA

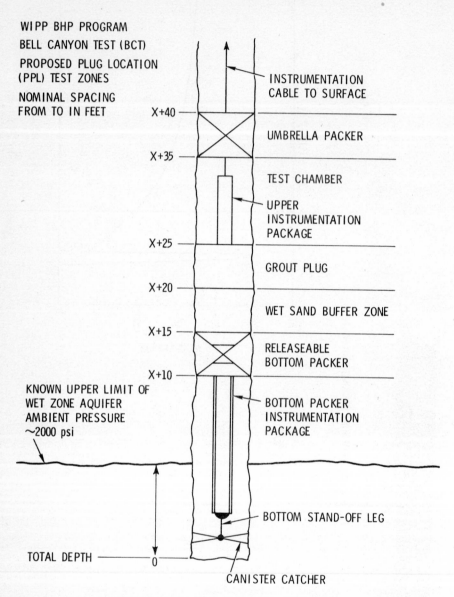

FIGURE 12. PROPOSED BCT CONFIGURATION

A more reliable approach was to run a shut-in test. To conduct such a test, a packer is installed above the plug with a digital memory recorder (DMR) reading a pressure transducer which is mounted in the region between the packer and the plug. The data are recorded in a memory which is interrogated at the surface after recovering the DMR. If the pressure readings equilibrate at the known pressure of the Ramsey Sands, it is assumed that the test is valid. By using the time to reach equilibrium, the packed off volume, and the properties of the brine, a quantitative value for the permeability of the plugged region can be calculated. The primary problem with the shut-in test is that it is not known whether the packer was set properly until the data are read, typically 72 hours after the start of the test. Some success has been achieved in making the shut-in test data available on the surface in real time by using a modified packer and a CWL pressure transducer. Regardless of the readout system used, a shut-in test is subject to interpretation regarding its validity. Figure 11 shows the data from the first BCT shut-in test, which shows the expected behavior [2]. Initially the hydrostatic head in the borehole is read followed by a small increase due to setting the packer. The pressure drop is a reading of the hydrostatic head in the tubing (also acting on the plug) as the brine is swabbed out. After swabbing, a valve is set in the packer and the shut-in test is initiated. It should be noted that all shut-in tests do not give this classic response.

5.   INSTRUMENTATION DEVELOPED FOR "DRY" WELLBORES

As initially conceived, the BCT would be a configuration similar to that depicted in Figure 12 [1]. The umbrella packer would isolate the test chamber from the rest of the wellbore and only the fluids migrating past the plug would reach the test chamber. The upper instrumentation package was designed to monitor the conductivity, level, and temperature of the brine in the test chamber. Also provisions for sampling for tracer gases from the bottom packer instrumentation package were included. The components of two probes designed for "dry" wellbores and the probe assemblies are described in the following sections.

5.1 Conductivity probe

A conductivity probe was included in the instrumentation package because it was hypothesized that conductivity could provide a rough measure of dissolved salts. This would be useful in identifying the source of the water.

Since greater-than-4000-foot lines had to be used for the BCT, the conductivity probe was designed to operate as a four-terminal resistor to eliminate line effects. Carbon electrodes were used to minimize interaction between the probe and the liquid being measured. During the initial laboratory tests, it was determined that ac excitation was required to prevent the electrolyte from reacting to the excitation current. Excitation frequencies of 60 Hz to 20 kHz were found to be satisfactory. Considering line-loading problems at the higher frequencies and the possibility of 60 Hz interference from the on-site power, 1 kHz was selected for the probe excitation.

Figure 13 is a schematic of the conductivity probe. The voltage follower amplifiers (BB) and detectors (DET) were added to the signal conditioning to convert the ac signals to dc so that they could be recorded on the Data Logger. Current (I) through the liquid being measured is determined by dividing the voltage across the current-viewing resistor (10 Ω) by 10. The resistivity between the probe electrodes is simply V/I. Table II is a listing of results obtained from various mixtures. The Delaware

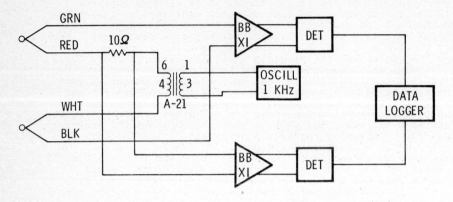

FIGURE 13. CONDUCTIVITY PROBE SCHEMATIC

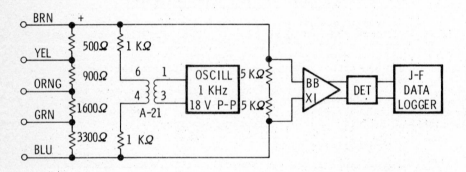

FIGURE 14. DISCRETE LIQUID LEVEL INDICATOR CIRCUIT DIAGRAM

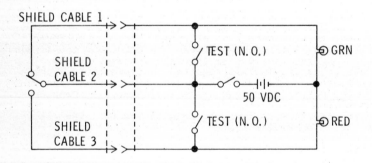

FIGURE 15. POSITION INDICATOR CIRCUIT

and Culebra are simulated samples representing these two aquifers. The values listed are for a particular electrode geometry. Different electrode geometries will give different apparent resistivities, but the different samples show similar ratios. These results indicate that the conductivity probe should be useful in identifying the source of the liquid being measured.

TABLE II - LABORATORY CONDUCTIVITY PROBE DATA

Sample	Apparent Resistivity (ohm)
Air*	920.8
Tap Water	5.7
Salt Water (77.5 g/L)	52.8
Culebra	143.5
Delaware	28.9

* In this case the reading represents the load afforded by 4923 feet of blue-quad cable.

5.2 Discrete water level indicator

The primary method for monitoring the change in liquid level above a borehole plug uses a commercial pressure transducer. However, since a pressure transducer has limited resolution and its output is dependent on liquid density, an independent discrete level indicator was designed. This indicator allows arbitrary signal levels for arbitrary level changes. It also provides an in situ calibration for a continuous reading pressure transducer.

The discrete liquid level indicator uses a resistance ladder and carbon contacts for the circuit changes. Figure 14 is the circuit diagram for this indicator. As the schematic shows, one pair of signal leads provides indications at four different levels. The distance between reading levels can be varied from 0.5cm to many metres. Here again 1 kHz excitation is used.

Table III lists the laboratory test results for the dc signals due to level changes that would be provided to a data logger with 4923 feet of blue-quad cable. The absolute voltage levels are not important, only the fact that a significant signal level change occurs. Voltage changes similar to those listed in Table III were obtained when Culebra and Delaware simulated aquifer samples were used to provide the short between contacts.

5.3 Probe for 7-to-8-inch borehole

A probe containing two discrete water-level circuits, one conductivity probe, and a Validyne Model DP15 + 2 psi differential pressure transducer was assembled for the B$\overline{C}$T. The discrete water level increments in inches were 1, 1, 1, 1, 2, 4, 6, and 8 for a total range of 24 inches. Two additional features were included in the probe: An air-inflated packer was included so that S^3 could sample the volume immediately above the plug for tracer gases as well as the remainder of the test chamber formed by the umbrella packer, Figure 12; also it was planned to attach the probe to the bottom of the umbrella packer and let the probe contact the plug. To prevent damaging the probe, a shock

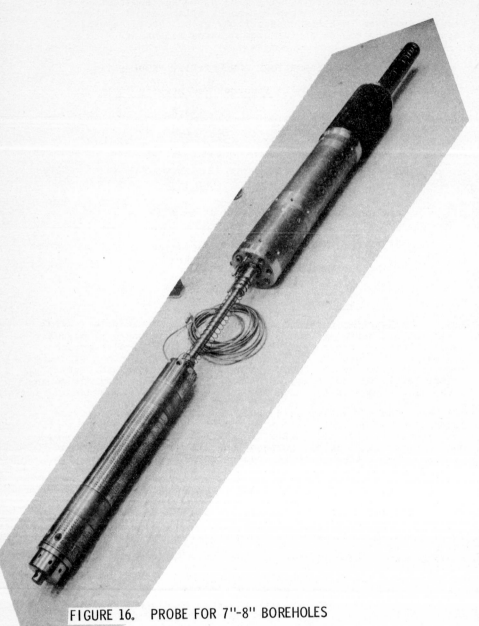

FIGURE 16. PROBE FOR 7"-8" BOREHOLES

absorber (spring-loaded collapsible section) was included in the probe. A position-indicator switch in the shock absorber permitted the operator installing the experiment to have a positive indication on the surface when the probe contacted the plug. Figure 15 is the circuit diagram used for the position indicator. The shields of the three blue-quads used in the probe provided the necessary conductors without affecting any of the other instrumentation. Figure 16 is a picture of the fully assembled probe for 7-to-8-in. boreholes.

TABLE III - DISCRETE LEVEL INDICATOR LABORATORY TEST RESULTS

Short*	Output Voltage (DC)	Incremental ΔV
None	10.00 volts	- volts
Blu-Grn	9.10	0.9
Blu-Orn	7.61	1.49
Blu-Yel	4.85	2.16
Blu-Brn	1.04	3.81

*See Figure 14.

5.4 Probe for 5-inch Casing

After completing the probe described in the previous section, it was determined that the production of fluids in the upper portion of the wellbore precluded following the initial test plan for the BCT.

An alternative approach was to case the wellbore with 5-in. ID casing down to the plugged region. To support this alternative a new probe was designed which would fit inside the casing.

The new probe included most of the features of the previous probe: two discrete water-level circuits, one conductivity probe, a position indicator switch, and a pressure transducer. In this probe the pressure transducer was changed to a 10 psi KULITE Model XTM-1-190 to permit a much higher water level to be read. Since water conductivity did not have to be a continuous reading, a two-active-arm, platinum-resistance thermometer was included to be read on the same blue-quad as the conductivity probe. A relay controlled through the shields of the blue-quads permitted switching between the desired measurands, conductivity or temperature. Figure 17 is a schematic of the resistance thermometer circuit. The completed probe for a 5-in. casing is shown in Figure 18. AEC-7 was never cased so this probe has not been fielded.

6. INSTRUMENTATION UNDER DEVELOPMENT FOR FUTURE APPLICATIONS

6.1 In-plug transducer development

Development has started on two probes that can be grouted in place to monitor the leakage of brine past the plug. The two probes are a Wenner resistivity probe and a monocoil inductance probe.

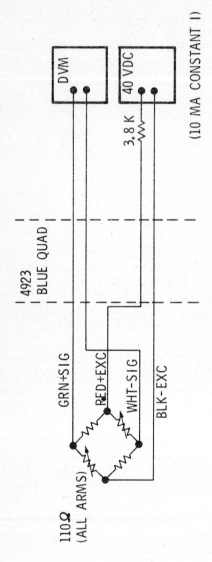

FIGURE 17. PLATINUM RESISTANCE THERMOMETER CIRCUIT

DVM

40 VDC

3.8 K

(10 MA CONSTANT I)

4923
BLUE QUAD

GRN÷SIG

RED+EXC

WHT-SIG

BLK-EXC

110Ω
(ALL ARMS)

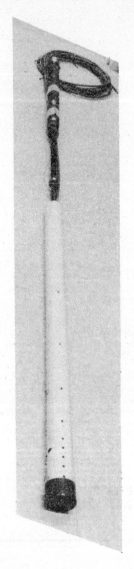

FIGURE 18. PROBE FOR 5" CASING

For a Wenner array, shown in Figure 19, the apparent resistivity is given by [3]

where
$$\rho_a = 2\ a\ \frac{\Delta V}{I}$$

ρ_a = apparent resistivity (ohms)

a   = probe separation

ΔV = voltage drop (volts)

I   = input current (amps)

The Wenner probe being developed uses a four-probe array to probe varying volumes that include the plug, interface, and region outside the plug. The four arrays have a = 4, 8, 12, and 24 inches. Its construction is carbon electrodes potted in a PVC pipe. One kHz excitation is used and the electrical signal conditioning circuitry is identical to that used for the conductivity probe discussed earlier. A prototype model of this probe has undergone preliminary laboratory tests.

The monocoil inductance probe, shown conceptually in Figure 20, is a backup approach that is also intended to monitor brine leakage past the plug. The primary advantage of the inductance coil is that it does not require good electrical contact with the plug material. Both probes will require further laboratory evaluations before they can be fielded.

6.2 Wire capabilities

Based on the BCT experience, the need for an in-house wireline capability has been demonstrated [4]. The wireline systems need the capability of real-time surface readout and recording. They also need to be designed to operate in a large hydrostatic environment (2500 psi). The following instrumentation will be procured, developed, and/or modified to meet the wireline requirements:

1. sensitive geophone

2. fluid surface detection

3. pressure sensing system, up to 3000 psi

4. wellbore fluid temperature and conductivity

5. wellbore closed circuit TV.

7.   SUMMARY

The instrumentation developed and the wireline capabilities in progress essentially satisfy the requirements for above-the-plug measurements. S^3's system for measuring permeability through the plugged region uses time-released tracer gases and has worked reliably. Some improvement in the sampling for the gases in brine-filled holes is needed.

Future instrumentation development will be influenced by the needs of the borehole plugging program as the program progresses.

References

1 Christensen, C. L., <u>Test Plan Bell Canyon Test WIPP Experi-mental Program Borehole Plugging</u>, SAND79-0739 (Albuquerque: Sandia Laboratories, August 1979).

2 Data obtained from C. L. Christensen.

3 Parasnis, D. S., <u>Mining Geophysics</u> (New York: Elsevier Sci-entific Publishing Company, 1975), pp. 177-178.

4 Christensen, C. L., Sandia Laboratories, Internal Memorandum, March 1980.

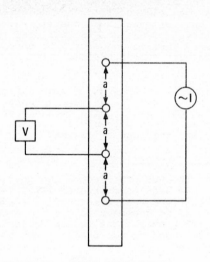

FIGURE 19. WENNER PROBE ARRAY

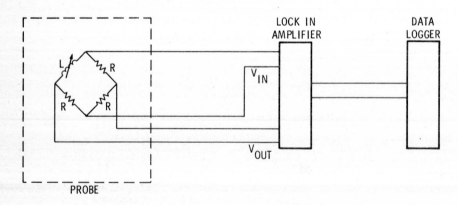

FIGURE 20. MONOCOIL INDUCTANCE PROBE

FILLING AND SEALING OF SHAFTS IN SALT MINES NECESSITATED BY AN INRUSH OF WATER OR BRINE INTO A SHAFT

H. E i c h m e y e r, H. S u k o w s k i, H. W o l f f

Technical University of Berlin
Federal Republic of Germany

ZUSAMMENFASSUNG

Im Rahmen der Untersuchungen zur Endlagerung radioaktiven Abfalls in
Bergwerken wurden auch die Möglichkeiten der Verfüllung und Abdich-
tung eines ersoffenen Schachtes betrachtet.
Falls es nicht möglich ist, das Bergwerk durch Pumpen vom Schacht-
sumpf oder durch Schließen eines großen Schiebers im Schacht vor dem
Ersaufen zu bewahren, muß der Schacht verfüllt und abgedichtet werden. Zu diesem Zweck werden mehrere Lagen unterschiedlicher Material-
zusammensetzung im Schacht eingebracht. Versuche zur Ermittlung ge-
eigneter Bindemittel für Pfropfen in Salzschichten wurden begonnen.
Einige Ergebnisse von Laborversuchen mit drei Zementarten werden
bekanntgegeben.

ABSTRACT

Within the investigations for the underground disposal of radioactive
waste in the Federal Republic of Germany also the possibilities of
sealing a flooded shaft have been studied.
If it seems not practical to safe the mine by pumping from the shaft
sump or by closing the shaft with a large valve, the shaft has to be
filled and sealed. For this purpose several layers of different materials are placed in the shaft. Tests have been started to find out
the most suitable binder for plugs in rock salt strata. Some results
of laboratory tests with three kinds of cement are published.

1) Introduction

In the Federal Republic of Germany (West) the concept of radioactive waste storage intends its deposition in a salt dome at a depth of approximately 850 meters within rock salt. Tests were begun several months ago at a suitable salt dome located near Gorleben. It is assumed that the Older Rock Salt Formation occurs in sufficient volume and in the necessary purity.

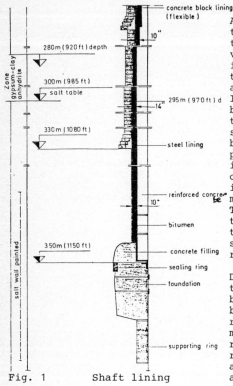

Fig. 1   Shaft lining

According to current mining operation planning, regardless of location, the mine is to be accessible via two vertical shafts 7.5 meters in diameter. The shafts are located at the center of the mine and approximately 500 meters apart. In the upper part down to 50 meters below the salt table (figure 1), the shafts are to be lined with steel-reinforced concrete with bitumenous joints. In the lower portion of the shaft lining only is planned in the presence of carnallite layers. A Koepe hoist is to be employed in the shafts to move the cages along rope guides. The sectional view (figure 2) of the shaft shows that in addition to the rope guides an emergency system will use tracks or guide rails.

During safety investigations in the preplanning phase, the possibility of an inrush of water or brine into the shaft was assumed, resulting in a flooding of the mine. This situation was assumed regardless of whether at the current state of mining technology a water or brine inrush could actually occur during the mine's approximately fifty year period of operation. Regardless of the cause of a water or brine inrush and of the probability of its occurance, when a shaft or mine is flooded, it is essential that the biosphere and storage areas do not come into contact or remain in contact with each other via the water or brine.

The following is the report on the work performed which was the object of the investigation.*

---

* The investigation was performed in order of the Physikalisch-Technische-Bundesanstalt (PTB), Braunschweig

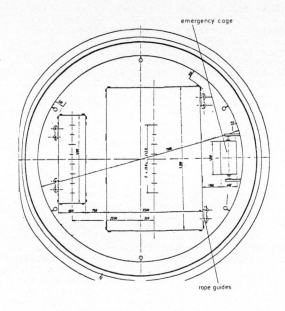

emergency cage

rope guides

Fig. 2   Shaft section

2) The reaction to an inrush of water or brine

2.1 The possibilties of preventing flooding of the mine.

The occurrance of a water or brine inrush into the shaft does not
necessarily lead to flooding of the mine. This fact must be emphasiyed
in view of the possibly serious consequences of a liquid leak in a
storage mine, and it is therefore departed briefly from the main
topic of this paper. On the one hand, the amount of liquid leaking
into the shaft could be of minor quantity, that it could be taken
care of by pumping while simultaneously carrying out sealing and
repair measures. On the other hand, it appears advisable, in the
case of large as well as small amounts of water or brine, to insert
a plug or a slide sealing off the lower portion of the shaft, in
order to prevent flooding of the mine. Among the many diverse technical solutions to this problem that are available, the use of a
hemispherical valve, also called a "shaft bell", seems to be the
most proven. In an emergency situation, the shaft bell would be
taken from a compartment adjacent to the shaft moved suspended along
two tracks, and lowered onto a circular steel-reinforced concrete
buttress at the perimeter of the shaft. A multi-layered rubber ring
would provide the sealing of the shaft as it is pressed into the
special profile of the buttress. After the shaft is closed, it can
be sealed additionally with a concrete plug and subsequently repai-
red at the site of the leak following a detailed step-by-step system.
Once this has been done successfully, it is possible to recommence
storage operations in the mine.

2.2 Procedures in case of flooding of the mine.

If it is not possible to close the mine with the help of the shaft
valve mentioned before, then the best must be made of the inevitable
flooding of the mine. Extensive measures must be taken immediately
to fill and seal the shafts. Since two shafts exist, and it can be
assumed that water is leaking into only one of them, there is a
chance that one of the shafts can be filled and sealed while still in
a dry state. The time available to do this depends on the rate of
flow. At a flow rate of approximately 15 m³/min., a rather high figure, based on experience, the dry shaft would have to be sufficiently
filled and sealed within 45 days. This is feasible technically and
organizationally.

The sequence of operations
can be seen in detail in the
following illustrations.
Figure 3, a general overview,
shows that during the first
45 days following the inrush
of brine, the filling and
sealing of the dry shaft will
be completed up to 84 meters
above the upper level of the
pit. In the second phase of
the filling and sealing, the
damaged shaft is first filled
and sealed temporarily, allowing shaft repairs to be
done at the site of the leak.
After successfull repair
the final filling and sealing
of the second shaft can be
done in the third phase.
The shaft is cleared up to
50 meters above the upper
level and then brought into
its final, filled and sealed
state in the same procedure
as the first shaft. The
filling and sealing is
carried out in a multi-step
operation which, for the
undamaged shaft (figure 4)
begins with the placing of
coarse basalt rock and grout
up to 30 meters above the
uppermost shaft station.
A layer of "Chemical seal"
is applied to the top of

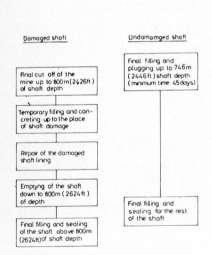

Fig. 3 Flowsheet of the shaft
filling and sealing
actions

this first plug-type seal to provide the complete additional seal of
the shaft. This is followed by a concrete plug 50 m high which serves
to withstand the pressure and simultaneously secures also the sealing
of the shaft. It is planned to fill the remaining part to the base
of the shaft lining with "earth concrete" (concrete with bentonite).
The lined part of the shaft will be filled with bitumen.
While standing under brine, the damaged shaft must first be filled
temporarily (figure 5). In the first step, coarse rock is poured in
from the surface and layer by layer injected with grout via pipes
located in the shaft, filling the shaft to 30 meters above the uppermost level. The following shaft areas up to the site of the leak are
filled with coarse rock only. In the area near the leak the coarse
rock is injected to create a temporary plug 50 meters high. While
protected by this plug, the leak can be sealed, from either the outside or the inside of the shaft. After this repair is completed, it

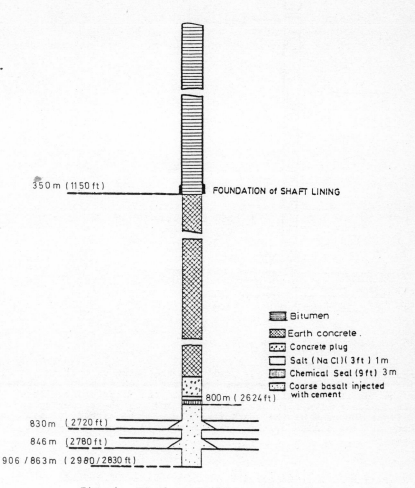

350 m (1150 ft) FOUNDATION of SHAFT LINING

Bitumen

Earth concrete .

Concrete plug

Salt (Na Cl)(3 ft) 1 m

Chemical Seal (9 ft) 3 m

Coarse basalt injected with cement

800 m (2624 ft)

830 m (2720 ft)

846 m (2780 ft)

906 / 863 m (2980 / 2830 ft)

Fig. 4   Final shaft filling and sealing

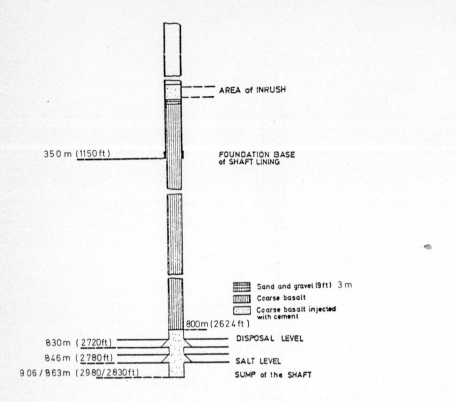

Fig. 5 Temporary shaft filling and sealing

is provided to remove the loose rock from the shaft so that the final
filling and sealing beginning from the first injected plug 30 meters
above the uppermost level can be done as in the first shaft. There
will be adequate time for this last phase and for the preceding repair, since the area of the shaft stations will be already closed
and more or less sealed by the first injected plug of coarse rock
and cement.

3)   Details of the final filling and sealing of the shafts

3.1   Initial findings and summary.

The particular problem of sealing off areas within a radioactive waste
storage mine is to provid a complete seal and to attain long-term
effectiveness. The filling and sealing materials must therefore be
high-reliability items, which by their nature require a great amount
of testing. It is also necessary to investigate all possible filler
materials, from natural substances like bitumen or clay to concrete-
type materials and synthetics. In this regard, natural substances
are especially attractive because of their known long-term stability.

Because of the diverse functional requirements, uniform sealing and

filling is not possible. Different functions have to be attributed
to the different parts of the shafts etc., resulting in a step-by-
step procedure for a flooded or flooding storage mine. This is described in detail as follows (figure 4):

1. The mine workings partially or totally filled with brine, must
   first be closed off from the shafts. In the lower part of the
   shaft, also a foundation for the subsequent filling section must
   be created simultaneously. This foundation is recommended since,
   although adhesion of the load-bearing concrete plug appears adequate, this cannot be absolutely guaranteed, based on experience.
   On the other hand, the initial sealing of a flooded mine is
   essential because subsequent filling and sealing work should be
   done in dry shafts to guarantee positive seal.

2. Particular positions of the shafts are designed to provide the
   complete seal of the shafts from the mine. These are best placed
   adjacent to the first portion serving the closing of the mine.

3. Since the column of filling material in the shaft may be subjected
   to stresses from the mine and from the upper shaft area, it is
   necessary to install a load-bearing, stress-resistant layer. This
   is best located directly above the sealing layer. The composition
   of this layer also has to take into consideration its sealing
   function. Its primary function, however is the bonding of the plug
   to the salt, i.e., the shaft wall.

4. The remaining shaft portions are also to be filled, since they
   may be subjected to long-term corrosion and rock stress resulting
   in collapse. In the shaft portion between the concrete plug and
   the lined area, it is recommended to use such material which can
   take the long-term load off the pressure-resistant superjacent
   portion and simultaneously assume a sealing function.

5. In the following lined shaft area, which also includes the salt
   table, the most important requirements are protection against
   corrosion and an effective seal. Therefore, it is best to use a
   filler material which not only above all protects the shaft from
   corrosion, but also intrudes and closes cracks carrying water or
   brine.

The detailed requirements described above for the filling and sealing
of shafts in a storage mine in massive rock salt, illustrate the
complexities of the problems and simultaneously summarize the solution to the problem, which were derived from investigations carried
out in the Federal Republic of Germany.

These suggestions can be advocated only in the special recognition of
circumstances dominating the discussions by the public and experts
concerning the storage of radioactive waste. There are two conclusions
to be drawn also from the solutions suggested:

1. To solve the general and special problems associated with the
   storage of radioactive wastes, the safety measures for the technical solutions should go far beyond normal safety precautions.
   This requirement is reflected by the suggestions presented in this
   report:

   - The closing off of the mine from the shafts already provides
     a substantial seal,

   - The pressure-resistant plug in the filling column functions simultaeously as a seal,

- The two following filling sections not only serve to protect the open space from pressure, but also provide a corrosion-resistant seal and a pressure-resistance, in addition to those provided by the underlying sections.

2. Economic arguments should play only a secondary roll in decision making because of the unusual measures required by the special problems associated with nuclear energy, and because of the small number of storage mines to be expected.

3.2 Technical installations and the technique of placing filling material in the shaft

Since there is adequate time to fill and seal the shaft, it is possible to dispense with the dumping of filling material into the shaft. Instead, transport systems used for shaft sinking can be employed while the normal hoisting installations are not suitable for moving the supplies needed to fill the shafts. Gravel, concrete, and other materials can be moved down into the shaft with buckets having a capacity of up to 5 m^3, allowing up to 40 m^3 of materials to be transported each hour, at a speed of 8 m/s. In addition, several pipes are located in the shaft and can be used to supply the grout to form the buttress at the shaft stations.

3.3 The individual steps in filling and sealing (figure 4)

3.3.1 Sealing off the mine workings

In the lower portion of the shafts, the mine workings are closed off in the area near the shaft stations by placing coarse rock which is strengthened with a binder. This rock must meet special requirements. According to mine regulations, the size of the rock particles must lie between 20 mm and 250 mm. In addition, the filling material must be solid and impermeable as well as not have a tendency to expand. Coarse basalt rock with sharp edges best meets these requirements, as it permits construction of a filler column with high resistance against deformation and flow. The coarse rock is always to be placed to a hight 30 meters above the uppermost shaft station. According to the angle of internal friction the rock will slope down 45° at the shaft stations. In spite of the high internal friction of the rock, however, long-term stability cannot be guaranteed. Additional measures are necessary. This involves filling of the spaces between the rocks with cement suspension or grout. A saturated brine solution is used to make this mix which is transported in the pipelines still present in the shaft.
The advantage of using a bucket system to bring the coarse rock into the shaft is that the filling operation can be continuously monitored to prevent the possibility of arching.

The main task of the plug of coarse basalt rock which reaches up to 30 meters above the uppermost level, is to serve as a support for the subsequent sealing layers. By the use of brine-resistant cements, the plug also has a substantial sealing effect, but it would not suffice if it were used as the sole sealer.

3.3.2 The sealing layer above the closed off portion of the shaft

A new synthetic sealing material (chemical seal), manufactured by
the Dow Chemical Company in Tulsa, Oklahoma, is applied on top of
the coarse basalt rock plug. Its purpose it to provide a complete
seal of the total shaft cross-section. This material has been used
in Canada in shaft sinking operations at potash mines to seal the
space between lining and surrounding rock, and has proven to have
excellent sealing charakteristics. When this material comes into
contact with liquids, it polymerizes and expands to fill a volume six
times its original size. At present, there are no data from which to
determine the optimal thickness of this layer for providing the best
shaft seal. In any case, the initially assumed thickness of 3 meters
requires additional testing. After the chemical seal has solidified,
it is first covered with a plastic foil and then with a layer of rock
salt approximately 1 meter thick. This is to prevent moisture of the
following layer of concrete from seeping into this sealing layer.

3.3.3 The concrete plug above the sealing layer

As shown in figure 4, a 50 meter high concrete plug is designed to
withstand the hydrostatic pressure on the filler column over a long
period of time. At present it is not possible to compute the required
dimensions of the plug. The equations derived by several authors have
lead to such greatly differing results that they could not be used
for solving practical problems. Plug heights between 9.21 and 32.2
meters have been computed, based on the dimensions of the shafts of
the storage mine. Even using the method of finite elements has not
resulted in any practical values, because the input parameters for
the computation are not yet exactly known. In addition, these equations only give information about the stability of the plug. Although
this is indeed an important parameter in dealing with sealing problems in the shafts, the sealing factor itself cannot be computed,
so that one must rely on experience. If available data on the effectiveness of shaft plugs, are compared, it can be seen that plugs
with heights of 40 meters and longer sufficiently meet the requirements of sealing and stability. For safety reasons, the height of the
plug has been increased to 50 meters. With this dimension, stability
is ensured and an enlarging of the shaft diameter ist not necessary
in the area of the plug to allow for the placement of additional
support. Since the concrete is transported with buckets, its flowabi-
lity dos not have to taken into consideration. An optimal water-
cement ratio can be chosen that results in the least permeability for
water. A relatively small amount of water, approximately 40 % of the
cement by weight, is necessary for the concrete to set. With a higher
water content, the excess water is forced to the surface, resulting
in flow channels which remain open after the concrete has set and
causing the concrete to be more permeable.

In addition to keeping to the optimal water-cement ratio, it is important to tamp or shake the concrete in order to reduce the air
bubble content. The choice of the best type of cement will require
further investigations. The following cements have been proven to be
suitable for use in rock salt depending of the composition of the
brine:

1. In rock salt brines without sulfate - Portland Cement or blast
   furnace cement

2. In sulfate brines with little magnesium chloride - blast furnace
   cement

3. In brines with much magnesium chloride - magnesium cement

1. Portland cement PZ 35 L - NW/ltS „Aquafirm". Cement Works Heidelberg. Water-tight, sulfate stable, low shrinking rate.

2. Poxmix 80. Deep hole drilling cement. Dyckerhoff Cement Works. 60% Basis Cement, 38% Tuff, 2% Diatomaceous earth sulfate stable, C 3A - free

3. Sakret SD1. Dry mortar. Sakret GmbH, Gießen. The mortar contains electro filter ash with limestone, Portland Cement and additives. The mortar is occasionally used as material for dam construction in german collieries and salt mines.

Fig. 6   Sorts of binding agents used for the tests

Although concretes made with these cements have been proven to be suitable for use, one cannot rule out, for example, that a mixture of granulate material, "Blitz-dämmer", and saturated brine may also be well suitable for use as a filler material. Another interesting material could be salt concrete of such a composition that after setting it will have most of the characteristics of the rock salt. Even synthetics are worthy of consideration. Without a doubt, further investigations in this area are necessary to determine the optimum materials for plugs in storage mines. Initial tests have already begun to find out what materials are best suited for a binder. For reasons of time and resources, only three types of cement were subjected to laboratory testing (figure 6). The cement in the illustration listed under 1., Portland Cement PZ 35 L-NW/HS Aquafirm, belongs to the readily available cements that can be procured in large quantities. The use of a "normal cement" in the tests provides a direct control for comparing the special cements. The other two binders used are special cements employed in deep hole drilling and mining. Pozmix 80 is used for the closing of boreholes and consists of 60 % base cement, 38 % trass (volcanic-tuff) and 2 % diatomaceous earth. This cement withstands sulfate and is C_3A free. In West German collieries and salt mines, dry mortar Sakret SD 1 has been used as a dam building material. It contains electro filter ash, chalk, portland cement and other ingredients.

The test mortars were mixed with both fresh water and a saturated brine. They were bedded in water (21° C/68° F) and in air, water and saturated brine at 37° C (98.6° F). For each mortar type there were three series of tests whose compositions are tabulated in figure 7. The mix ratio of sand and cement is uniformly 3:1, and the water-cement ratio between 0.45 and 0.5; for the Sakret dry mortar, the mix ratio of water to mortar is 0.3. At 36 %, the salt content of the mixing brine corresponds to the degree of saturation when using rock salt. Normal salt used in salting roads was used. It has a 95 % NaCl content. This simplification was necessary for initial testing. In future investigations other important salts will also be included.

In order to give the tested mortars the proper designation, the following parameters were checked at standard testing intervals of 1,2,7,28,90,200 and 400 days.

Cement or mortar	Portland Cement PZ 35			Deephole drilling cement Poxmix 80			Dry mortar Sakret SD1		
Series	I	II	III	I	II	III	I	II	III
Sand (g)	1350	1350	1350	1350	1350	1350	—	—	—
Cement or dry mortar (g)	450	450	450	450	450	450	1500	1500	1500
Water (g)	225	225	203	225	225	203	450	450	450
Water cement ratio	0,50	0,50	0,40	0,50	0,50	0,45	0,30[1]	0,30[1]	0,30[1]
Na Cl (g)	—	—	73	—	—	73	—	—	162
Na Cl (%)	—	—	36[2]	—	—	36[2]	—	—	36[2]

Fig. 7   Compound of the Mortars

     1) Ratio water/dry mortar

     2) Related to the water portion

1. Manufacturing abilities by measuring the spreading power and the workability.
2. Hydration curve by measurements during the first two days with an adiabatic calorimeter.
3. Change in length at each test date.
4. Strength development by measuring the compressive strength and the transverse strength at each test date.

The results of the 90th day testing are already available. Only a few basic statements about the results may be given here:

1. Hydration Curve (figure 8)

As the curve shows, cement A and P have an even temperature increase during the period 6 - 12 hours (A) and 4 - 10 hours (P). After an initial even temperature increase, mortar S shows a very strong thermal development after 10 hours, but drops off after 12 hours. The maximum temperatures developed were approximately 34° C (P), 36° C (A), and 39° C (S). The addition of salt in all cases caused a reduction of the maximum temperature to around 30° C, as well as a definite delay in the thermal reaction by 5 - 6 hours (P) and 10 - 12 hours (A,S), and also resulted in a slower temperature increase.

2. Change in length

The changes in length measured after 90 days are depicted as the mean values in figure 9. Mortars kept in water at room tempera-

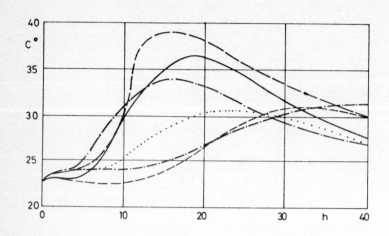

Fig. 8   Course of temperature of mortars

───── = Aquafirm tempered with water, ‒‒‒‒ = Aquafirm tempered with brine
──·── = Pozmix   "   "   "   ,········ = Pozmix   "   "   "
──── = Sakret   "   "   "   ,──·──·── = Sakret   "   "   "

Sample of mortar			Bedding in			
			Water (21°C)	Water (37°C)	Air (37°C)	Brine (37°C)
Aquafirm	(A)	Tempered with water	20	25	-70	40
Pozmix	(P)		20	15	-80	5
Sakret	(S)		35	20	-280	-25
Aquafirm	(A)	Tempered with brine	───	70	-210	50
Pozmix	(P)		───	95	-330	50
Sakret	(S)		───	80	-550	50

Fig. 9   Change in length at samples of mortars
          (Length measured after 90 days, average value in μm)

ture expanded only slightly. Mortars made with brine expanded three to four times as much as those mixed only with water and attained factors of around 0.5 ‰. Somewhat less expansion was found in mortars with brine which were bedded in brine. Different influences of the binding materials are apparent in salt-free mortars. While cement A expands 0.25 ‰ when bedded in brine, samples P expanded only slightly; mortar S shrunk somewhat. Bedding in an air lead to strong shrinkage, especially in the case of mortar S. The mixture with brine doubled or tripled the degree of shrinkage, and in the case of mortar S, resulted in a shrinkage factor of 3.5 ‰.

3. Strength development

With regard to the development of strength in the tested mortars, the first comment is directed to normal conditions, i.e., at a temperature of 21° C and setting of the mortar samples which were mixed with water, in a water environment. As shown in figure 10, a continuous development in the compressive and transv. strengths of all three binders (mortars) was observed. Mortars A and P behaved quite similarly, while mortar S attained somewhat unsatisfactory results.

The subsurface temperature of 37° C is assumed under the given conditions. The strengths were measured on each test date for mortars which set under water, in a saturated salt solution and in air. The numerous data will not be reviewed here, since they were derived from initial testing which has not yet been completed and much more testing remains to be done. The following preliminary statements can, however, be made:

1. The mortars mixed with saturated brines, as opposed to mortars mixed with water only, have quite strongly differing compressive strengths when allowed to set under water or brine (figure 11). None of the strength values falls short of the estimate for the required minimum value. A similar result was attained with transverse strengths (figure 12).

2. The relations for mortars bedded in air are not so uniform, albeit with a definite trend towards increased strengths with mortars mixed with saturated brines and special cements (figures 13 and 14).

3. All mortars set in air have substantially lower strengths than those set in liquid, regardless of the composition used (figures 15 and 16).

4. The strengths of the mortars set under brine, especially the special mortars, do not deviate substantially from the mortars set under water (figures 11 and 12).

5. Only a few mixtures of the special mortars (P and S) have more favorable strengths than the mortars made of standard cement (A). The latter can therefore be considered to be equal to any of the special mortars used in the testing (figures 15 and 16).

Further testing is necessary to check the initial test results obtained with three types of cement and, if required, to widen the scope of testing to find a suitable binder. Additional ingredients should furthermore be included in the investigations, as well as other suitable binders, whereby synthetics should also be considered. The question of stability under changing temperature conditions will also have to be investigated. Other important problems like water perme-

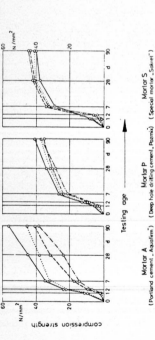
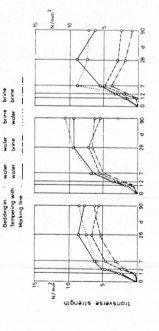

Fig.10 Development of compression strength and transverse strength for several mortars under „ normal conditions" (21°C bedded in water)

‒‒‒‒‒ = Mortar A (Portland cement „ Aquafirm")
‒·‒·‒ = Mortar P (Deep hole drilling cement „ Pozmix")
‒ ‒ ‒ = Mortar S (Special mortar „ Sakret")

Fig.11 Development of the compression strength for several mortars bedded in water or brine

Mortar A   Mortar P   Mortar S
(Portland cement „ Aquafirm")   (Deep hole drilling cement „ Pozmix")   (Special mortar „ Sakret")

Bedding in : water   water   brine   brine
Tempering with : water   brine   water   brine
Marking line   ‒‒‒‒   ‒‒‒‒   ‒‒‒‒   ‒‒‒‒

Fig.12 Development of the transverse strength for several mortars bedded in water or brine

Mortar A   Mortar P   Mortar S
(Portland cement „ Aquafirm")   (Deep hole drilling cement „ Pozmix")   (Special mortar „ Sakret")

Bedding in : water   water   brine   brine
Tempering with : water   brine   water   brine
Marking line   ‒‒‒‒   ‒‒‒‒   ‒‒‒‒   ‒‒‒‒

- 334 -

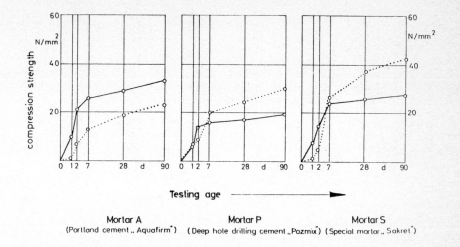

Mortar A
(Portland cement „Aquafirm")

Mortar P
(Deep hole drilling cement „Pozmix")

Mortar S
(Special mortar „Sakret")

Fig.13 Development of the compression strength for several mortars bedded in air

Bedding in	:	air	air
Tempering with	:	water	brine
Marking line	:	_____	....

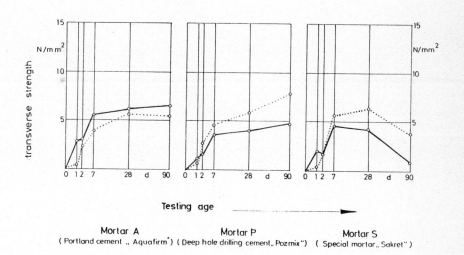

Mortar A
(Portland cement „Aquafirm')

Mortar P
(Deep hole drilling cement „Pozmix")

Mortar S
(Special mortar „Sakret")

Fig.14 Development of the transverse strength several mortars bedded in air

Bedding in	:	air	air
Tempering with	:	water	brine
Marking line		_____	......

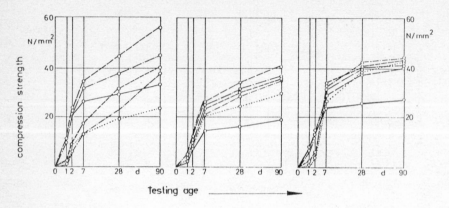

Mortar A
(Portland cement „ Aquafirm")

Mortar P
(Deep hole drilling cement „Pozmix")

Mortar S
(Special mortar „ Sakret")

Fig 15 Development of the compression strength for the several mortars bedded in water, air

Bedding in	:	water	water	air	air	brine	brine or brine
Tempering with	:	water	brine	water	brine	water	brine
Marking line	:						

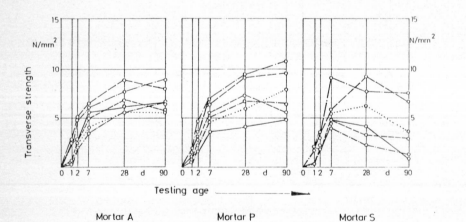

Mortar A
(Portland cement „ Aquafirm")

Mortar P
(Deep hole drilling cement „Pozmix")

Mortar S
(Special mortar „ Sakret")

Fig.16 Development of the transverse strength for several mortars bedded in water, air or brine

Bedding in	:	water	water	air	air	brine	brine
Tempering with	:	water	brine	water	brine	water	brine
Marking line	:						

ability corrosion resistance, and adhesion to rock salt should also be the object of further investigations.

3.3.4 Filler column of bentonite concrete (earth concrete)

After the shaft is sealed off from the rest of the mine using the measures described in the preceding paragraphs, it is necessary to fill the rest of the empty spaces in the shaft. For this purposes, it is important to use a material that simultaneously assumes other important functions, e.g., to provide an effective seal and protection against corrosion. It is suggested to fill the shaft portion at 746 - 350 m depth which is in rock salt and unlined, with bentonite concrete, which is hoisted down with buckets. Due to the bentonite admixure, this concrete has the essential attribute of swelling under moisture, this improving the sealing of the shaft. Wether or not concrete with bentonite will be suitable for this purpose, in the final analysis will have to be determined in additional tests. Of interest will be the optimal mixing ratios, the composition of the mixing liquids, and the behavior of the mortar under increased temperature. If concrete with bentonite should prove unsuitable, other materials will have to be considered. Bitumen, as planned for the lined portion of the shaft would be suitable in any case, but is not a first choice for filling of unlined shafts because of its extremely high costs.

3.3.5 Filler column of bitumen

The lined portion of the shaft is particulary endangered because it is located in liquid-carrying strata and subject to corrosion. Under this aspect, bitumen is recommended as a filler material to provide corrosion protection. Many reasons speak for the choice of this material. Bitumen consists of multi-molecular hydrocarbons and is not water soluble. Absorbtion of liquids by osmosis, can also be ruled out for all intents and purposes. After being subjected to water for 10 years, a 5 mm thick layer of bitumen absorbed 1 - 3 % water, depending on the hardness of the bitumen, and only superficialy. In addition, bitumen is resistant to most inorganic acids, brines and salts. Because of its large molecules, bitumen has no distinct melting point, but instead it softens slowly over a more or less wide temperature range. For distilled bitumen, which would be used to fill the shaft, this thermoplastic range is between 27 - 72° C. When subjected to mechanical stress, this material behaves elastic, plastic or viscous depending on temperature. With increasing temperature and pressure the liquid attributes become more pronounced. This is the main reason for choosing bitumen. The hydrostatic pressure of the filler column can also push water from the shaft back into the surrounding rock, and fill the cracks that open up. Additionally, bitumen can protect the shaft from corrosion, so that an almost unlimited stability can be attained in the lined portion of the shaft.

In order for the bitumen to fulfill these requirements its density must be increased by the addition of filler ingredients. Powdered rock has proven to be usefull, raising the density of bitumen from 1 to 1.3 g/cm³. Bitumen types B 300 or B 400/500 are proposed. A mixture used in lining shafts and also suitable for the filling of storage mine shafts, consists for example of 63 % weight primary bitumen "Spramex 400/500" and 37 % powdered limestone of an extremely fine grade.

The bitumen is to be brought into the shafts via special pipes that can be shortened. The bitumen is delivered to the mine using tanker trucks. Immediately before the filling, the bitumen is mixed with the powdered stone in heated vessels. It is expected that it will be possible to mix and fill 6 m³/hour.

3.4   Concluding remarks

The concept proposed for the filling and sealing of shafts in a radio-active waste storage mine located in a salt dome, is the result of first rough investigations into the difficult matter of shutting down such a storage mine.

The concept contains several redudant safeguards to assure that both a watertight and pressureproof seal in the shafts is maintained. This proposal is therefore relatively complex and requires further investigations and accurate testings to find optimal solutions.

Discussion

<u>T.O. HUNTER</u>, United States

      Is the "earth concrete" similar to saltcrete used at Asse ?

<u>H. EICHMEYER</u>, Federal Republic of Germany

      No, the earth concrete is a mixture of sand, gravel and bentonite with a certain amount of cement to stabilize it.

<u>D.M. ROY</u>, United States

      I am interested in the aggregate contained in your earth concrete. You indicated the contents were bentonite, cement, sand (quartz ?) and gravel. Was the "gravel" from basalt, or other type rock ? This may be important in determining the long-term reactivity.

<u>E. EICHMEYER</u>, Federal Republic of Germany

      The gravel will be of a conventional kind, the same which is used in the construction industry. But, for our purpose, we will have to investigate the problem of the composition of the earth concrete and the relative abundance of the different components.

<u>R.D. STATLER</u>, United States

      How do you emplace the "earth concrete" in the shaft and how do you emplace the bitumen ?

<u>E. EICHMEYER</u>, Federal Republic of Germany

      The "earth concrete" is hoisted down in buckets. For this purpose a hoisting equipment will be installed for the filling and sealing operation. A capacity of about 40 m^3 per hour will be possible. The bitumen, however, is intended to be emplaced by pumping ; the pipes will be progressively shortened during the filling operation. It is expected that about 6 m^3 per hour could be emplaced.

<u>T. YAKIRO</u>, Japan

      What kinds of rock are suitable as components of earth concrete to be used in shaft sealing ?

<u>H. EICHMEYER</u>, Federal Republic of Germany

      We will have to investigate the optimum composition of earth concrete for shaft sealing. For normal uses earth concrete is made by mixing sand and normal gravel with cement and bentonite.

SESSION 4

Chairman - Président

N.A. CHAPMAN

(United Kingdom)

SEANCE 4

A STRATEGY FOR EVALUATING THE LONG-TERM STABILITY OF HOLE-PLUGGING MATERIALS IN THEIR GEOLOGICAL ENVIRONMENTS

S. J. Lambert
Nuclear Waste Technology, Division 4511
Sandia National Laboratories
Albuquerque, New Mexico   87185

ABSTRACT

Material used to plug boreholes will not in general be in chemical equilibrium with its host rock. Adverse long-term performance of a plug can involve changes in phase assemblage in the plug/ rock system which are difficult to observe at low temperatures in real time. The thermodynamics of multiphase equilibria provides a technique of predicting what phase changes might occur. The thermodynamic treatment of plug/rock systems utilizes (1) a formulation of possible chemical reactions among phases in the system and (2) de-terminations of changes in values of Gibbs' free energies for the hypothetical reactions, to identify the theroretically-permitted reactions which could degrade plug performance. Time-dependent pre-diction of phase changes requires a knowledge of rate laws and constants for specific reactions whose mechanisms are well known.

INTRODUCTION

It is generally recognized that boreholes and shafts, if left open in the vicinity of a radioactive waste repository, could serve as a conduit for water from the local hydrologic system to enter the repository, and perhaps for waste radionuclides to escape. In spite of the fact that the consequence of an unsealed borehole or shaft penetrating a geologic repository can be numerically evaluated [1] , prudence dictates that reasonable state-of-the-art attempts be made to inhibit the flow of material through man-made openings in and around the repository. Since the time interval required for radio-active waste "isolation" is estimated to be several human lifetimes, materials used to seal shafts and boreholes will be expected to have no less long-term integrity than the rock system containing the waste. The long-term behavior of rock during the time a repository is effectively isolating waste can be neither precisely predicted nor directly observed. It is similarly difficult to predict the behavior of materials used to seal the man-made openings in rocks.

An alternative to direct observation of phenomena which take place at "geologic" time scales is the approach adopted by Bingham and Barr [2] . In essence, events and processes that might disturb the buried waste are identified. Such identification is a necessary step in preparing assessments of "consequences that would attend releases of buried waste." In addition, the identification would point out the events and processes that should receive further study in "experimental and developmental programs, which may produce methods of preventing them or of mitigating their effects." In the application of the above approach to evaluation of long-term stability of hole-plugging materials in their geological environ-ments, the "events and processes" leading to adverse plug perform-ance are identified as mineralogical phase changes in the system of plug-rock-groundwater. "Adverse plug performance" due to long-term changes is here taken to be (1) formation of soluble, friable or permeable phases in the plug or nearby rock, or (2) the opening of the interface between plug and rock due to shrinkage or degradation of adhesion.

Generally, hole-plugging materials are engineered so as to have a desired set of physical properties which can be tested in a laboratory. In the case of cementitious grouts, such properties include various moduli, porosity, permeability, push-out strength and expansivity. It cannot be assumed a priori that properties of a certain grout mix measured at one time will remain within desired specifications for hundreds or thousands of years. The alternative to direct observation of changes in properties is to assume that phase changes will take place and to evaluate their consequences. Certain phase changes will be thermodynamically permitted, whereas others will not. The identification and treatment of the phase changes and the reactions leading to them proceed as follows:

1. Phases in the system are identified.

2. All possible reactions among components of rock, grout and groundwater are formulated.

3. Thermodynamic data are acquired for all hypothetical reactants and products at their initial and final states.

4. Reactions determined by thermodynamics to be possible are assumed to take place.

5. The consequences of the reactions and their products are evaluated for their potential severity in compromising the integrity of a borehole or shaft seal.

The thermodynamic model developed and used to evaluate the reactions predicts what reactions can occur. In a system of even a few chemical components, a multitude of reactions can be formulated. Some reactions, although theoretically permitted by the thermodynamics, may defer to others; some may be kinetically slow or chemically inhibited. The thermodynamic model does not and cannot predict the progress of any reaction as a function of time; the model in effect predicts what reactions will occur in infinite time, without regard to activation energy barriers.

LIMITATIONS OF DIRECT EXPERIMENTATION

Let us assume that a material used to plug a hole is chemically unlike its host rock. Chemical thermodynamics dictates that mineralogical phase changes will occur, involving the interaction of components in rock and plug at ambient subsurface temperature, until chemical equilibria prevail. The interacting system of plug and rock (and perhaps groundwater) components can be treated much the same as a phase assemblage undergoing metamorphism.

The expression "chemical equilibria" implies a set of instantaneous, reversible reactions among components. In actual fact, a reaction may be neither instantaneous nor reversible, but may require a non-infinitesimal amount of time to occur. Turner [3] has described factors which tend to make many metamorphic reactions exceedingly slow:

1. The required liberation of components from reactants, diffusion of components to sites where regrouping into a transient species of high free energy (the activated complex) can occur, and nucleation of product phases.

2. The dependence of reaction rate upon temperature, which for some reactions might be approximated by the Arrhenius equation, once the activation energy barrier is overcome:

$$k = Ae^{-E/RT} \tag{1}$$

where k = reaction rate constant
A = "constant" representative of the instantaneous abundance of the activated complex
E = activation energy (an intermediate stage barrier to the reaction)
R = gas constant
T = temperature.

Notice that lower temperatures indicate lower reaction rates.

3. The exponential increase of reaction rate with reaction entropy.

4. The extremely slow rates of diffusion in the solid state.

5. The instability of a small nucleation site relative to a larger crystal.

6. The high activation energy required to disrupt crystal lattices.

7. The presence of water (or other fluid phases) which may act as a catalyst or as a stoichiometric reaction constituent to raise or lower the reaction rate.

In a general sense, a <u>reversible</u> reaction:

$$dD + bB + \ldots \rightleftharpoons pP + qQ + \ldots \qquad (2)$$

has associated with it an expression for the rate of change in concentration for one of the constituents:

$$\frac{dC_D}{dt} = K\, C_D^{\alpha}\, C_B^{\beta}\, \ldots \qquad (3)$$

The instantaneous rate of change with time in the concentration of D is directly proportional to the product of the concentrations of all the constituents each raised to a power. The sum of the powers is called the order of the reaction, and is commonly 0, 1, or 2. Some constituents, such as catalysts, may not appear in (2), but may have a non-zero exponent in (3) [4]. It must be emphasized that rate laws such as (3) and constants contained therein apply to specific reactions (or sets of serial or parallel reactions) and specific reaction paths having unique activated complexes. It may be diffi-cult to identify the interrelationships among reactions, inert constituents and catalysts. For example, in a dry system, the melting of orthoclase is incongruent at approximatey $1150^{\circ}C$ [5] :

$$\underset{\text{orthoclase}}{KAlSi_3O_8} \rightleftharpoons \underset{\text{leucite}}{KAlSi_2O_6} + \text{melt} \begin{array}{l} \text{(about 28 wt \% } SiO_2 \\ \text{72 wt \% } KAlSi_2O_6) \end{array} \qquad (4)$$

The reaction becomes congruent at 5000 bars water pressure, yet water does not appear in the reaction as written [6]. This has implied that water has taken part in an intermediate reaction (activated complex?) perhaps depolymerizing the tectosilicate structure.

Several simultaneous reactions may share one or more reactants or products, giving rise to competition, illustrated generically as follows:

$$cC + dD \rightleftharpoons mM + bB \qquad (5a)$$
$$pP + qQ \rightleftharpoons xM + rR \qquad (5b)$$

The rate law governing the disappearance of reactant Q in (5b) might entail the instantaneous concentration of product M, in the system, regardless of the source of M:

$$\frac{dC_Q}{dt} = KC_M^{\alpha}\, \ldots \qquad (5c)$$

For example, several metamorphic dehydration reactions may liberate water from several different minerals simultaneously. Any rate law in which water's exponent is non-zero signifies the dependence of reaction rate upon water from <u>all</u> sources in the system. Unless all sources are identified (and their rate laws deduced), prediction of degree of production of a given component as a function of time exceeds the limitations of the data. Experimental determination and verification of reactions, rate laws and constants are carried out in real time. Such an endeavor would consume an immense amount of time and resources for the complete kinetic characterization of even one system of rock and plug of a given bulk composition.

PREDICTIVE THERMODYNAMICS

The application of chemical thermodynamics as a tool to predict the likelihood of occurrence of reactions requires a knowledge of values of thermodynamic functions of the initial and (hypothetical)

final states of the plug/rock system. There is no consideration of
intermediate, transient or metastable phases. The power of thermodynamics lies in its ability to determine which of the hypothetical
final states can be achieved for a specific set of thermodynamic and
stoichiometric variables. Ultimately it is possible to deduce the
phase assemblage with the lowest free energy -- the "most stable"
assemblage.

The data required for predictive thermodynamics are in principle
simple to acquire. Many of the data for the reactant and product
phases of interest will have already been published. Previously
unavailable data for some phases will have to be generated by a
moderately large amount of relatively straightforward laboratory
experimentation.

Another advantage of predictive thermodynamics is its
independence of time. Once a reaction has occurred, if no lower
free energy states exist, no further reaction will take place in an
initially closed system that remains closed. Additional reactions
will be initiated only by a change in an intensive variable (bulk
composition, temperature, electromotive force, etc.).

THERMODYNAMIC MODELING

As with any modeling endeavor, the use of thermodynamics to
predict the phase assemblage of "ultimate" stability is iterative.
The following discrete steps have been specifically identified:

1. Identify potential reactant phases in plug and rock (and
    groundwater, if any).

2. Formulate all possible reactions among components of rock,
    plug and groundwater. For simplicity, the reaction
    products thus hypothesized should be phases actually known
    to exist.

3. Obtain thermodynamic data, particularly values of Gibbs'
    free energies of formation (ΔG^0) for the reactant and
    product phases.

4. Calculate ΔG^0 values for the reactions, identifying
    those for which ΔG^0 is less than 0 as being able to
    take place.

5. Evaluate reaction products which are soluble, friable,
    permeable or of decreased volume. Based on this,
    re-evaluate the constituency of the plug for long-term
    compatability with rock.

6. Formulate new reactions using products obtained in
    previous formulations and repeat data collection,
    calculation and evaluation steps until a phase assemblage
    is found in which no further changes can take place.

We must recognize that with each successive iteration of the above
steps, the confidence in the prediction diminishes.

MODIFICATIONS TO THE MODEL

The reader who is familiar with chemical thermodynamics will
recognize that the ΔG^0 value for an overall reaction is an
oversimplification. For any generalized reaction (2), the free
energy expression is:

$$\Delta G = \Delta G^O + RT \ln \frac{a_P^p \; a_Q^q \ldots}{a_D^a \; a_B^b \ldots} \tag{6}$$

In (6) the ΔG^O value refers to the standard state of all products and reactants (the pure phases). The ratio is called the activity ratio, and is unity (thus, $\Delta G = \Delta G^O$) at the standard state. At equilibrium $\Delta G = 0$, and the ratio becomes the equilibrium constant, K_{eq} so that

$$\Delta G^O = -RT \ln K_{eq} = -RT \ln \frac{[P]^p \; [Q]^q \ldots}{[D]^d \; [B]^b \ldots} \qquad \frac{\gamma_P^p \; \gamma_Q^q \ldots}{\gamma_D^d \; \gamma_B^b \ldots} \tag{7}$$

The thermodynamic activity ("a") of a component consists of the product of some measurable property (concentration, partial pressure, mole fraction, molality, etc.) and γ, the activity coefficient (a measure of the non-ideal behavior of the component in the system). Ideally, $\gamma = \underline{1}$.

There are a number of factors in a system which offset it from its standard state. First, thermodynamic functions vary with temperature, as illustrated by the definition

$$\Delta G^O = \Delta H^O - T \Delta S^O \tag{8}$$

in which ΔH^O and ΔS^O are temperature-dependent. If changing pressure is a factor, a change in volume must be accounted for, which requires a knowledge of molar volumes, thermal expansions and isothermal compressibilities for the phases. The existence of solid solutions among product and reactant phases will cause deviations from the standard (pure) state, and perhaps deviations from ideal ($\gamma = 1$) solution behavior as well. Similar deviations will occur in liquid solutions (i.e., groundwater), and deviations from ideality will be especially relevant in concentrated solutions of strong electrolytes (i.e, brines).

METHODS

Phase identification is best accomplished by a combination of x-ray diffraction, optical petrography, and electron probe micro-analysis. In this activity, an energy-dispersive x-ray spectrometer attached to a scanning electron microscope will probably not afford a sufficiently quantitative basis for the unambiguous identification of mineral phases and solid solutions.

The formulation of all the possible reactions among phases is largely an exercise in "bookkeeping." It is essential not to overlook any potentially detrimental reactions in this step of the procedure. It might be appropriate to "turn a computer loose" to generate various combinations of reactants (initial phases identified above) and hypothetical product phases. In such a "random walk," a large number of non-existant phases are likely to arise. Conversely, a number of products of interest may not appear among the reactions so generated. Any use of computer-generated reaction formulations will need to be supplemented by an inspection to (1) eliminate non-existent phases, and (2) ensure the incorporation of certain phases of interest. In the latter case, the computer may be given some reactants and products for which it must generate a reaction relating them. In any case, the total number of phases in a system is limited by the mineralogical phase rule, and use of the relationship can be made to reasonably limit the number of phases in the model.

In its most general statement, Goldschmidt's mineralogical phase rule states that

$$f = c + n - \phi \qquad (9)$$

for a given bulk composition, where

ϕ = number of phases in system
c = number of components in systems
n = number of intensive variables which may
    assume arbitrary values
f = number of degrees of freedom

If the number of phases is greater than that allowed by the phase rule, disequilibrium is indicated. Disequilibrium allows for the possibility of components in a system of grout, rock and groundwater recombining to seek a phase assemblage with a lower value of Gibbs' free energy.

Acquisition of thermodynamic data for reactant and product phases begins with a search of the literature to determine the existence of data. For some phases, calorimetric measurements will be necessary to establish values of standard-state functions. The entropy will be readily determined from the heat-capacity as a function of temperature:

$$S^O_T = \int_O^T C_p(T) \, d\ln T \qquad (10)$$

Heat capacity is measured at constant pressure over as much of the temperature range (between 0^OK and ambient) as practicable. Approximations at very low temperatures can be made by the application of the theoretical relationship of Debye. Numerous discussions of the measurement of activity and activity coefficient are given by Lewis and Randall [7] .

The most painstaking part of collecting thermodynamic data will involve the measurement of ΔH^O values (standard enthalpy of formation from the elements). This will entail a substantial amount of calorimetry to be performed, probably by solution. A detailed discussion of various kinds of calorimetry is beyond the scope of this paper.

The arithmetic calculations of free energies for reactions, even taking into consideration the effects of varying pressure, temperature, activities, activity coefficients and bulk composition, are straightforward, if time-consuming.

KINETICS OF INTERACTIONS BETWEEN PLUG AND ROCK

As argued in a foregoing discussion, the determination of time-dependence of a reaction is not to be taken lightly. It may be useful, however, to determine rates for potentially-detrimental reactions identified by thermodynamic modeling. Since it is anticipated that plug-rock interactions will be quite slow at the temperatures of interest (30 to 40^OC), it should be expected that individual kinetics experiments may require "runs" of several months to several years before measurable changes will be observed. In addition, there are certain pitfalls in the standard kineticist's "bag of tricks" which could yield misleading results.

There are a number of popular "accelerated aging" techniques which are intended either to raise the rate constant for a reaction, or circumvent a reaction of unmitigateably small rate constant. One such technique is raising the temperature, since rate constants have

approximately the Arrhenius dependence (1). Unfortunately, values
of thermodynamic functions have a temperature dependence as well.
An entirely different reaction may predominate in the system at
higher temperatures, or the reaction under study may be inhibited at
higher temperatures.

Another technique involves varying the amount of certain
reactants or products to favor the reaction. In actual fact, the
concentration dependence of a reaction (3) is generally not known
until the rate law is deduced. Variation of all the relevant
components requires a respectably large number of experiments (run
in parallel, if time is of the essence). Even with this optimiza-
tion, reactions largely in the solid state will still probably
require several months or longer for measurable changes.

In some experiments, reactants might be ground finely to reduce
the grain size, increase surface area, and encourage the reaction.
A system which is too fine-grained may be difficult to analyze
quantitatively for changes in reactant concentration, especially by
the various kinds of microscopy. In addition, there is a free energy
dependence on grain size (i.e., surface area) but more importantly,
some reactants might undergo phase changes during ultrapulverization.
For example, some silicate minerals are known to be more easily
hydrolyzeable at extremely fine grain sizes, and the pH of a coexist-
ing aqueous solution can skyrocket.

A non-trivial problem confronting any kineticist is the
monitoring of reaction progress. Benson [4] has suggested several
experimental methods, all of which have their limitations. A few
additional methods for a solid-state matrix are given here as
examples for consideration.

X-ray diffraction can be used to identify phases in an
aggregate, and the relative intensities of diffraction peaks could
be used to semi-quantitatively determine abundance changes as a
function of time for certain phases. Preparation of x-ray diffraction specimens will require either an intact slice of plug/rock
system for repeated (non-destructive)analysis, or a freshly-powdered
sample for each analysis (essentially destructive). The latent
textural and mineralogical inhomogeneities of the first preparation
will probably degrade the sensitivity of the technique for some
phases. Local inhomogeneity could be alleviated in the second
preparation, but gross inhomogeneities would require new calibration
of peak intensities for each analysis. The precision required
exceeds that which is offered by the instrument in most cases.
Regardless of the preparation, a multiphase matrix will generate
overlapping diffraction peaks, which in general cannot be resolved.

Electron and photon (e.g., x-ray) optics probably have the
precision required for determining phase abundances as a function of
time. Special requirements for specimen preparation make this
essentially a destructive test. In addition, sample homogeneity
will make difficult the comparison of analyses at different times,
since each specimen will be unique.

Resonance spectroscopy (nuclear magnetic, electron paramagnetic,
etc.) offers a reasonable prospect for required precision and sensitivity. Specimen preparation and analysis can be non-destructive.
Hydrogen NMR can be especially useful for observing the increase and
decrease in abundance of various hydrated phases in cementitious
grout, for example. A chief disadvantage, as with x-ray diffraction,
is overlap of peaks. In addition, NMR studies of minerals are in
their infancy, and a larger data base for minerals would have to be
established. Furthermore, an instrument sensitive to more than one
element could be quite costly.

SUMMARY

The long-term stability of plugging material in geological environments cannot be directly observed in a human lifetime. Consequently, an approach similar to that of safety assessment is proposed.

The plug will in general not be in chemical equilibrium with its host rock or with natural groundwater solutions which may accumulate in the rocks. Chemical reactions can be formulated from various combinations of components in grout, rock and groundwater, under assumptions made for future postulated juxtaposition of the three. The reactions will be sluggish at the ambient temperatures (less than 40°C), but calculations using thermodynamic data for postulated products and reactants can be used to identify the reactions that may ultimately take place.

ACKNOWLEDGEMENT

This work was supported by the United States Department of Energy.

References

[1] Draft Environmental Impact Statement, Waste Isolation Pilot Plant, DOE/EIS-0026-D, 1979, United States Department of Energy.

[2] Bingham, F. W. and G. E. Barr, 1979, Scenarios for long-term release of radionuclides from a nuclear-waste repository in the Los Medanos region of New Mexico: Sandia National Laboratories, Albuquerque, New Mexico, SAND78-1730.

[3] Turner, F. J., 1968, Metamorphic petrology; mineralogical and field aspects: McGraw-Hill Book Company, New York.

[4] Benson, S. W., 1960, The foundations of chemical kinetics: McGraw-Hill Book Company, New York.

[5] Bowen, N. L., 1928, The evolution of the igneous rocks: (reprinted 1956, Dover Publications, Inc., New York).

[6] Turner, F. J., and J. Verhoogen, 1960, Igneous and metamorphic petrology: McGraw-Hill Book Company, New York.

[7] Lewis, G. N. and M. Randall, 1961, Thermodynamics (revised by K. S. Pitzer and L. Brewer): McGraw-Hill Book Company, New York.

Discussion

J.C. WRIGHT, United States

     I agree with your model and feel that it is needed.
However, the interplay between physical and geochemical programs
appears to be missing. What kind of time frame are you talking
about to complete the geochemical model for a single, largely
anomalous, interaction system between cement plug and host rock ?

S.J. LAMBERT, United States

     There must be an interaction. Unfortunately, the geochemist
is left with the materials to study long after the field operations
have ended, and commonly the geochemist's results lag far behind
field results, because geochemical investigations require more time.
The interaction is no less important for the people interested in
physical properties than for those interested in chemical properties.

     The time required to thermodynamically model a single
plug-rock-ground water system, including the acquisition of thermodynamic data, is on the order of 5 years ; variations may depend
upon the complexity of the system and the resources available.

N.A. CHAPMAN, United Kingdom

     Do you think sufficient thermodynamic data are presently
available for this type of modelling and, if not, how much effort
do you think ought to go into pure research to obtain them. Also,
what is your opinion of the usefulness of existing thermal models
such as PATHCALC and EQ 3/6 ?

S.J. LAMBERT, United States

     The effort required for obtaining new thermodynamic data
is dictated by the choice of reactions to be modeled. One could
conceivably constrain the formulation of hypothetical reactions to
produce simple products for which data are easily acquired.

     I am not prepared to evaluate the various existing
computer codes. The actual numerical calculations may be the easiest
and most satisfying part of this endeavor. The non-trivial parts
determine which reactions to model, and anticipate which reactions
will be of concern or consequence.

GEOCHEMICAL FACTORS IN BOREHOLE AND SHAFT PLUGGING MATERIALS STABILITY

D. M. Roy
Materials Research Laboratory
The Pennsylvania State University
University Park, PA 16802, U.S.A.

ABSTRACT

Factors affecting the performance and potential longevity of cementitious borehole and shaft plugging materials are being examined. Materials being studied are those which have high potential of approaching thermodynamic stability or, if not achieving full stability, then undergoing changes with time, which would not compromise their properties. Laboratory studies of materials include studies of their mechanical, physical, thermal, microstructual, and rheological properties. Thermodynamic studies to determine the most stable states, and kinetic and diffusion studies are in progress, which include detailed characterization of products from experiments. Insights are also sought from analyses of cements and concretes.

1.0 Introduction

The development of adequate technology for repository sealing requires the availability of sealing materials which fulfill certain requirements.[1]   1) The plug or seal materials should be mechanically adequate, bond well to the surrounding rock, and should not undergo significant volume change with time, maintaining very low permeability; 2) they should be tailored for compatibility with the surrounding rock; 3) they should be resistant to destructive expansion and contraction, assuring low permeability to circulating fluids to minimize possible transport of radioactive (RW) species; 4) they should be stable and chemically durable in a selected geochemical environment over the time frame deemed necessary for effective isolation; and 5) they should maintain desirable retardation to radionuclide transport over long time periods.

It is likely that cement-based materials either alone or in combination with other compatible plug component materials, will ultimately satisfy these requirements. To achieve these objectives, the research described here deals with such materials, first by a materials selection and evaluation process, which addresses the materials characteristics relevant to placing and hardening; early stage properties which include mechanical, physical, and thermal; and the necessary characterization of starting materials and products. Then the geochemically related factors that impinge upon longevity of plugging and sealing materials, over the extraordinarily long time period required, on the order of thousands of years and longer are treated. The latter studies include detailed characterization of some selected potential plugging materials; evaluation of thermodynamic stability; determining kinetic factors; selection criteria using knowledge of natural mineral associations and from studies of ancient materials; and determining potential stability of material host rock/environment combinations.

2.0 Technical Studies

The interrelationships among the types of information needed for the seal materials evaluation and selection, and to address also the problems concerning longevity are further categorized in Figure 1.

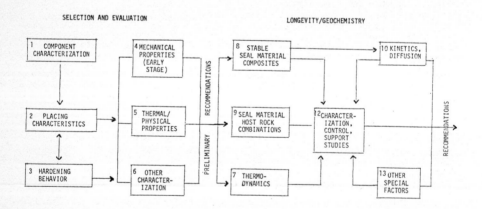

Figure 1

Interrelation of factors concerned with borehole and shaft plugging materials selection and geochemical stability.

2.1. Component Characterization

These studies, which are discussed in greater detail elsewhere [23], involve the characterization necessary to assure that the materials (typically commercial) which will constitute components of an actual field mix, may be verified with respect to all significant factors which affect the reproducibility of their performance. They include chemical composition, particle size distribution, variability within a component, and sensitivity to changes taking place during storage. In addition to the usual standards and characterization methods, these studies will draw from a broad range of techniques to define precisely the permissible materials component variables, based upon an understanding of how these affect the early stage materials properties and potential longevity. Thus, they will draw as well upon information gained through the studies reported in the following sections.

2.2. Placing Characteristics

Optimal borehole plugging and shaft sealing materials will normally be those having lowest porosity and negligible permeability. In order to assure the ability to remotely place such materials having optimal properties, and even to assure superior properties of such materials when placed by normal methods, it is necessary to define the factors which affect early stage rheological properties. Use of chemical admixtures which enable reduction of the water content of mixes and therefore generate lower porosities has been part of these investigations, carried out to generate empirical data as well as understanding of the operating mechanisms.

Systematic studies of rheological properties of various mixes have been carried out. [2-4]. Figure 2 illustrates the increase in workability of of cementitious mixes, all having the same water/solids ratios, upon addition of small concentrations of water-reducing chemical admixtures, in this instance a sulfonated napthalene formaldehyde condensate. Its use generates lower porosities and permeabilities in the final products formed from mixes which are still workable or pumpable. From such viscometric relationships one is able to derive two principal features: the yield stress on the horizontal axis, that is the basic stress applied before movement takes place in the mix, and the viscosity. The latter was usually calculated from the following relationship: $\eta = \tau - \tau_0 / \dot{\gamma}$ where η = plastic viscosity, τ = measured stress, τ_0 = yield stress and $\dot{\gamma}$ = shear rate. It is noted that the yield stress becomes minimal with addition of sufficient concentration of admixture, and that the viscosity behavior may become almost Newtonian.

Since the goal is generating materials having optimal properties and assuring reliable emplacement, it is useful to express the results of such studies in terms of concentration of cement solids, as shown in Figure 3. A

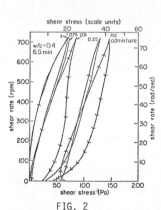

FIG. 2

Shear stress-shear rate relationship as a function of admixture A4 concentration, w/c = 0.4.

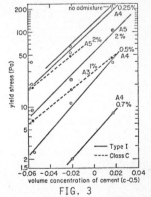

FIG. 3

Changes in yield stress with volume concentration and type of cement, and different admixtures; increasing shear rate, beginning from time 6.5 minutes.

basic mathematical relationship between yield stress and volume concentration of the solid phase of the following type: $\tau_0 = A_0 (V_c - 0.5)$ was verified in the experimental data of Figure 3. This illustrates that the use of increasing concentrations of a chemical admixture, by decreasing the yield stress makes the mix behave as though it contained a lower volume concentration of solid. This phenomenon is reflected in the shifting of the curves to form essentially parallel lines. Such relationships are used to design proper mixes. A comparable relationship is shown in Figure 4, where viscosity and volume concentration of cement are similarly related. This was mathematically expressed as follows: $\eta = B_0 \exp (K_1 V_c + K_2)$.

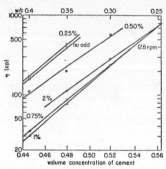

Additional knowledge of the mechanisms operating in controlling flow properties of cementitious mixes was gained from the study of electrokinetic phenomena [5,6]. The dispersion of particles to generate fluid mixes is dependent upon surface effects. The generation of a highly negative zeta potential upon cement particles through the addition of superplasticizing admixtures, was found to cause this dispersion, and to produce changes which parallel those more empirically reflected in flow properties. Figure 5 shows the effect of admixture concentration upon the zeta potential of a particular cement: while parallel effects of admixture concentration upon mortar flow were shown in Figure 6. The importance of the character of the mixing water is also illustrated in

FIG. 4

Effect of volume concentration on viscosity; measurements after 6.5 minutes.

Figure 5, which shows the type of variability which may be expected due to impurities. Thus, the dispersing effect of admixtures is apparent in both zeta potential and flow measurement.

2.3. Hardening Behavior

The hardening behavior of candidate mixes is investigated in sufficient detail to assure that neither unanticipated premature setting nor excessive delay in hardening takes place which would harmfully affect the performance of the hardened materials. The relevant knowledge may be generated through the viscometric studies [2-4] such as described in the previous section 2.2, especially when they are carried out as a function of time and temperature. Vicat setting time determinations have also been made, [7] as shown in Figure 7.

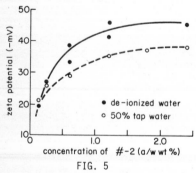

FIG. 5

Change in zeta potential of Type I portland cement with increasing concentration of admixture

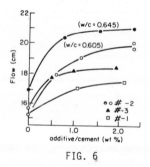

FIG. 6

Change in Mortar Flow with Admixture Type and Concentration.

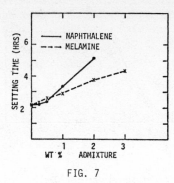

FIG. 7

Effect of type of water-reducing agent upon initial setting times of Type I cement as a function of admixture concentration.

In this particular case, contrast is shown between the setting time in mixes utilizing two different types of chemical admixtures.

Other methods used include investigation of the early heat evolution characteristics of cement slurries or mixes, with measurements which are highly quantitative. The results with a particular mix are shown in Figure 8. Characteristically the calorimetric curves show two heat evolution peaks, one strong peak during the first minutes, and the second after some hours. This particular set of data illustrates the strong effect of temperature in accelerating reaction, the consequent heat evolution and hence hardening. (Initial setting corresponds to the onset of rapid heat evolution with the start of the second peak). Such measurements are also useful for detecting changes in manufacturing which might not otherwise be evident, but having potential for significantly modifying material properties.

2.4 Mechanical Properties, Early Stage

Studies of the earliest stage volume change in cementitious mixes, prior to, during setting and just post-setting have been carried out, for they not only generate initial knowledge of the consolidation process but are important in predicting later stage mechanical properties. Excess shrinkage or expansion at this early stage may have deleterious effects upon the longer term mechanical properties. Bulk chemical-conditioned volume change has been measured by a technique in which partial removal of early bleeding water may be carried out [8]. Under such conditions, volume shrinkage of pastes was found to range from about 0.6 to 1.0% at 20 hours age. These conditions may simulate placement of a column of material from which there is some segregation of water, which cannot be completely recombined. Figures 9-11 illustrate some of the types of difference measured. Figure 9 shows the differences in magnitude of cumulative volume change observed with mixes having different water-to-cement (w/c) ratios. Figure 10 contrasts the rates of volume change for two mixes. One mix incorporated a superplasticizing admixture and the other was without. The effect of the admixture was seen in the delay in the realization of maximum shrinkage. Finally, Figure 11 illustrates the difference between pure cement paste and those incorporating two different filler materials.

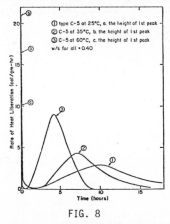

FIG. 8

Rate of Heat Liberation vs. Time for Class C (C-5) Cement Slurries.

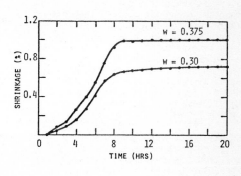

FIG. 9

Bulk Chemical Shrinkage of Cement Paste with Different w/c Ratio

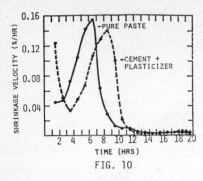

FIG. 10

Shrinkage Rate of Cement Paste With
Without Plasticizer

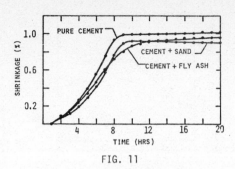

FIG. 11

Bulk Chemical Shrinkage for Pastes
Containing a Filler (w/c = 0.375).

The volume change becomes smaller with increasing proportions of inert filler.
Currently studies are in progress with an apparatus to study similar changes at
elevated temperatures and pressures, more closely simulating expected repository
conditions.

2.5. Thermal/Physical Properties

Thermal properties of seal materials are important in a number of ways:
at depth the materials will be subjected to temperature elevation from the geothermal conditions, plus at least local temperature variations due to waste loading of the repository. When subjected to temperature elevation, cementitious materials undergo thermal expansion which is characteristic of most materials [9],
upon which is superimposed the complexities of creep and shrinkage due to moisture
movement, and additional complexities resultant from the nature of the materials
in which aggregate and matrix possess different thermal behavior. Thermal expansion studies are being conducted in order to examine effects of variables such as
proportions of components, curing time or age, and curing temprature. A typical
thermal expansion curve of a cement paste is shown in Figure 12; also superimposed
is a thermogravimetric analysis curve showing the weight loss over a similar interval. The thermal expansion curves are characterized by an expansion up to approximately 100°C (depending upon the particular mix and curing procedures) followed by a usual contraction due to water loss. The details and magnitudes may
be modified; however, mixes containing large proportions of aggregate assume more
nearly the characteristics of the aggregate and exhibit more steady expansion.

2.6 Other Characterization: Permeability

The permeabilities of seal
materials constitute one of their most
important properties, since the most
likely path for release of radioactive
materials from the repository is by
fluid transport. Thus is it necessary
to minimize seal material permeabilities, and to use the measurement as an
essential routine characterization
method. Permeabilities to both liquid
and gas have been measured [10], and
for the latter the Klinkenberg extrapolation [11] has been used to convert
the data from measurements at several
different gas pressures to an equivalent liquid permeability. Some typical
extrapolated liquid equivalent gas
permeabilities are given in Table I
for a series of different fly ash-
containing mixes, with fly ash contents, curing times, temperatures, and

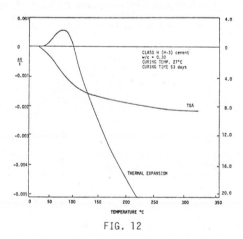

FIG. 12

Thermal Expansion and TGA - Class H
Cement

Table I. Mixes containing fly ash

sample	14 days	28 days	2 months
gas permeability			
1A-C-1-0.73-FA-0.33-OT-Ca(OH)$_2$-RT	1.30×10^{-5}	8.50×10^{-6}	9.50×10^{-6}
1B-C-1-0.73-FA-0.33-OT-Ca(OH)$_2$-60	1.50×10^{-8}	3.30×10^{-6}	2.45×10^{-6}
2A-C-1-0.73-0.33-1T-Ca(OH)$_2$-RT	2.00×10^{-5}	8.25×10^{-6}	2.30×10^{-6}
2B-C-1-0.73-FA-0.33-1T-Ca(OH)$_2$-60	6.00×10^{-7}	2.30×10^{-6}	8.20×10^{-6}
3A-C-1-1-FA-0.33-OT-Ca(OH)$_2$-RT	1.75×10^{-5}	1.05×10^{-5}	1.15×10^{-5}
3B-C-1-1-0.33-OT-Ca(OH)$_2$-60	3.30×10^{-6}	2.50×10^{-6}	2.75×10^{-6}
4A-C-1-1-FA-0.33-1T-Ca(OH)$_2$-RT	1.35×10^{-5}	1.15×10^{-5}	7.00×10^{-6}
4B-C-1-1-FA-0.33-1T-Ca(OH)$_2$-60	1.35×10^{-6}	3.40×10^{-6}	4.60×10^{-6}

conditions as variables. The results for most of the samples fell within the desired permeability range ($<10^{-6}$ Darcy). The gas measurement method is somewhat limited because of some increase in permeability due to drying of the samples. Thus, for evaluating materials potentially used in contact with aqueous solutions, liquid permeability measurements are more useful, and the values thus obtained are usually lower, as will be discussed later.

2.7 Thermodynamic Properties

Up to this point we have examined factors important in designing seal materials and evaluating their overall performance. The following section considers those factors more directly concerned with longevity, of which a major consideration involves the relative thermodynamic stabilities of the materials and their components. Portland cement-based materials hydrate relatively slowly at room temperature and atmospheric pressure, their short-term reaction products are poorly crystalline, and the changes that take place under normal conditions to increase the degree of crystallinity are usually slow. A rigorous method for predicting long-term behavior of cement-based materials in a geologic environment involves considering thermodynamic equilibria, and calculating the lowest free-energy (most stable) state for a given compositional system or subsystem, under the expected environmental conditions. Such studies are in progress which utilize existing data and approximations in performing these calculations, and also involve generating new data to increase the accuracy of the results. These have begun with room temperature conditions, but are extending to consider elevated temperatures and pressures.

The free energy of formation for each candidate phase, or free energy of reaction between sets of phases is the principal quantity required to assess their relative stabilities. The Gibbs free energy of formation (at $T_0 = 298.15K$, 1 atm pressure, standard state) is given by the following relationship:

$$\Delta G^o_{f,T_0} = \Delta H^o_{f,T_0} - T_0 \Delta S^o_{T_0}$$

If the free energies of formation (derived from the enthalpies of formation $\left(\Delta H^o_{f,T_0}\right)$ and entropies of formation $\left(\Delta S^o_{T_0}\right)$ are known for each candidate phase, then the most stable individual phase or combination of phases (from a given total chemistry) can be determined (i.e., that combination which has the most negative free energy of reaction).

One example is given in the following comparisons, beginning with the major portland cement component, C_3S[1] to follow the most favorable hydration reactions:

---

[1] Cement abbreviations: $C=CaO$, $S=SiO_2$, $A=Al_2O_3$, $F=Fe_2O_3$, $H=H_2O$

$$C_3S + 3H_2O = \frac{1}{2}\,(Ca_3Si_2O_7 \cdot 3H_2O) + \frac{3}{2}\,Ca(OH)_2$$

$$\text{(gel, non-crystalline)} \tag{1}$$

$$\Delta G = -16.0 \text{ Kcal/mole } C_3S \text{ [12]}$$

Optional hydration reactions to generate crystalline products, and hence more negative ΔG's include:

$$C_3S + 2.17H_2O \rightarrow 2CaO \cdot SiO_2 \cdot 1.17H_2O + Ca(OH)_2$$

$$\text{(hillebrandite, crystalline)} \tag{2}$$

$$\Delta G = -25.87 \text{ Kcal/mole } C_3S$$

$$C_3S + \frac{18.5}{6}\,H_2O = \frac{1}{6}\,(5CaO \cdot 6SiO_2 \cdot 5.5H_2O) + \frac{13}{6}\,Ca(OH)_2$$

$$\text{(tobermorite, crystalline)} \tag{3}$$

$$\Delta G = -24.14 \text{ Kcal/mole } C_3S$$

In addition, if silica (quartz)[2] is added to the system to react with excess calcium hydroxide, the following reaction applies:

$$5Ca(OH)_2 + 6SiO_2 + 0.5H_2O \rightarrow 5CaO \cdot 6SiO_2 \cdot 5.5H_2O$$

$$\text{(tobermorite, crystalline)} \tag{4}$$

$$\Delta G = -36.30 \text{ Kcal/mole tobermorite} \tag{4.1}$$

or

$$\Delta G = -15.73 \text{ Kcal/}\frac{13}{6} \text{ mole } Ca(OH)_2 \tag{4.2}$$

Adding (3) and (4.2), for the overall reaction in which all the calcium hydroxide formed is consumed:

$$C_3S + \frac{13}{5}\,SiO_2 + \frac{16.5}{5}\,H_2O = \frac{3}{5}\,(5CaO \cdot 6SiO_2 \cdot 5.5H_2O)$$

$$\Delta G = -39.92 \text{ Kcal/mole } C_3S \tag{5}$$

The values listed in Table II were used for the calculations.

The series of calculations in Table II therefore show that the addition of excess silica to cement composition is likely to generate a more thermodynamically stable state in the products; hence, this factor has been incorporated in the design of experimental seal material compositions.

2.8 Stable Seal Material Composites

The major objective of these studies is to generate seal materials which approach, or have the potential of approaching thermodynamic stability, and at the same time have suitable physical properties. Two different sets of materials were examined with this objective in mind: one group has been designed for field tests in an evaporite formation[3]

Table II. Gibbs Free Energies of Formation of Relevant Phases (Kcal/mole)

$Ca(OH)_2$	-213.77 [13]
H_2O	-56.68 [13]
SiO_2 (quartz)	-204.66 [13]
β-C_2S	-516.6 [12]
C_3S	-657.8 [12]
$5CaO \cdot 6SiO_2 \cdot 5.5H_2O$	-2361.45 [14]

[2] The ΔG of reaction becomes more negative if amorphous silica, rather than quartz is added as a compound.

[3] USAE Waterways Experiment Station and Sandia personnel worked jointly to design this mix for use in the Bell Canyon Test [15].

and the second set was part of a preliminary exploratory group which was more generically designed. Viscometric measurements were used as a basis for setting lower water content limits that could be used and still produce workable mixes.

Two of the mixes are given in Table III. Number 2 is a mix formulated with the addition of slag and quartz to generate higher-silica compositions, with the goal of attaining an ultimate lower free energy state, while the other, number 1, was the field test expansive cement mix optimized rheologically with the addition of fly ash and also possessed added silica, but less than in number 2. Mix 2 was optimized from chemical considerations (the use of the finer Class C cement was arbitrary; the coarser H would be more favorable rheologically). Continuing studies are in progress to optimize both chemical and early stage mixing parameters. The typically high compressive strengths are evident in the Table IV data. The longer-time higher-temperature samples showed an increase in strength even in the higher w/c ratio (0.45) slag formulation.

Table III. Mix Formulations

Component % in Mix	BCT-IFF 1	Slag + cement + quartz 2
Class C (C-5)	–	35.15
Class H (H-4)	68.0	–
Fly Ash (B15)	22.9	–
Slag (B19)	–	29.7
Quartz (B5)	–	35.15
Expansive Additive A (A29)	9.1	–
Deionized Water (E1)	30.0	45.0
Dispersant (A28)	0.29	–

Table IV. Compressive Strengths of Mixes from Table III

Mix No.	W/S	Curing Temp °C	Time (da) [Strength (MPa)] 7	28
1	0.30	27°	[76]	[92]
1	0.30	90°	[52]	[53]
2	0.45	27°	[45]	[64]
2	0.45	90°	[57]	[94]

Detailed properties were determined for these samples, along with similar data for a larger number of mixes, which are described in a separate paper: [16] permeabilities, microstructures, x-ray diffraction characterization of solids to determine phases present, volume change, and microhardness are discussed. Among the additional mixes studied are compositions utilizing finer grained amorphous silica sources to react with much of the calcium hydroxide as it is generated from C3S hydration, different water-reducing admixtures and other additives selected for the purpose of determining the effects that different particle size distributions in the components have on the properties of the products.

2.9 Seal Material/Rock Combinations

In addition to the properties of the seal materials as such, knowledge of the relative thermodynamic stabilities of seal materials in probable geologic repository environments is important. These materials should be tailored to the host rock to the extent necessary to assure that any changes engendered in the seal material/rock system with time will not compromise their plugging performance. Thus, both thermodynamic stabilities and physical properties evaluations should be considered in determining very long term seal material performance.

If the chemistries of candidate materials are relatively compatible with those of the host rock and fluids, the likelihood of achieving adequate longevity is high. Table V shows in general terms gross chemical comparisons between plug/seal material constituents (portland cement-based) and typical host rocks being considered for waste repository sites. Fortunately, the major oxide components of hydraulic cements are shared with the constituents of basalts, granites and shales, although there are obviously differences in each. This fortuitous compositional coincidence suggests that it should be feasible by compositional adjustments to assure general compatibility of cement-based sealants with a variety of rock types. The comparison with the chemistries of evaporites is not as straight forward, for the major cement components are combined in minor silicate constituents of evaporites, while the major components of evaporites are halides, sulfates and carbonates, the anions of which combine with the cations

Table V. Cross Chemical Comparisons: Constituents of Potential Plug Material
Components and of Typical Host Rocks

	Major Components				Minor Components
		Hydraulic Cements			
Oxides	Ca	Al	Si	Fe^{3+}	Mg, SO_4, Na, K
			Basalt		
"	"	"	"	Fe^{2+}	Na K
			Granite	Mg	
"	"	"	-	K, Na	Mg Fe^{3+}
			Shale		
"	"	"	"		Mg, Na, K, S, carbonate

	Minor Components				Major Components
		Salt/Evaporites			
Silicates	Ca	Al	Si	Fe Mg	(Na K Mg (Ca)) (Halides, Sulfides, Carbonates)

which are minor in the cements. Nevertheless, it should be possible to design
compatible combinations, as there is strong evidence from examining cementitious
mixes aged in place where they plugged boreholes in evaporites, that the phases
have not changed substantially or degraded their integrity with time (see Section
2.11, and Ref. [17]). In addition, they are likely to be strong candidates for
sealing formations adjacent to evaporite horizons.

Studies are in progress to evaluate the effectiveness of a selected
cement composite/rock system. Measurements have been made of permeabilities of
cored rock cylinder samples containing the seal material cast as the core. Table
VI gives water permeability data for two sets of samples (one, BCT-IFF is mix #1
of Table III). The data include permeabilities measured of the seal material itself,
the rock alone, and finally the seal material/rock couple after different curing
conditions and durations. Although the permeability of a "couple" generally has
been found to be greater than either of the individual phases, in the measurements
shown here the cement/quartzitic sandstone couple has a low permeability comparable
to that of the rock itself. The cement/anhydrite couples similarly possess low
permeabilities which, if not quite as low as the rock, nevertheless are very low.

The bond strength values (also reported in Table VI, measured from push-
out tests) increase with curing time for both rock types, and are considered to be
very adequate. In the case of the quartitic rock, the relatively high bond strength
may be due to chemical interaction at the interface; while with the anhydrite, the
bond strength is likely to be the result of increased formation of the expansive
component, ettringite in the cement, causing a greater stress to be exerted by the
cement against the rock wall. In either event, if the bond is strong and the
permeability very low, satisfactory retardation of fluid flow is to be expected.

2.10 Reaction Rates, Diffusion, Kinetics

Processes taking place at the
interface or zone of contact between a
plugging material and the host rock, or
through interconnecting pores in the
solid are important considerations for
borehole plugging. The former have been
discussed in some detail separately [18,
19], while the latter are the subject of
a recent report [20]. It is necessary to
know not only the processes taking place,
but also the rates at which they happen,
and the factors which cause variations
in these rates, in order to project future
plug performance.

Table VI. Water Permeabilities and Bond Strengths of Cement/Rock Couples

	Permeabilities (Darcys)	Bond Strength (MPa)
BCT-IFF 4 weeks, 35°C	$<2.7 \times 10^{-8}$	
Anhydrite core	$<2 \times 10^{-10}$	
Cement/Anhydrite 2 weeks, 35°C 4 weeks, 35°C	$<7 \times 10^{-6}$ $<2.5 \times 10^{-7}$	7.1 14.2
Quartzite	1.6×10^{-7}	
Class C cement + slag + fly ash 2 weeks, 60°C	1.7×10^{-6}	
Cement (Quartzite) 2 weeks, 60°C 4 weeks, 60°C	---- 2.2×10^{-7}	6.3 10.0

The factors which affect diffusion of ions through a solid include the following: chemical composition, electrical potential, chemical reaction, porosity (including texture), solution concentration, pressure and temperature. A cementitious composite typically behaves as an electronegative semi-permeable membrane, the permeability depending upon the initial preparation parameters; but the electronegative character suggests that anions should diffuse faster than cations. Investigations are currently in progress to determine diffusion rates of salts, keeping certain of the above parameters constant while others are being varied. The results may be used to help define the effectiveness of cementitious plug materials utilized in contact with salts and brines.

Typical examples of the data obtained for room temperature atmospheric pressure diffusion of sodium and chloride ions under the potential of a concentration gradient through a relatively porous (7-day cured, w/c=0.45 paste) cementitious material are shown in Figure 13. It is seen that the diffusivity of the sodium is approximately one and one-half orders of magnitude smaller than that of the chloride ion. The diffusion rate in $cm^2 sec^{-1}$ is calculated from the following relationship:

$$D = \frac{\Delta Q_{mt} (mol \cdot cm^{-2})}{\Delta t (sec)} \cdot \frac{\xi (cm)}{\Delta C (mol \cdot cm^{-3})}$$

where Q is the total amount of diffusing ion, Δt = time, ΔC is differential concentration and ξ is sample thickness.

In other studies, electron microprobe profiles of interfacial zones across cement/rock boundaries [18] and detailed scanning electron microscopic studies of interfaces [19] have been carried out to determine extent and type of reaction.

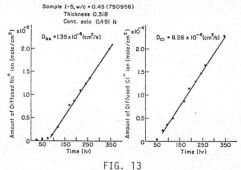

FIG. 13

Diffusivity of NaCl through Type I Cement (I-5), w/c = 0.45.
Modified (degassed for 15 min.) API mixing procedure.

2.11 Other Special Studies

Within this category are the 1) examination of ancient cements, mortars and concretes, ranging from specimens from Roman and Greek antiquity to more modern representatives to discern the reasons for their superior performance through the years; and 2) the examination of the effect of microorganisms upon cementitious materials. The latter involves a current literature study, while the former is a continuing effort.

A segment of a cementitious plug emplaced some 17+ years ago in salt strata of the Salado formation, in a potash mine near Carlsbad, NM was made available to us by Sandia Laboratories. Although the cement itself was a rather routine material, it was found to retain its integrity in a reasonable manner. The objectives of the study of this material were to 1) determine the mineralogy of the cementitious plug, the immediately adjacent host rock and the interfacial region; 2) determine mechanical and physical properties of the hardened plug material and 3) characterize the interface surface, and determine if possible the nature of the chemical and/or physical interaction across this surface.[17] A summary of some of the information gained is given in Tables 7 and 8, the former giving the mineralogical, petrographic and scanning electron microscopic characterization data for the plug and rock, and the latter physical, and mechanical properties, and chemical analytical data.

The plug itself was a rather routine, high water-to-cement ratio salt-saturated plug which by means of its prior saturation was apparently able to maintain its state of equilibrium with the surrounding rock. The mineralogy of the plug indicated that the pores were saturated with salts, and the presence of Friedel's salt is characteristic of cements containing significant amounts of alumina, which also contain large concentrations of chloride. Although, because of its high w/c ratio the (gas) permeability of the plug and interfacial region were higher than that of the salt itself, the physical integrity appeared adequate. Thus the results suggested that the plug, while adequate, could be considerably improved using new technology.

3.0 Discussion and Conclusions

As part of the development of potential repository borehole plugging and shaft sealing cementitious materials, new material mixes have been formulated and investigated, partly in collaboration with other workers.[5] The use of chemical admixtures has been investigated in detail, for the purpose of generating lower water content and hence lower porosity and minimal permeability materials, expected to be durable. Rheological properties in the fresh state, of these materials have been investigated for use in designing mixes, assuring that various combinations of additives are compatible and will perform in an

TABLE VII. Minerals Identified in 1. Host Rock (Bedded Evaporites) and 2. Cementitious Borehole Plugs, and Methods of Identification

1. Rock

mineral name	chemical composition	methods of identification
gypsum	$CaSO_4 \cdot 2H_2O$	(?) DTA
halite	NaCl	SEM-EDX, op.mic., XRD, DTA
illite (group)	$KAl_4Si_7AlO_{20}(OH)_4$	SEM-EDX, XRD, DTA
kainite	$KMgClSO_4 \cdot 11H_2O$	SEM-EDX, op.mic.
langbeinite	$K_2Mg_2(SO_4)_3$	SEM-EDX, XRD, DTA
magnesite	$MgCO_3$	SEM-EDX, XRD, op.mic., DTA
polyhalite	$Ca_2K_2Mg(SO_4)_4 \cdot 2H_2O$	SEM-EDX, op.mic., DTA
syngenite	$K_2Ca(SO_4) \cdot 2H_2O$	SEM-EDX, op.mic.

2. Cementitious Plug

(calcium silicates and chlorides)		SEM-EDX
C-S-H	$(CaO)_{varies}SiO_2(H_2O)_x$	XRD
Friedel's salt	$[Ca_2Al(OH)_6][Cl \cdot 2H_2O]$	XRD
halite	NaCl	XRD
portlandite	$Ca(OH)_2$	XRD, DTA
sylvite	KCl	SEM-EDX, XRD

op.mic. -- optical microscopy, transmitted polarized light
XRD -- x-ray diffraction
SEM-EDX -- scanning electron microscopy, energy-dispersive x-ray analyses

TABLE VIII. Physical and Mechanical Properties of Borehole Plug, and Permeability Values for Host Rock and Interface

A.	Bulk density	1.45 g/cm^3
B.	Helium density	1.78 g/cm^3
C.	Porosity	18.5%
D.	Compressive strength	2.59 MPa (375 psi)
E.	Solubility	∼2 gm/l in deionized water @ room temperature

F. Elemental analysis of filtrate: (ppm)

Cl*	575	CaO	58
Na_2O	310	Al	< 1
K_2O	216	Mg	< 1
SiO_2	139	*gravimetric; all others atomic abs.	

G. Elemental and XRD analysis of filtrate solids:

Elemental (SEM/EDX)	XRD
Cl	NaCl
Ca	KCl
Na	$CaCO_3$
Si	Friedel's salt $[Ca_2Al(OH)_6][Cl \cdot 2H_2O]$
S	C-S-H
K	
Fe	

H.	Permeability: (gas)	salt	∼2 x 10^{-6} (darcies)
		cement	∼1 x 10^{-3}
		interface	∼3 x 10^{-3}

---

[5] Some mixes were either initiated by USAE Waterways Experiment Station and Sandia Laboratories personnel or modified from such formulas. Many others were new experimental mixes.

optimal manner during placement. Early volume change of the materials has been investigated, and determination of degree of sensitivity to proportioning of materials, water contents, and additives has been of particular concern. Calorimetric measurements of heat evolution during hardening have been made, both for quality assurance of materials components, and as the data are indicative of early hardening behavior of the mixes. Thermal expansion studies have been carried out to determine the magnitude of effects anticipated upon variation in other materials parameters. Water (and gas) permeabilities are determined as a major diagnostic measurement indicative of the effectiveness of the seal materials to minimize fluid transport, and hence restrict possible radionuclide transport.

Thermodynamic properties of cementitious materials and mineral phases from host rocks are under investigation, and current calculations have been used to predict the potential for thermodynamic stability of some cementitious seal materials in the projected environments. This information has been used in the design of potentially stable seal materials; and examples have been given of some of the materials prepared, and their properties evaluations relevant to long-term performance. Effects of temperature and pressure on both rate of phase formation and upon materials properties are studied, as they are relevant to seal materials longevity. Compressive strengths are determined as a major diagnostic measurement, in conjunction with microhardnesses of small samples or portions of samples. Nondestructive mechanical properties evaluations are also carried out using resonance method elastic moduli determinations, which can be repeated on the same samples at successive time intervals. X-ray diffraction characterizations of crystalline components of seal materials and host rock components are made, along with microstructural studies (including chemical composition determinations) of detailed regions such as interfaces between seal materials and host rock. These are used in combination with mechanical and physical properties in evaluation of likely changes in the seal material-rock combinations with time. Detailed investigations are made of ancient cements, mortars and concretes as available, and more recent specimens of plug materials, for diagnosis of their long-term performance. Ionic diffusion measurements through seal materials are carried out in order to determine the rate of possible transport of certain ionic species, as affected by different conditions, including ionic concentrations, temperature, and seal material parameters.

The overall results point to the versatility of cementitious seal materials, and the likelihood that suitable materials design modifications based upon sound experimental and theoretical considerations will enable their adequate performance in repository penetration sealing, either alone or in combination with other seal materials.[24]

References

[1] Roy, D.M., M.W. Grutzeck, and P.H. Licastro, Evaluation of Cement Borehole Plug Longevity, A Topical Report, ONWI-30, Feb. 1979 (35 pp.).

[2] Roy, D.M. and K. Asaga, Rheological Properties of Cement Mixes. III. The Effects of Mixing Procedures on Viscometric Properties of Mixes Containing Superplasticizers, Cem. Concr. Res. 9(6) 731-40 (1979).

[3] Asaga, K. and D.M. Roy, Rheological Properties of Cement Mixes: IV. Effects of Superplasticizers on Viscosity and Yield Stress, Cem. Concr. Res. 10, 287-295 (1980).

[4] Roy, D.M. and K. Asaga, Rheological properties of Cement Mixes: V. The Effects of Time on Viscometric Properties of Mixes Containing Superplasticizers; Conclusions, Cem. Concr. Res. 10, 387-394 (1980).

[5] Daimon, M., D.M. Roy. Rheological Properties of Cement Mixes: II. Zeta Potential and Preliminary Viscosity Studies. Cem. Concr. Res. 9, 103-110 (1979).

[6] Roy, D.M., M. Daimon and K. Asaga, Effects of Admixtures upon Electrokinetic Phenomena during Hydration of C_3S, C_3A and Cement, 7th Intl. Congress Chem. Cement, Paris, 1980 (in press).

[7] Roy, D.M., M. Daimon, B.E. Scheetz, D. Wolfe-Confer, K. Asaga. Role of Admixtures in Preparing Dense Cements for Radioactive Waste Isolation, pp. 461-467, in Scientific Basis for Nuclear Waste Management, Vol. 1; Ed. G. J. McCarthy. Proceedings of the Symposium on "Science Underlying Radioactive Waste Management," Materials Research Society Annual Meeting, Boston, MA, November 28-December 1, 1978.

[8] Roy, D.M. and N. Setter. Chemical Shrinkage in Cement Pastes. Cem. Concr. Res. 8, 621-632 (1978).

[9] Sullivan, P.J.E., The Effects of Temperature on Concrete, Chap. I in Developments in Concrete Technology I, F. D. Lydon, Ed., Appl. Sci. Publ, Ltd., London (1979).

[10] Scheetz, B.E., D.M. Roy, E.L. White and D. Wolfe-Confer, Comparison of Tailored Cement Formulation for Borehole Plugging in Crystalline Silicate Rocks and Evaporite Mineral Sequences, Proc. Symp. on "Science Underlying Radioactive Waste Management," Mat. Res. Soc. Annual Meeting, Boston, MA, November 1979 (in press), Plenum Press.

[11] Klinkenberg, L.J. The Permeability of Porous Media to Liquids and Gases, Drilling and Production Practice 200-213 (1941).

[12] Brunauer, S. and Greenberg, S.A. The Hydration of Tricalcium Silicate and β-Dicalcium Silicate at Room Temperature, Vol. I, U.S. National Bureau of Standards, Chemistry of Cement, IVth Intl. Symp. 135-165, Washington, DC (1960).

[13] Robie, R.A., B.S. Hemingway and J.R. Fisher, U.S.G.S. Bulletin 1452 (1978).

[14] Babushkin, V.I., G.N. Matveev and O.P. Mchedlov-Petrosyan, Thermodynamics of Silicates (in Russian) (1972).

[15] Christensen, C.L., Borehole Plugging Program Status Report, Oct. 1, 1978-Sept. 30, 1978, Sandia 79-2141, Jan. 1980.

[16] Grutzeck, M.W., B.E. Scheetz, E.L. White and D.M. Roy. Modified Cement-Based Borehole Plugging Materials: Properties and Longevity, Submitted to Workshop on Borehole and Shaft Plugging, May 1980, OECD and USDOE.

[17] Scheetz, B.E., M.W. Grutzeck, L.D. Wakely and D.M. Roy, Characterization of Samples of a Cement-Borehole Plug in Bedded Evaporites from Southeastern New Mexico, ONWI-30, July 1979, A Topical Report for ONWI.

[18] Langton, C.A., M.W. Grutzeck and D.M. Roy, Chemical and Physical Properties of the Interfacial Region Formed Between Unreactive Aggregate and Hydrothermally Cured Cement, Cem. Concr. Res. 10, 449-454 (1980).

[19] Langton, C.A. and D.M. Roy, Morphology and Microstructure of Cement Paste/Rock Interfacial Regions, 7th Intl. Congress Chem. Cement, Paris, 1980 (in press).

[20] Roy, D.M. et al. Borehole Plugging and Shaft Sealing Systems, ONWI/SUB/78/E512-04200-3, Quarterly Report, Chap 5.0 by S. Goto, Jan. 15, 1980.

[21] Roy, D.M. and W.B. White, Borehole Plugging by Hydrothermal Transport, A Feasibility Report, ORNL-Sub-4091-1, May 30, 1975 (250 pp.).

[22] Roy, D.M. et al., Geochemistry of Cement-Based Borehole Plugging and Shaft-Sealing Systems, ONWI/SUB/78/E512-04200-2, Interim Progress Report, Nov. 15, 1979.

[23] Roy, D.M. et al., Borehole Cement and Rock Properties Studies, ONWI/SUB/78/E512-00500-4 (Task I), Annual Progress Report, Nov. 30, 1979.

[24] This research was supported by the U.S.D.O.E. under and R & D Subcontract with the Office of Nuclear Waste Isolation, Battelle Memorial Institute.

Discussion

J.C. WRIGHT, United States

You are studying materials longevity or stability. Are you also considering material property or engineering function longevity in your studies ?

D.M. ROY, United States

Our longevity studies attempt to consider both original phases present, transition phases and potential ultimate products. We have data on physical and mechanical properties for both initial and ultimate products of some of the materials I discussed. If we apply the principle of constant volume and conservation of mass, then we can predict that the change will not cause deterioration in physical properties to a significant extent. This needs to be done for all materials that might be used.

J. HAMSTRA, Netherlands

The permeability of 2×10^{-6} cm sec^{-1} measured for the rock salt around a concrete plug was established in the laboratory. What were the loading conditions ? If no loading was applied, do you agree that this permeability should not be assumed to be representative of the in situ situation ?

D.M. ROY, United States

The "loaded" situation permeabilities should be lower by at least an order of magnitude. Waterways Experiment Station data on halite from the site demonstrated substantial decreases in permeability when the samples were triaxially loaded in the laboratory. Obviously more "loaded" laboratory tests need to be made. Some of our current tests are radially restrained.

W. FISCHLE, Federal Republic of Germany

In some tests you found "Friedel salt".

Did you determine if this is good or bad for a borehole plug ?

Are you planning more tests to investigate this problem ?

D.M. ROY, United States

We do not have specific evidence that Friedel's salt formation is harmful in itself. If one is designing an expansive cement, then the expansiveness will be diminished by replacement of sulfate by chloride to form Friedel's salt. We are continuing to make tests and our diffusion studies will be coupled with phase characterization and properties measurements.

MODIFIED CEMENT-BASED BOREHOLE PLUGGING MTERIALS:
PROPERTIES AND POTENTIAL LONGEVITY

M. W. Grutzeck, B. E. Scheetz, E. L. White, and D. M. Roy
Materials Research Laboratory
The Pennsylvania State University
University Park, PA 16802, USA

Abstract

Cementitious materials modified to increase geochemical compatibility with the plugging environment have been investigated. Extensive mechanical and physical property investigations have been conducted on six formulations. The compositionally adjusted cements were modified with the addition of a variety of pozzolanic materials including fly ash, slag, ferrosilicon dust, and fine quartz. One of these formulations was designed as an expansive cement specifically for field emplacement in an evaporite sequence; some included brine in mixing water; while the remainder were generic formulations designed to develop long-term stability in geochemical environments.

Typical of the 28-day physical properties of these formulations are water permeabilities of ten to one thousand nanodarcies and compressive strengths of 50 to 120 MPa. All of these measurements were recorded for formulations whose rheologic characteristics have been determined as pumpable under field emplacement conditions. X-ray diffraction and SEM microstructural characterization studies were also performed.

Introduction

The demonstrable ability to safely seal a nuclear waste repository for countless generations is of high importance if nuclear energy is to become a viable source of energy. Preliminary studies [1-6] have shown that, to date, there are no better materials for borehole or shaft sealing than common portland cement or portland cement derivatives. The greatest uncertainty for many sealing materials is the required time frame for sealing, where half-lives of some radionuclides range into tens to hundreds of thousands of years [7]. It is with this aspect of the concept of demonstrable plugging, that the present paper is concerned.

Present studies at The Pennsylvania State University, Materials Research Laboratory are concerned with developing and optimizing cement based sealing materials. Realizing that cements, mortars, and concretes are some of the few materials for which there are substantial historical data from archaeologic and engineering records, studies are concerned with composites of portland cement and derivatives whose compositions have been modified for certain purposes.

In addressing longevity, a series of commercially available cements were taken as a base and their compositions were adjusted by addition of various reactive high silica components. Additives studied include dehydrated Ludox, dehydrated silicic acid, ferrosilicon dust, micrometer-sized quartz, slag, and fly ash. The additives were used to either react with the available free calcium hydroxide liberated from cement hydration, or to remain available for slower reaction over long time periods. Calcium hydroxide itself has relatively inferior mechanical properties, and because of its relatively higher solubility, might prove a pathway for leaching, in contrast to calcium silicate hydrates which are more inert.

The favorable thermodynamic stability of probable cement hydration products from compositions adjusted with higher silica contents has already been discussed in our previous paper [8]. Not only is longevity necessary, but always of paramount concern are the accompanying physical performance factors. The borehole plugging material must not only be stable in the borehole environment, it must also have adequate bond strength to host rock, low permeability both through the plug material and along the interfacial zone between the rock and plug (which is often of different chemistry from the plug material itself), and have appropriate physical and mechanical properties to be able to withstand tens of thousands of years of geologic processes (earth movements, weathering, etc.).

Method of Approach

Conceptually, initial mix design was based on the previously mentioned mineral duplication longevity concept. To date, four chemically adjusted experimental portland cement based materials and one field test recipe have been tested. A schematic/ approximate representation of the relative locations of the cementitious mixtures on a three-component diagram is given in Fig. 1. Accompanying compositions for the appropriate phases are given in Table I. The text of the paper deals with the experimental results which include conventional and non-conventional physical

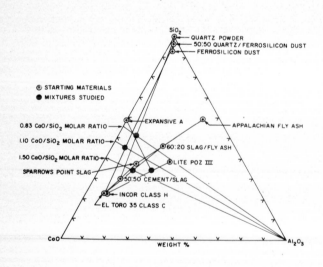

Fig. 1. Simplified "ternary" diagram showing location of starting materials and mixtures studied.

Table I. Composition of Starting Materials

	El Toro 35 (C-5)	Incor H (H-4)	Quartz Silicic Acid Ludox	Lite Poz III Fly Ash	Appalachian Fly Ash	Slag	Ferrosilicon Dust
SiO_2	21.2	23.0	100[1]	36.9	58.2	35.7	89.0-90.0
Al_2O_3	2.72	3.0		23.3	29.8	10.52	0.95
TiO_2	0.13	--		--	1.2		0.02-0.05
Fe_2O_3	5.17	5.2		5.5	3.6	0.68	1.83
MgO	0.90	2.3		4.5	1.1	11.64	1.06
CaO	65.7	64.8		25.5	1.3	40.42	0.52
MnO	0.22					0.18	0.17
SrO	0.09						
Na_2O	0.21				0.3	0.20	0.81
K_2O	0.18				3.2	0.37	1.86
P_2O_5	0.05						0.07-0.08
SO_3	2.62	1.3			0.3	0.27	0.40
LOI	0.76	0.97		0.11	1.3	0.23	2.67
CO_2	0.26						
Other						0.41[2]	0.30-0.90[3] 1.66[4] 0.12[5]
Total	100.21				100.3	100.21	variable
Blaine		4580		6711		5500	

[1]Assumed to be 100% SiO_2
[2]S
[3]SiC
[4]C
[5]heavy metal oxides and carbides

and mechanical test data and other experimental observations. The water content of the adjusted experimental portland based cements was chosen so that mixes would have relatively the same rheological characteristics after a given period of time. The limit set was at the upper viscosity limit, as thick as possible, but yet pumpable, using the "minimum" water concept as a guideline [9]. The w/s (water/cementitious solids) ratio was allowed to vary, fixing consistency such that the w/s ratio chosen would assure pumpability after two to three hours of hydration in the slurry stage, thus matching laboratory studies with field emplacement capabilities.

Viscosity Studies Used to Adjust w/s Ratio

All adjusted mixes fell into a broadly defined viscosity range, such that the apparent viscosity of the mix after two to three hours was between 18-22 poise at 117 sec^{-1} (150 rpm) shear rate. Consistency measurements were made using a Haake Rotovisco RV III viscometer. Measurements were made using a serrated rotor and cup (SVIIP), and both increasing and decreasing rate of shear of 78 sec^{-1}/min. (100 rpm/min) at various elapsed times (6.5, 30, 60, 120 minutes) after initial blender mixing. The temperature of 35°C was chosen for the viscosity measurements to represent typical downhole conditions. Rheologic data were collected starting from rest, increasing to 507 sec^{-1} (650 rpm), and immediately changing direction and decreasing to rest. The time necessary for a run was approximately 13 minutes. Therefore, the elapsed times after mixing which are presented are "nominal" since hydration continues even as the mixture is undergoing testing. The mixes undergoing testing were approximately Bingham plastic in nature, and the "down" portion of the rheologic curves (i.e., 650 → 0 rpm) often resulted in better Power law plots than did the equivalent "up" curve data (0-650 rpm) which were slightly curved. With the addition of water reducing additives, the trend in some cases seems to be reversed. The up curve seems to give better power law plots, while the down curve is now slightly curved. Effects of admixtures on viscometric characteristics have been discussed in more detail elsewhere [10] and are continuing.

Since different mixing procedures produced different rheologic characteristics of the mixes [11], it was necessary to use a fixed procedure to enable reproducible comparisons. Figure 2 is a plot of shear stress in Pascals vs. shear rate in sec^{-1} for a neat cement paste, w/c = 0.40, without the presence of water reducing additives. These are typical plots for the Haake Rotovisco system; both "up" and "down" data are given for purposes of illustration. As can be seen, the mixture receiving the smallest amount of mixing (upper curve) had the highest apparent viscosity [(shear stress/shear rate) x 10 = poise]. When mixing was more severe (20 minutes @ 140 rpm--middle curve), or extremely intense (blender), apparent viscosities drop. These preliminary experiments also showed that for a given cement at a

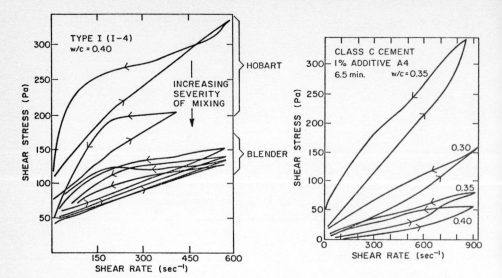

Fig. 2. Changes in Shear Stress-Shear
Rate Relationships with Variations in Mixing Procedure.

Fig. 3. Changes in Shear Stress-
Shear Rate Relationships
with Varying w/c Ratio.

given state of hydration, the lower the w/c ratio, the higher was the apparent
viscosity (Fig. 3), or if a water reducer/ superplasticizer was added to the mixture, the apparent viscosity of the mix would drop as a function of increasing
admixture concentration (Fig. 4).

Rheologic determinations were made on cements with and without the use of
water reducing agents. Representative data for cement numbers 1 and 5 (Table II)
are summarized as a series of Power law plots in Figs. 5 and 6. Although these
plots are not very accurate, they do demonstrate the major differences between
cements formulated with and without water reducing superplasticizer additives. The
basic BCT-IFF mixture has low K' (Y intercept) values, and the n' (slope) values
approach unity (Newtonian behavior). Thicker mixtures without water reducers, have
characteristically higher K'
values and are more susceptible
to shear thinning, as reflected
by their smaller n' values.
The w/s ratio chosen for the
adjusted cements (#1-6) and
given in Table II were selected
by using similar Power law plots
for each cement. The BCT formulations were not adjusted in
this manner, since they are expansive mixes designed to be
very fluid, as apparent from
comparison of Figs. 5 and 6.

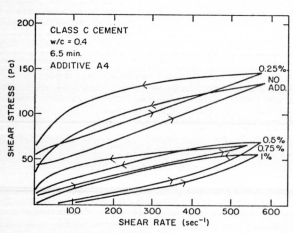

Fig. 4. Changes in Shear Stress-Shear Rate Relationships with Admixture Concentration.

Compressive Strength Methodology

Compressive strengths
of the mixes based upon Table II
formulations were evaluated
after curing at four temperatures (27°, 35° or 45°, 60°,
90°C) for 7, 14, and 28 days.
Samples were mixed according to
a modified API method based
upon a blender-mixing procedure.

Table II. Cement Mix Formulations Used for Mechanical Properties and Permeability Measurements

Formulation	ID and/or Source	Class C + Slag + Fly Ash 1	+ Quartz 2	FeSi-Dust + Quartz + Class C 3	+ Class H 4	"BCT-IFF" 5	BCT-IF 6
C/S	--	1.5	0.83	1.1	1.1	--	--
w/s	--	0.35	0.45	0.375	0.325	0.25	0.28
% Class C (C-5)	Southwestern Portland Cement Co.	51.0	35.15	69.4	--	--	--
% Class H (H-4)	Lone Star Industries	--	--	--	69.4	68.0	62.4
% Fly Ash (B15)	Lite poz III	--	--	--	--	23.0	21.1
% Fly Ash (B3)	Appalachian	19.6	--	--	--	--	--
% Slag (B19)	Bethlehem Steel	29.4	35.15	--	--	--	--
% Quartz (B5)	5 or 30 micrometer Min-U-sil	--	29.7	15.3	15.3	--	--
% FeSi Dust (B17)	Iceland	--	--	15.3	15.3	--	--
% Expansive Additive A (A29)	Dowell	--	--	--	--	9.0	8.4
% Salt (B-16)		--	--	--	--	--	8.1
Deionized Water (E1)		35.0	45.0	37.5	32.5	24.75	27.6
Dispersant	D-65 (A28)					0.23	0.25
Defoamer	D-47 (A27)					0.02	0.02

Cylindrical samples, nominally 2.2 cm diameter x 4.5 cm high, were cast in Teflon coated glass vials, vibrated for 60 seconds, and then cured for 24 hours at temperature in a water saturated air environment. After 24 hours, the samples were removed from the glass vial and allowed to continue curing in saturated $Ca(OH)_2$ or WIPP-A brine/$Ca(OH)_2$ solution for the designated length of time. Testing was carried out on right circular cylinders (ends were ground on a disk grinder) with a Tinius Olsen testing machine and a 0.025 inch/min. rate of loading.

Permeability and Microstructure

Permeabilities were measured using pressurized deionized water (0.1 to 4.8 MPa) as the permeating fluid. Two distinct types of measurements are reported. The first used solid materials which were either cast directly (cementitious mixes) or epoxied into 5 cm I.D. steel rings (rock cores). These data established a baseline with which to evaluate the merits of various cement/

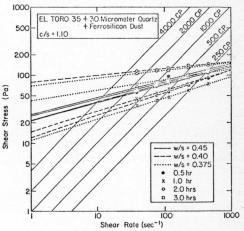

Fig. 5. Power Law Plot for Class C + Ferrosilicon Dust + 30 μm Quartz Mix (C/S = 1.10) Showing the Effects of Varying w/s Ratio upon Rheology.

rock pairs which comprise the second type of permeability data reported. Since it was hypothesized that the interfacial region between borehole or wall rock and plug may prove to be the weakest link in plugging boreholes and shafts, a series of "mini-boreholes" filled with the experimental (Table II) cement mixes were evaluated. Three rock types were evaluated: a high silica quartzite, a high carbonate limestone, and anhydrite obtained from the Castile formation, AEC-7 drilling site near Carlsbad, New Mexico. Cylinders 5 cm diameter x 5 cm high having a 2 cm I.D. central hole were processed with water or kerosene cooled diamond saws. All samples were cleaned with carbon tetrachloride and acetone prior to filling with the cementitious mix. Rock cores were filled with cement, puddled, and fitted with Teflon end

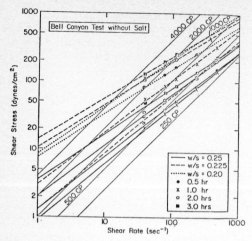

Fig. 6.   Power Law Plot for the Basic BCT-IFF
Formulation Without Salt, Showing
the Effects of Varying w/s Ratio
upon Rheology.

Table III. Compressive Strength (MPa) of
Class C + Slag + Fly Ash
C/S = 1.5, w/s = 0.35

$Ca(OH)_2$ Curing	Temperature			
	27°C	45°C	60°C	90°C
7 days	56.4	64.6	65.9	76.9
14 days	70.8	71.0	87.6	72.3
28 days	64.2	73.1	63.3	79.5

caps. Samples were allowed to cure
at temperature in jars of either saturated $Ca(OH)_2$ solution or WIPP-A
brine/$Ca(OH)_2$ solution (BCT formulations).

Microstructures of cured
cement mixes as a function of time
and temperature were sometimes
studied using a scanning electron
microscope. Fractured inner surfaces
of samples after compressive strength
tests were gold coated and imaged at
approximately 1000X and 3000X. Representative data are presented in the
text of this paper at appropriate
times.

Results

The compressive strengths of the Class C + fly ash + slag (#1, Table II)
mixes as a function of curing time and temperature are presented in Table III. The
effect of temperature on the acceleration of the reactivity of the slag is obvious
with the very high strengths apparent even at 7 days, at 90°C, and relatively little
increase between 7 and 28 days. Strengths of 45°C and 60°C samples were also higher
at 7 days. Generally, this type of mix appears very favorable from the standpoint
of strength development, although studies should be continued for longer times.

X-ray diffraction data for this matrix of samples verified the increase in
starting material reactivity with temperature. Residual cement phases (β-C_2S)
tended to decrease in intensity, as did the $Ca(OH)_2$ produced during the reaction.
Small amounts of calcium silicate hydrate (C-S-H) (broad peak at 29-30°2θ) were
detected at 60° and 90°C in the 7-day samples. In addition, at 14 and 28 days,
C-S-H was also present at lower temperatures (27, 45°C).

Water permeabilities of this mix for samples cast in 5 cm I.D. steel rings
after different curing intervals and temperatures are summarized in Table IVa (solid
samples) and IVb (rock/cement combination samples). The 2 week cured solid samples
described in Table IVa indicate that a lower permeability is achieved at higher
temperatures. At 4 weeks, the bonding to steel was not very good.

A correlation seems to exist between increasing bond strength (60°C to 90°C)
and decreasing permeability. Perhaps the more crystalline nature of the product at
higher temperatures also leads to lessened paths of permeability.

Table IVb data deal with the permeability of this same cement cast in quartzite rock cores. The water permeability of the quartzite used was approximately 2 x
10^{-7} darcy. It is obvious that the lower limit of the cement/rock couple is the
approximate permeability of the rock itself. Therefore, values approaching 1 x 10^{-7}
would seem excellent for a particular cement/quartzite rock pair. In all cases,
flow rates were linear with no associated bonding problems as was the case with
solids cast in steel rings. Permeability was found to be dependent on the curing
temperature, never being quite as low in the 90°C samples as in the 60°C samples.
Permeability also decreased with curing time, reaching average values of 5.0 x 10^{-7}
and 8.0 x 10^{-7} after 4 weeks at 60°C and 4 weeks at 90°C, respectively. Perme-
bility values in this range are approaching the quartzite and as a whole, the Class

C + slag + fly ash mix seems very promising and definitely is a candidate for further study.

Compressive strengths of the Class C + slag + 5 μm quartz (#2, Table II) formulation are given in Table V. Seven-day compressive strengths tend to remain nearly constant as a function of temperature, except for the 90°C sample which was substantially higher. Fourteen and 28 day strength measurements are not as constant, showing 60° and 90°C variability when compared to the 7-day trend.

Table IVa. Water Permeability (Darcys) of Class C + Fly Ash + Slag Mixes Cast in 5 cm Steel Rings and Cured in Ca(OH)$_2$ Solution

2 weeks @ 60°C	2 weeks @ 90°C
solid	solid
1.7×10^{-6}	6.0×10^{-7}
4 weeks @ 60°C	4 weeks @ 90°C
solid	solid
no bond	no bond

Table IVb. Deionized Water Permeability of Class C Cement + Slag + Fly Ash Cast in Quartzite Cores* (w/s = 0.35)

identification	length cm	curing time weeks	temp. °C	elapsed RT time between cure and test, d	water pressure MPa	viscosity of water cP	volume H$_2$O, cc	time hrs.	K_{H_2O} darcys	comments
Class C (C-5) cement + slag (B19) + FA (B3) (C/S = 1.5)	3.86	2	90	3	1.6	0.9579	6.87	36	4.39×10^{-6}	very linear flow rate
				5**	3.1	0.9358	8.28	47	2.03×10^{-6}	very linear flow rate
	3.85	2	90	25	4.4	1.0175	7.27	69	9.60×10^{-7}	very linear flow rate
	3.54	2	60	7	3.0	0.9579	0.71	22	3.64×10^{-7}	linear flow rate
	3.77	2	60	28	4.3	0.9579	13.53	74	1.5×10^{-6}	very linear flow rate
	4.06	4	90	29	3.0	0.9250	4.44	48	1.13×10^{-6}	linear flow rate
	3.86	4	90	31	3.1	0.8737	3.03	71	4.6×10^{-7}	linear flow rate
	3.96	4	60	27	3.1	0.9810	3.03	48	7.8×10^{-7}	linear flow rate
	3.79	4	60	39	3.1	0.9469	1.01	52	2.2×10^{-7}	linear flow rate

X-ray diffraction studies of these cured mixes duplicate previous trends of decreasing quartz, residual cement phases (β-C$_2$S) and Ca(OH)$_2$ with increasing temperature. Unique to this cement, however, is the fact that the Ca(OH)$_2$ had reacted completely and was not evident in the 7- and 14-day samples cured at 90°C. At 28 days, the Ca(OH)$_2$ had reacted, and was not found in either the 60° or 90°C samples. Reactivity in these mixes is generally high, resulting in rapid development of high compressive strengths. This cement appears to have considerable potential for longevity/durability since at least at higher temperatures, the calcium hydroxide reacts nearly as rapidly as it is formed. In addition, significant C-S-H is present at all temperatures and times.

Permeabilities determined for solid samples cast in 5 cm I.D. steel rings and cement/rock pairs are given in Table VIa and VIb, spectively. Permeabilities for solid solid samples cast in steel ranged from 3.9×10^{-6} to no detected flow, with water pressures of 2.76 to

Table VIa. Water Permeability (Darcys) of Class C + Slag + 5 μm Quartz Mixes Cast in 5 cm Steel Rings and Cured in Ca(OH)$_2$ Solution

2 weeks @ 60°C	2 weeks @ 90°C
3.9×10^{-6}	no flow - 120 hrs. @ 4.1 MPa
4 weeks @ 60°C	4 weeks @ 90°C
no flow - 48 hrs. @ 2.8 MPa	no bond

Table V. Compressive Strength (MPa) of Class C + Slag + 5 μm Quartz C/S = 0.83, w/s = 0.45

Ca(OH)$_2$ Curing	Temperature			
	27°C	45°C	60°C	90°C
7 days	42.9	45.5	45.1	57.9
14 days	82.5	61.4	62.2	87.0
28 days	65.2	55.3	50.7	92.5

Table VIb. Water Permeability (Darcys) of Class C + Slag + 5 μm Quartz Mixes Cast in Limestone Cores

2 weeks @ 60°C	2 weeks @ 90°C
1.9×10^{-6}	3.5×10^{-7}
4 weeks @ 60°C	4 weeks @ 90°C
no bond	no bond

Table VII. Compressive Strength (MPa) of Class C + Ferrosilicon Dust + 30 μm Quartz
C/S = 1.1, w/s = 0.375

Ca(OH)₂ Curing	Temperature			
	27°C	45°C	60°C	90°C
7 days	51.3	43.3	44.1	41.4
14 days	62.3	67.1	67.0	40.6
28 days	57.8	60.3	61.6	57.5

4.13 MPa. With limited data, this mix cast in limestone cores did not perform as well. Whereas 2-week permeabilities of limestone/cement pairs are adequate and decrease with increasing temperature, both of the 4-week samples showed inadequate bonding to limestone--with actual separation occurring along the interfacial zone. The reason for this annular cracking is not positively known, but is probably related to the high w/s ratio of the mix.

Compressive strengths for Class C + ferrosilicon dust + 30 μm quartz (#3, Table II) are given in Table VII. Seven-day sample compressive strengths generally decrease with increasing temperature; while at 28 days, curing temperatures caused relatively little difference in strength.

The major trends in phase changes determined from x-ray diffraction data were the general decreases in amount of Ca(OH)₂ with increasing temperature; it was

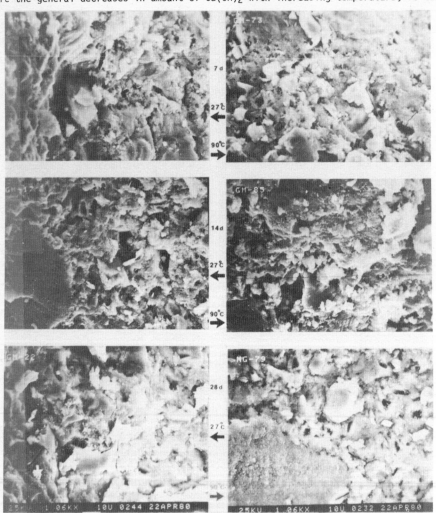

Fig. 7. Scanning Electron Microscope Images of 7, 14, and 28 day Cured Samples of Class C + Ferrosilicon Dust + 30 μm Quartz as a Function of Temperature. 1000X.

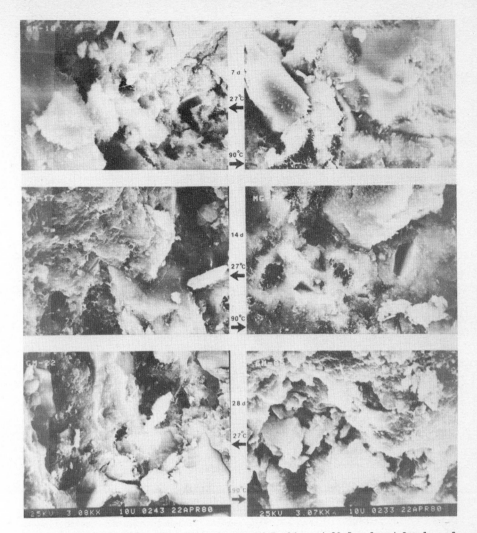

Fig. 8. Scanning Electron Microscope Images of 7, 14, and 28 Day Cured Samples of Class C + Ferrosilicon Dust + 30 μm Quartz as a Function of Temperature. 3000X.

absent from samples cured at 90°C at all curing times (7, 14, 28 days). C-S-H material was present at all times and temperatures, apparently increasing with length of cure. Permeability values for this mix ranged from 10^{-5} to 10^{-8}. Work is continuing on this mixture.

Figures 7 and 8 depict a series of scanning electron microscope images for 7, 14, and 28 day curing intervals at both 27° and 90°C. Figure 7 represents a relatively low (1000X) magnification image of fractured surfaces for these samples, while Figure 8 represents a 3000X image of similar features.

Both figures contain representative examples of the presence of quartz grains. Alteration was observed in many samples. However, the cements that were cured at 90°C begin to show

Table VIII. Compressive Strength (MPa) of Class H + Ferrosilicon Dust + 30 μm Quartz C/S = 1.1, w/s = 0.325

$Ca(OH)_2$ Curing	Temperature			
	27°C	45°C	60°C	90°C
7 days	52.1	49.2	42.1	34.9
14 days	63.3	65.4	51.8	41.9
28 days	71.4	72.1	59.7	64.0

Table IXa. Water Permeability (Darcys) of Class H + Ferrosilicon Dust + 30 μm Quartz Cast in 5 cm Steel Rings and Cured in $Ca(OH)_2$ Solution

2 weeks @ 60°C	2 weeks @ 90°C
no flow - 119 hrs @ 4.1 MPa max	$<4.4 \times 10^{-6}$

4 weeks @ 60°C	4 weeks @ 90°C
no flow - 65 hrs @ 4.1 MPa max	$<2.6 \times 10^{-6}$

Table IXb. Water Permeability (Darcys) of Class H + Ferrosilicon Dust + 30 μm Quartz Cast in Quartzite and Limestone Cores and Cured in $Ca(OH)_2$ Solution

2 weeks @ 60°C		2 weeks @ 90°C	
quartzite	limestone	quartzite	limestone
2.5×10^{-6}	1.5×10^{-7}	4.7×10^{-6}	no flow
	no flow - 4.8 MPa max	1.7×10^{-6}	$<3.4 \times 10^{-6}$

4 weeks @ 60°C		4 weeks @ 90°C	
quartzite	limestone	quartzite	limestone
1.1×10^{-6}	3.6×10^{-7}	1.1×10^{-6}	$<1.5 \times 10^{-5}$
7.3×10^{-6}	6.0×10^{-7}		$<1.3 \times 10^{-6}$

evidence of the initial stages of chemical reaction across the cement-paste/quartz boundaries. Samples GM-73 and MG-79 contain grains of quartz that contain etch or growth features altering the surfaces of the initial grains. Similar features were not observed in any of the 27°C samples.

Both sets of samples at 27° and 90°C exhibit pits apparently generated by the plucking of grains from the cementitious matrix during the fracturing process. Sample GM-22 and GM-73 contain fractures generated during the sample preparation.

The cements represented in these figures are generally very dense products. Examination of the 27°C cured samples revealed some 1 to 5 μm sized pores typically containing growths of elongate hydrated calcium silicate phases. At 90°C, the pores contain growth of hydrated calcium silicate phases but the mean length of the filaments appears to be smaller and the filaments appear more densely packed, perhaps indicating more extensive hydration, i.e., Fig. 7, GM-79.

Compressive strength determinations for a lower w/c ratio mixture, but conducted with a Class H cement (Class H + ferrosilicon dust + 30 μm quartz, (#4, Table II) were determined as a function of time and temperature. Data are presented in Table VIII. Compressive strengths decrease with temperature, but increase with time.

X-ray diffraction data once again indicate a relatively constant, slightly decreasing quantity of quartz and residual cement phases with increasing temperature. Once again the $Ca(OH)_2$ decreases with temperature in each of the 7-, 14-, and 28-day determinations and is absent in all of the 90°C samples. C-S-H was present at all temperatures except 27°C at 7- and 14-days curing.

Permeability values for solid samples cast in 5 cm I.D. steel rings and rock/cement pairs are given in Table IXa and IXb, respectively. The 60°C solid samples (Table IXa) both were essentially impermeable, having no measured flow after 119 and 65 hours, respectively. The 90°C samples tended to have decreasing flow rates and progressively lower apparent permeabilities with length of time the permeability data were collected. The data given are 15- and 18-hour values, respectively. Permeability of the 2- and 4-week samples seem comparable.

Table IXb data deal, not only with similar cement/rock pair permeabilities as before, but also with the effects of rock type on the nature of the observed permeability. Two- and four-week cure data at 60°C indicate very little difference in magnitudes of observed permeabilities for a given rock type as a function of time. The cement/quartzite values are approximately one order of magnitude higher than the comparable cement/limestone values. Both solid 60°C samples (Table IXa) were essentially impermeable and the quartzite itself was 2×10^{-7} darcys. Therefore, it is apparent that the one order of magnitude greater permeability could be attributable to interfacial flow. Cement/limestone values are at least one order of magnitude lower than the quartzite at 60°C (in one case it was low enough not to produce any flow at all). This indicates a permeability approaching both the cementitious material and the limestone itself (although the limestone is softer, it is more dense than the quartzite and presumably has lower permeability).

The 90°C data are more variable, and both the solid and cement/rock permeabilities are higher than their 60°C counterparts. At four weeks, the limestone couple is more permeable.

Compressive strengths for the BCT-IFF formulation (#5, Table II) are given in Table X. Samples were cured from 7 to 180 days at 27°, 35°, 60°, and 90°C. As a whole the observed strengths are fairly consistent, all behave in a similar fashion and have similar trends (except for 27°C-90 days) as a function of temperature. At a stage of hydration, up to 180 days, the strengths generally increase, reaching maximum values at about 60°C. Above 60°C, however, strengths fall off radically, often reaching values which are lower than their 27°C counterparts. Physical deterioration of samples cured in WIPP-A brine at 90°C for 90 days has been observed. Buildups of $Mg(OH)_2$ and Friedel's Salt deposits on the surface of the sample have been observed. Since the BCT-IFF formulation is an expanding cement and the samples were cured without restraint, the deterioration may be explainable as an artifact of curing procedure.

Table X. Compressive Strength (MPa) of BCT-IFF

WIPP-A Brine Curing	Temperature			
	27°C	35°C	60°C	90°C
7 days	78.2	90.1	92.1	51.3
14 days	60.8	122.3	75.4	42.4
28 days	91.9	122.9	107.2	52.3
56 days	85.5	117.9	128.5	53.9
90 days	126.2	85.2	95.2	46.0

X-ray diffraction results generally duplicate previous trends to greater reaction at higher temperatures at a specific length of cure. Ettringite and gypsum are present at all times. In addition, at 90 days a trace of Friedel's salt is also present. At 90 days, there seems to be more C-S-H present in the 27°C sample than in the higher temperature samples.

Permeability data for BCT-IFF mixes cast in steel rings and anhydrite core (AEC 7 drill hole) are presented in Table XIa and XIb, respectively. The permeability of anhydrite to gas was essentially nil. It was therefore assumed that the anhydrite was similarly impermeable to water in the present time frame. The BCT-IFF solid samples (Table XIa) had very low permeabilities, therefore the anhydrite/cement samples should likewise be very low unless interfacial permeability was an important feature. Table XIb data on anhydrite/cement core samples indicate that early permeabilities on the order of 1-3 weeks at 35°C fall in a 10^{-6} to 10^{-5} darcy range. Thus it would seem that the interfacial region is

Table XIa. Water Permeability (Darcys) of BCT-IFF Mixes Cast in 5 cm Steel Rings and Cured in WIPP-A* Brine

1 week @ 35°C
no flow - 26 to 115 hrs @ 450 psi

2 weeks @ 35°C
2×10^{-7}

4 weeks @ 35°C
no flow - 58 hrs @ 225 psi

*NaCl   8.98 wt %
MgSO4   0.36 wt %
KCl   4.77 wt %
MgCl2 11.14 wt %
CaCl2   0.14 wt %

Table XIb. Deionized Water Permeability of Bell Canyon Test Formulations Cast in Anhydrite Cores*

identifi- tion	w/s	length cm	curing time (weeks)	temp. °C	elapsed RT time between cure and test, d	water pressure MPa	viscosity of water cP	volume H_2O cc	time hrs.	K_{H_2O} (darcys)	comments
BCT-IFF 4349.45	0.3	3.34	1	35	1	2.8	0.9579	6.5	41	1.9×10^{-6}	linear flow rate
BCT-IFF 4365.?	0.3	3.31	2	35	3	2.8	0.9358	1.6	8	2.2×10^{-6}	increasing flow rate value @ 20 hours
BCT-IFF	0.3	3.30	2	35	5	2.8	0.9358	7.1	9	7.0×10^{-6}	increasing flow rate value @ 112 hours
"BCT-IFF" 4488.1	0.25	3.56	3	35	1	2.8	0.8937	40	21	2.2×10^{-5}	decreasing flow rate value @ 7 hours
"BCT-IFF" 4492.65	0.25	3.52	3	35	2	2.8	0.9142	18	90	2.3×10^{-6}	linear flow rate
BCT-IFF 790727	0.30	3.64	4	35	12	4.4	0.8937	no flow	29	$<2.5 \times 10^{-7}$	maximum value
BCT-IFF 790729	0.30	3.64	4	35	31	4.2	0.9250	no flow	50	$<1.6 \times 10^{-7}$	maximum value

*Specimens prepared at PSU using modified API mixing procedures, vibrated for 60 seconds, and then cured at 35°C in a water saturated air environment ("BCT-IFF") or in WIPP-A brine (BCT-IFF).

Table XII. Compressive Strength (MPa) of BCT-IF

Saturated NaCl + Ca(OH)$_2$ Curing	Temperature		
	27°C	60°C	90°C
14 days	49.1	63.2	46.2
28 days	63.2	65.8	59.8
56 days	60.5	48.5	34.2
90 days	79.4	70.7	71.0
180 days	81.0	77.3	58.6

indeed allowing the passage of water in a preferential manner. However, by 4 weeks, no flow was detected in two replicate samples after 29 to 50 hours at 610 to 640 psi.

Compressive strengths for the BCT-IF with NaCl MPa (#6, Table II) formulation cured in a saturated solution of Ca(OH)$_2$ and NaCl as a function of time and temperature are given in Table XII. The 14 day strengths of 60°C samples were higher, but at 28 days the values were independent of temperature.

X-ray diffraction data for samples cured up to 6 months in a saturated solution of Ca(OH)$_2$ and NaCl once again depict a greater reactivity at higher temperatures. Residual cement mineral peaks (β-C$_2$S) and Ca(OH)$_2$ tend to decrease with temperature. Ettringite was found at 27°C at all times, but was generally absent at 60° and 90°C. Friedel's salt was generally present throughout.

Conclusions

Six cementitious formulations (four experimental and two field mixes) have been investigated. The field mixes were studied in some detail, adjunct to field emplacement. the other four mixes were preliminary experiments initiating a materials longevity study. For the experimental mixes, a temperature matrix was also introduced to provide information on changes in properties as a function of possible repository heating; and also to generate data to be used in kinetic studies. The compositions of the mixes were designed primarily for their estimated thermodynamic properties, within limitations of feasible pumpability. Viscosity, compressive strength, rock and steel bonding, water permeability, cement/rock combination permeability, phase identification (by x-ray diffraction) and microstructural characterization were the primary measurements reported herein.

The absence of free calcium hydroxide in some of the silica adjusted cement formulations indicates that the silica adjustment design is working. It is possible to minimize the content of a potentially soluble phase, and generate a more stable phase assemblage. The data on comparative permeabilities for samples hydrated up to 28 days, containing fly ash and free of Ca(OH)$_2$, are insufficient yet to make generalizations and definitive correlations other than those such as w/c ratio. However, effects that would be dependent on differential solution in a very low permeability solid, would take longer than the duration of the permeability tests, to generate significant measurable permeability differences. Experiments using longer term permeability/flow measurements, changing the permeating fluid, and also with effects of longer curing and aging of the cementitious materials, have been initiated. Experiments of this type should help in assigning relative figures of merit, and establishing the most stable cement-based formulations.

The data for samples cured at 90°C (compressive strength, x-ray phase identification, and permeability) have shown the greatest variability among the samples studied. The 27°-60°C data showed more gradual changes in properties with time and temperature. Therefore, experiments 90°C or higher are also among the continuing studies [12].

References

[1] Borehole Sealing--Subcontract 78X-15966C-Final Report; Dowell.

[2] Rennick, G. E., J. Pasini, R. F. Armstrong, and J. R. Abrams, Plugging Abandoned Gas and Oil Wells, Mining Congress Journal, Dec. 1972, pp. 37-42.

[3] Roy, D. M., and W. B. White, Borehole Plugging by Hydrothermal Transport, A Feasibility Report, ORNL/Sub-4091/1, May 31, 1975.

[4] Black, D. L. (Westinghouse Astronuclear Laboratory), A Study of Borehole Plugging in Bedded Salt Domes by Earth Melting Technology, WANL-TME-2870, June 1975 (prepared for Union Carbide, Oak Ridge, TN).

[5] a) Martin, R. T., Feasibility of Sealing Boreholes with Compacted Natural Earthen Material (Vol. 1), June 1965, Prepared for Oak Ridge National Labora-

    tory, Contract ORNL-Sub-3960 (MIT Res. Rept. 75-28).

    b) Fernandez, R., Borehole Plugging by Compaction Process, Final Report Y-OWI-Sub-7087-1, August 1976 (Draper Lab., Cambridge, MA).

[6] Herndon, J., and D. Smith, Plugging Wells for Abandonment, Y/OWI/Sub-76/99068 (Sept. 1976).

[7] Scheetz, B. E., M. W. Grutzeck, L. D. Wakeley, and D. M. Roy, Characterization of Samples of a Cement-Borehole Plug in Bedded Evaporites from Southeastern New Mexico, ONWI-30, July 1979, A Topical Report for ONWI.

[8] Roy, D. M., Geochemical Factors in Borehole and Shaft Plugging Materials Stability, Submitted to Workshop on Borehole and Shaft Plugging, May 1980, OECD and USDOE.

[9] API Recommended Practice for Testing Oil-Well Cements and Cement Additives, American Petroleum Institute, Washington, DC.

[10] Roy, D. M., and K. Asaga, Rheological Properties of Cement Mixes: V. The Effects of Time on Viscometric Properties of Mixes Containing Superplasticizers; Conclusions, Cem. Concr. Res. 10, 387-394 (1980).

[11] Roy, D. M., and K. Asaga, Rheological Properties of Cement Mixes. III. The Effects of Mixing Procedure on Viscometric Properties of Mixes Containing Superplasticizers, Cem. Concr. Res. 9, 731-740 (1979).

[12] This research was supported by the U.S.D.O.E. under an R&D subcontract with the Office of Nuclear Waste Isolation, Battelle Memorial Institute.

Discussion

K. MATHER, United States

I have two questions and a comment on Mr. Grutzeck's very interesting presentation. The questions are : (1) What is the ferro-silicon dust from Iceland ? (2) Was the simulated borehole specimen from which you got lower permeability the one in limestone or the one in quartzite ? I would expect to find the lower permeability in the specimen in limestone, at earlier ages.

Now the suggestion : the low strengths characteristic of the materials cured at 90°C may represent a silica starvation, and the strengths would perhaps be substantially improved by larger doses of pozzolanic silica.

M.W. GRUTZECK, United States

I suspect that the ferrosilicon dust which is high in SiO_2 (~ 96 %) is not amorphous due to the traditional melt mechanism found in fly ashes, but is a silica material, so finely crystalline to appear amorphous to X-rays.

The same adjusted cement was cast in both a limestone and quartzite core ; the resulting permeabilities were an order of magnitude lower for the limestone than for the quartzite at similar times and at 60°C. I suspect that permeability might be reduced by carbonation of the interface.

E.W. PETERSON, United States

What type of water did you use when making the measurements which showed a decrease of permeability with time ?

M.W. GRUTZECK, United States

Deionized.

S.J. LAMBERT, United States

Naturally occurring tobermorite, with which I am familiar, is so badly weathered that it requires pick and shovel to expose. Perhaps before invoking the existance of a mineral as a demonstration of longevity or stability, we should qualify the use of such minerals by considering the conditions of stability in which they are found.

M.W. GRUTZECK, United States

There are different deposits of tobermorite throughout the world. Some are better than others. I agree with your statement and we are using conservative approaches when integrating thinking of this type into the total geochemical evaluation.

G.M. IDORN, Denmark

Have you tried to experiment with alkalies in your systems ?

M.W. GRUTZECK, United States

Not yet ; we have started with the simple system.

DOWN-HOLE TELEVISION (DHTV) APPLICATIONS IN BOREHOLE PLUGGING

C. L. Christensen and Robert D. Statler
Sandia National Laboratories
Albuquerque, New Mexico   87185
United States of America

and

E. W. Peterson
Systems, Science and Software
La Jolla, California
United States of America

ABSTRACT

The Borehole Plugging (BHP) Program is a part of the
Sandia experimental program to support the Waste Isolation Pilot
Plant (WIPP).[1]   The Sandia BHP program is an Office of Nuclear
Waste Isolation (ONWI)-funded program designed to provide inputs to
the generic plugging program while simultaneously acquiring WIPP-
specific data.   For this reason a close liaison is maintained
between the Sandia WIPP project and the ONWI generic program.
Useful technology developed within the Sandia BHP to support WIPP is
made available and considered for further development and application to the generic Borehole Plugging and Repository Sealing Program
at ONWI.[2]   The purpose of this report is to illustrate the usefulness of downhole television (DHTV) observations of a borehole to
plan plugging operations.

BACKGROUND

The Sandia BHP includes a field test program for testing candidate plugs in situ. One such field test now in progress is the Bell Canyon Test.[3] While conducting permeability measurements during this test, a need arose to locate competent packer seats in a borehole. The nature of this testing required that locations having a minimum of five linear feet of undisturbed wellbore wall be identified. The accepted procedure to locate these seats is to identify them by use of a caliper log; however, in our case, after several failures to obtain satsifactory seats, it was decided that some technique other than a caliper log selection process was required. Accordingly, a local water-well service company, which uses a closed-circuit TV system, was hired to assist in selecting the desired packer seats. (A more detailed description of the rationale in selecting a packer seat location for a precision permeability measurement is presented in Appendix A.) The results of this service are presented here as information for those unfamiliar with the appearance of a wellbore at depth. This report is an attempt to make these results generally available.

The photographs included herein were randomly selected from a video tape of the actual wellbore DHTV survey to illustrate the types of wellbore conditions that may be encountered, especially in formations subject to dissolution effects.

EQUIPMENT

As noted, the DHTV rig was part of a hired service, generally used for logging of local water wells. Application to a field-test program requiring accurate and repeatable positioning within inches was not intended. The system consisted of a van with a 3000-foot cable reel, drive and controls mounted in the rear; a nominal 6-inch OD video camera, including two lighting arrangements--one for dry holes and one for underwater operations; video monitor and a video recorder capable of superimposing the cable reel footage readings. The rig included a well-head pulley arrangement to lower the assembly into the wellbore, and the entire operation was conducted by one operator and a helper. Assembly time required about one hour.

RESULTS

The accompanying illustrations provide an indication of the wellbore conditions observed. Figures 1 and 2 illustrate the equipment and setup procedure used in the evaluation of AEC-7 for the Bell Canyon test series. Figures 3-8 present a sequence of pictures at various depths as the DHTV rig is lowered through the wellbore. Sample photographs taken with both dry and underwater lamps for illumination are included. The caliper logs for the same depth are included for comparison. Again, the reader is cautioned not to try to make depth correlations closer than 12 to 24 inches because of the uncalibrated nature of the TV cable system. General comments are provided on the illustrations.

CONCLUSIONS

The use of downhole television provides additional knowledge of wellbore conditions. While only qualitative at this stage, the potential usefulness of this technique is obvious. While in some cases the correlation to the caliper log is good, there is evidence that a one-arm caliper will miss some of the cavities picked up by the camera. Extension to three- and six-arm calipers can reduce this risk but adds complexity and expense. In those cases where knowledge of the wellbore condition is required, as in a field test program for borehole plugging, the DHTV system can be invaluable.

Sandia National Laboratories is in the process of assembling a downhole system which will provide greater depth capability for DHTV coverage and will also include a limited wire line logging capability for monitoring downhole pressures, temperatures, and other selected borehole conditions.

REFERENCES

1. Christensen, C. L. and T. O. Hunter, Waste Isolation Pilot Plant (WIPP), Borehole Plugging Program Description, January 1, 1979, SAND79-0640, Sandia National Laboratories, Albuquerque, NM 87185, August 1979.

2. Office of Nuclear Waste Isolation, Program Plan and Current Efforts in Repository Sealing for the NWTS Program, ONWI-54, p. 12, Battelle Memorial Institute, Columbus, OH, October 1979.

3. Christensen, C. L., Bell Canyon Test, WIPP Experimental Program, Borehole Plugging, SAND79-0739, Sandia National Laboratories, Albuquerque, NM 87185, June 1979.

APPENDIX A
PACKER ELEMENT/BOREHOLE INTERFACE SEALING

E. W. Peterson
Systems, Science and Software

The measurement of liquid and gas flow through geologic media possessing microdarcy-range permeabilities is recogized as an important technical consideration in the design of nuclear waste repositories and associated barrier zones. Conventional oil-field technology developed and successfully used for testing millidarcy-range formations has not addressed certain problems inherent in performing reliable, reproducible, quantitative measurements in the microdarcy range. One unresolved problem is that of forming a sufficiently tight seal along the packer element/borehole interface so that conventional pressure buildup, fluid production, and injectivity tests can be performed. While three-arm caliper logs are beneficial in identifying borehole intervals possibly suitable for packer placement, the added resolution provided by a closed-circuit TV log becomes invaluable when testing microdarcy-range media. Examples of packer/borehole sealing requirements for such testing are given below.

In conjunction with the borehole plugging program at the WIPP site, in situ testing has been performed to determine the salt bed permeability at the proposed repository level, the formation permeability at the plug emplacement position, and the capability of the borehole plug to block flow from a high-pressure aquifer. Each of these tests presents different requirements for packer/borehole sealing if quantitative data are to be obtained.

Permeabilities of the salt beds were evaluated using the guarded straddle packer system shown in Figure A-1. The protected volumes afford a means of quantitatively assessing both the horizontal and vertical flow components. This latter quantity, which is not measured using conventional systems, must be measured when testing formations with microdarcy-range permeabilities to insure that any flow bypassing packers closing off the test interval is small compared to that entering or leaving the formation. Test intervals range from 30 to 60 meters, and their surface area is large compared to the wellbore cross-section area. As a result, this measurement has the least severe packer/borehole seal requirements of those discussed. Even so, working with air at a control-zone pressure of .8 MPa requires the leak rate past the enclosing packers be < 10 SCCM (e.g., equivalent water flow $\approx 3 \times 10^{-4}$ STB/D), if formation permeabilities in the order of a microdarcy are to be measured.

The most stringent measurement requirements occur when performing plugzone integrity tests or when evaluating plug performance. The standard configuration for performance evaluation is shown on Figure A-2. Tests performed to date evaluated either the fluid buildup or pressure buildup. With the configuration shown, a 16.0 MPa pressure is maintained in the annulus above the umbrella packer (i.e., the borehole makes water). In order to measure a flow corresponding to that occurring through a cement plug having a micro-darcy permeability, the flow around the umbrella packer must be on the order of $< 10^{-5}$ STB/D.

The preceding examples illustrate some requirements that must be satisfied when performing in situ testing supporting nuclear waste isolation programs. Packer element/borehole interface seal requirements are so critical that they obviously cannot be assured pretest even by visual examination of the borehole, since gap or fracture dimensions (see insert on Figure A-2) must be $< 10^{-5}$ cm.

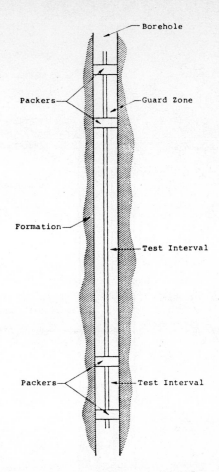

Figure A-1.   Sketch Showing the S-Cubed
Guarded Straddle Packer System.

Final evaluation of the seal integrity can, many times, be done only
through post-test data evaluation. However, it must be emphasized
that visual examination of the borehole provides improved wellbore
surface definition and would certainly supplement three-arm caliper
logs. In that respect, use of the DHTV saves rig time by greatly
increasing the chance of setting the packer in a competent zone.

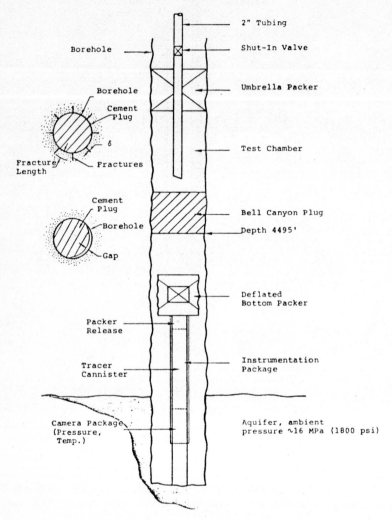

Figure A-2. Sketch Showing Configuration Used for Evaluation of Bell Canyon Plug Performance.

GENERAL DISCUSSION

N.A. CHAPMAN, United Kingdom

We can now move on to a general discussion of the papers in this Session. The basic themes have been the physical and chemical properties of various cementing compounds, and the thermo-dynamic approach to modeling rock-ground water interactions. We have also heard about the use of a wide variety of these materials to seal shafts in rock salt. The thrust of many of the papers was to determine how to estimate the long-term stability of these materials ; perhaps we could begin by discussing the longevity of potential cements ; how much value we ought to put on archeological evidence and whether 17 years is really a long enough record.

K. MATHER, United States

I wish somebody would go back to India and investigate the length of time during which pozzolanic materials in the form of brown clay bricks have been used there. These traditional materials may have been in use for thousands of years and they might provide the best example of stability in a surface environment. Of course, there are also the Roman concretes that can be traced back about 2000 years. I am not pessimistic about the longevity of systems based on the kinds of reactions we have in Portland cements. Additional components may further improve the properties of the cement, for example increasing the resistance to high temperatures or to various kinds of chemical attack. Pozzolanic materials are quite interesting and silica particles may also be an excellent additive. I think that there has been some use of silica in concrete in Norway and Iceland.

N.A. CHAPMAN, United Kingdom

Does anybody else want to comment on the issue of longevity ?

T.G. CLENDENNING, Canada

Domestic hot water tanks lined with a proprietary Portland cement coating have been in use about 30 years both in the United States and Canada. The interesting condition is that the material is exposed to elevated temperature. I think that these tanks might be a useful source of information.

N.A. CHAPMAN, United Kingdom

Of course, this is a very short time period. Does anybody have any feeling about the required longevity of seals and plugs ? Or is it something that will be variable in different repository systems or in different parts of the same repository ? Are we talking in terms of hundreds or thousands of years ?

J.C. WRIGHT, United States

Everyone is analysing materials longevity ; but when we put down a plug we are really interested in the longevity of the function. In relation to concrete we are primarily concerned about the durability of the plugging function. If we look at ancient concretes we can say that they are stable, but we do not know what the properties were at the beginning. On the other hand if we are interested in the properties of a mixture as a function of time we need to assess the mineralogical changes which will take place.

G.M. IDORN, Denmark

Some experiments carried out today approach hydrothermal conditions. I am not aware how hydrothermal conditions could be produced in a borehole.

S.J. LAMBERT, United States

The occurrence of particular minerals has been used to demonstrate the long-term stability of natural materials. As Mrs Mather pointed out it is quite conceivable that by means of various reactions one could actually obtain products with properties even more desirable then those that were originally intended for the mixture. My quarrel with tobermorite is only due to the fact that sometimes we tend to use a mineral as demonstration of longevity without regard to the conditions of stability.

K. MATHER, United States

What we need to look out for in formulations for boreholes and shafts is that the temperature, later on, will get elevated. We should be looking for high creep capacity in the material we put down there so that some movement can be accomodated and we should also be looking for a formulation which is higher in silica than a normal Portland cement.

N.A. CHAPMAN, United Kingdom

Can I just play devil's advocate for a moment ? It does not seem to me as though we really know what we are trying to achieve. It seems like a good idea to seal boreholes ; but do we really want to seal them or do we just want to retard ground water flow ? If we want to seal them, do we want to seal for a long time or a short time ?

J. HAMSTRA, Netherlands

I want to go one step further. You should identify the host rock first because the situation for crystalline rock, for example, is completely different compared to rock salt or argillaceous rocks. So if you want answers to your questions you should first identify your objective, which will vary in relation to the host rock. Since rock salt is impermeable the objective is to produce equally impermeable seals. M. Eichmeyer's proposal to use bitumen as the main sealing material is on the right track because that is a natural material and we know its behavior and longevity. Whether pure bitumen or a mixture of bitumen and rock fragments is used makes very little difference ; the only difference is the cost since bitumen is quite expensive. If the problem is to seal a shaft in granite it will not be necessary to make a seal more impermeable than the host rock.

N.A. CHAPMAN, United Kingdom

Has anybody else got any comments on the difference between achieving a complete seal and restoring the hydraulic properties of a rock before any boreholes were drilled ?

T.O. HUNTER, United States

        I scanned the literature briefly looking for safety
assessments in which doses to people resulting from borehole or
shaft failure had been evaluated. I have not found any and I would
like to ask if anybody knows of any such studies.

J. HAMSTRA, Netherlands

        We have calculated the consequences for the transport of
radionuclides in open disposal boreholes in a flooded mine. In this
case you have convective cells in the open boreholes and the rates
of radionuclide release become too high. This assumes of course,
that the open pathway extends all the way to the biosphere. That is
why, as I stated in my paper, we try to design plugs that will ensure
that diffusion is the only operating transport mechanism. That would
give us the time to reenter the flooded mine and to take remedial
actions because diffusion would take many years.

D.M. ROY, United States

        There is a good deal of redundancy in the isolation system.
We design waste forms for very low release rates and we have overpacks
and backfilling and then we have borehole plugging, and shaft sealing.
If each one of these barriers does its job well, then maybe the system
is overdesigned. Nevertheless I think that one should try to make
progress in the development of all barriers.

R.D. ELLISON, United States

        If the criteria are based on doses due to radionuclide
release then the need for sealing should come directly out of the
calculations. As an alternative, sealing could become an institutional
requirement. A seal, particularly in a salt environment, has two
main functions, the first is to restrict the migration of radionuclides and the second is to keep the water out of the repository.
Possibly the second function is the most important one.

J. HAMSTRA, Netherlands

        I disagree about this. Getting water into a disposal area
is one thing but then getting it out is another story. It is very
difficult to imagine a mechanism to get brine out of a flooded mine
other than by convergence of the excavation itself. If you calculate
the resulting rate of release of the brine you find out that there
is little cause for concern. It is an unpleasant idea to have a
flooded mine but the consequences are minor, and this should be
the only important consideration.

R.D. ELLISON, United States

        I do not think that we disagree, since basically we said
the same thing. We calculate the consequences and design accordingly.
In relation to the water inflow, if you have two shafts and you
leave them open, it is at least possible for water to go in one
shaft and out the other. I think that that would not be desirable.

T.O. HUNTER, United States

        Most release scenarios assume that flow is established,
one way or another. If you look critically, in fact you discover

that flow would not be established, because, for example, the density of saturated brine would make, a U tube effect impossible. Nevertheless people go ahead and assume that radionuclides are released, even if that violates the laws of physics. On the other hand you have to do that or you would not have any consequences. I would like to ask the people here what they think about the public's perception of this problem. I am always amazed by the attention that plugging gets in both technical and lay circles. My view is that in most cases borehole plugging and shaft sealing would be a redundant barrier. But I fear that the public and maybe even regulatory authorities see it as a crucial and fundamentally weak link in the system.

J.C. WRIGHT, United States

The concept of isolation in geological formations is based on several assumptions ; one of them is that you have considered everything and then that you have built-in redundancy and that you got the best possible product. From the point of view of communicating with the public the problem is to prove that you have considered everything. You can carry out a consequence analysis and show that plugging is not important, but you have no way of showing that you have considered everything. On the other hand I think that the philosophy of doing the best we can is the right one.

T.O. HUNTER, United States

I agree with the philosophy, I just wonder about the public's perception.

K. MATHER, United States

Mr. Statler told us that in Nevada somebody fell down a 24" hole. Now it is clear that the public is going to be much happier about 24" holes with fixed caps over them than it would be about the same holes left open. The same philosophy would apply to penetrations around a waste repository.

N.A. CHAPMAN, United Kingdom

Radiological assessments of radioactive waste disposal generally end up with some sort of risk committment as a result of a particular model or release scenario. Are we saying that releases from a repository with plugged boreholes should be no worse than releases that would take place in case there were no boreholes or are we willing to accept an incremental release due to the boreholes ? We already have models and release scenarios assuming that there are no boreholes.

P.H. LICASTRO, United States

The approach in most engineering projects is to leave the site in the same situation that existed before drilling and mining. So if you look at borehole plugging from that standpoint, it would follow that you have to produce the best seal in order to show that it is better than what surrounds it. Basically that is the philosophy that NRC will pursue.

SESSION 5

Chairman - Président

T.O. HUNTER
(United States)

SEANCE 5

T.O. HUNTER, United States

   Basically our goal today is to summarize what we have discussed in the last two days and try to arrive at some condensations of our views ; in addition everyone should go away feeling that on some specific topics he gathered new insight. In order to do that I have prepared a list of topics which I thought would be appropriate to discuss. For each of these topics, I have asked an individual within the group to prepare a brief summary of the situation as an introduction to the discussion.

   The first subject on my list is : what are the requirements or criteria for penetration sealing ? At least an assessment of where we are with respect to that subject. Another subject is the characterization of the borehole or shaft environment in terms of what is important. There was discussion of the disturbed zone and of how the deformation of the rock and the compatibility of sealing materials with the rock might be important. We had several papers that discussed a methodology for arriving at borehole plug designs. In those discussions issues were mentioned but not everybody agreed on their relative importance. Several people have been concentrating on concrete as a sealing material ; it would be good if we could summarize our views on that particular subject. Another important question is that of the long-term stability of the plugs. Also we have seen in several papers, different methods of analysis for assessing the performance of borehole seals. Several people mentioned that we are in the process of either undertaking or planning field tests in different media. I would also like to summarize that situation. Last, but not least, we should have some discussion on what opportunities there are for further international cooperation in the area of borehole and shaft sealing. So what I did was to ask eight people in the audience to think about these subjects and to ask themselves the questions : where are we today, where do we need to go and then finally, what is the most immediate thing we should do ? The first topic is that of requirements and criteria and I asked Dr. Smith to provide some thoughts on this.

M.J. SMITH, United States

   I would like to start by describing what I think I have seen so far in terms of NRC requirements for borehole plugging. If we try to categorise it we have the site characterization report, the environmental report with the application for the construction licence, then we have the updated licence for operation and, finally, the licence application for closure. One of the assessments that we have made in the program office is that by the time the site characterization report is completed we have to provide some assurance that the exploratory shaft, if one was required, could be sealed. Then by the time you get to the licence application for construction, you need design considerations that relate to the decommissioning of the facility. By the time you get to the licence application for operation you are looking for questions that weren't resolved during the construction authorization. That is the point where we ought to provide reasonable assurance that the repository can be sealed. Then by the time you get to the closure stage you are looking at the disturbance that the repository might have caused. In other words before construction, you try to address constructability and/or design functions, and then after the start of operation you look at verification of design. With that in mind I thought about a way I might provoke people into making some comment. The borehole plugging system is part of the overall system of multiple barriers that provide waste isolation. The system includes both engineered and natural components. There is little you can do about the natural barriers other than to site the repository at the appropriate depth, and at the appropriate distance from points of ground water discharge.

But there are quite a few things that you can do to the engineered components of the system. My concern is that we are attempting to decide about the engineered components without making an assessment of the desirable performance for those components or the system. The normal procedure is to begin by describing the functions that the component must perform. From the functions you then move on to an assessment of required performance. In looking at the things that affect the required performance you find that it is strongly affected by the host environment, by the allowable dose to man, and by the interaction with other engineered components. Now the next step is to develop, from that required performance, the performance criteria that you are going to place on the system. And finally, once you have the criteria, you then can proceed with the design of the system.

The point is that the required performance of the engineered components in a repository or in a multiple barrier system, is intimately related to the host environment. The geologic conditions of the host rock determine what the borehole plugging system should be.

So I see a potential pitfall here ; that is, if the criteria define something more than the overall performance of the barriers system, of which the borehole plug is one part, if we get to the point where someone tells us how each of the components of that system has to perform, then constraints are placed on the actual design of the system. I see this as two points that can be joined by a great number of different paths ; if we have too many restrictions some of the paths may be eliminated. Therefore the criteria should not define anything else than the performance of the overall system.

J. HAMSTRA, Netherlands

I believe we are all looking for some sort of multi-barrier solution ; we will have to define for ourselves the quality criteria for each barrier in order to achieve an overall system performance that will comply with the criteria imposed by the regulatory authority. If you do not succeed in getting the level of quality for one of the barriers then you might revise the design criteria and place more weight on other parts of the isolation system.

M.J. SMITH, United States

That is exactly my point. There is a myriad of combinations and permutations of components of a multiple barrier system. If someone defines for us something else other than the overall system performance criteria, then our ability to choose a system that will work could be restricted.

T.O. HUNTER, United States

Let me try to summarize the comments up to this point. Dr. Smith has presented us a framework for the discussion of plugging criteria. One specific point has been identified, that is the separation of overall performance criteria from the criteria for the design of specific components.

R.D. ELLISON, United States

The system you are talking about is the total system of barriers, of which the seal is only one component, as opposed to the system of seals.

L.F. HARTUNG, United States

I would make one suggestion : in a multiple barrier system the borehole/shaft seals are a function of the disturbance of the geologic barrier. I think Mr. Hamstra is correct, in that each one of these operates separately, and if you have to make trade-offs within the multiple system of, for instance, shaft seals, then they will only come up with the eventual criterion for that part of the multiple barrier system.

D.M. ROY, United States

In the discussion so far we seem to be treating just the repository, and I think that we shouldn't forget the waste, since it is also a component of the total system. Second I would say that whilst you work on the overall system you must begin with each component and you have a constant iterative process. You can't only work on the overall system without considering each component.

D.E. STEPHENSON, United States

You put forward two components of the system. When you define the criteria for the system, I'm not sure whether you are separating the performance criteria of the two. That is the engineered and the natural components.

M.J. SMITH, United States

If you look at 10 CFR 60 both are in there. I don't have any problem with people defining their expectations for the engineered components, or for how one expects the overall system to behave and that in a sense tells you what the criteria are. But my whole contention is that if you begin asking for performance figures for individual components of the system you tie people's hands. You no longer have the freedom of designing the system with the trade-offs that are necessary. The proposed technical criteria that are to be published do in places describe how to do the job ; I think that is a serious mistake.

T.O. HUNTER, United States

Are we at the point of specifying, as Mr. Hamstra suggested, some interim criteria for the components of borehole seals, and if so, what should they be ?

R.D. ELLISON, United States

Put in the position of providing a conceptual design for a site, I would suggest that in the USA you take the existing EPA standards and divide them by a factor of safety of 10 or 20 and make that your goal, in order to find out what is necessary to meet it. Alternatively I think we might take the suggestion we had the other day of looking at variations of background levels.

J. HAMSTRA, Netherlands

I hate safety factors. It is the approach that most licensing people use. I think that if you do not trust or you cannot prove your components and their performance, then you should go for a separate and independent back-up.

R.D. ELLISON, United States

        The safety factor is suggested because we don't know what the regulatory authorities will decide, and we must assume that the limits will be equal to or lower than existing standards. The other justification is that there is more than one mechanism for release, and seal failure is only one of them.

M. GYENGE, Canada

        I see here a very strong analogy with civil engineering design. For example if I wish to design a beam made out of concrete, the beam must be very large. I have also the alternative of using steel. If I put the two together to make reinforced concrete, I still have to know the properties of each component, but the whole design concept changes.

T.G. CLENDENNING, Canada

        I would suggest that the sealing is not part of the engineered barrier, but related to the penetrations, and I think that that would tend to put the thing in perspective because it would relate the sealing to the natural system.

F.P. MUZZI, Italy

        I am new to this circle and I would like some information in order to clarify my mind. My main work is with site selection and reviewing nuclear power plants from the safety point of view.

        This morning the first speaker said that the performance requirements of sealing are mainly derived from the host environments but I have not yet seen clearly which host environment factors can really condition design. I see that many approaches to sealing are based on stiff plugs, which are placed in host environments which are not compatible as far as modulus is concerned. I believe that due to the time scale with which we are working, an earthquake problem cannot be avoided. Two questions arise. First, how will the seal perform in case of large displacements and second, how it will perform when waste is released ? You have now two important changes : one is that the seal has become very weak, the other is that the boundary conditions of the geologic environment, existing when the shaft, tunnel, or borehole were drilled, have been modified. What is the interaction of these factors with seismic waves ? Recent literature has shown for instance that surface waves may be very effective when you are deep enough so that you may have interaction with shafts or tunnels at frequencies which can be compatible with the frequencies of your system. In conclusion, a proper approach to the problem could be first of all to define which environmental conditions should affect design, then, having that clear in your mind, you can proceed to the required development work.

T.O. HUNTER, United States

        I think that could be an important factor in the design of plugs. It has been identified as one to consider in this country also ; some work has been done on the effects of seismic activity in underground locations. I think it relates back to the comment that Mr. Clendenning just made about how the penetrations affect the natural system. I will ask Mr. Hartung to give us a closing comment.

L.F. HARTUNG, United States

I'm enthusiastic that the meeting is going along the same lines that the Nuclear Regulatory Commission generally supports ; that there is significant advance in materials development ; that participants are generally taking the approach of the multiple seal in order to better meet the performance criteria, and that intrusion into the geologic barrier is being considered as the controlling condition and not the design of the engineered barrier itself. This interaction between the geologic barrier and the engineered barrier is very important, especially in long-term performance assessment. There are practical points that I would like to hear discussed later on, such as drilling techniques designed to minimize disturbance of the geologic barriers. As for placement techniques, we can come up with some very fine studies, but the fact is that placement of the material is ultimately going to be the controlling condition. I didn't see mention of verification techniques yet. It would seem reasonable that the verification of design would occur prior to licensing or upgrading of the licence application for operation. It would be a difficult situation to licence for operation and find out that you couldn't seal the repository. Also I think a very good discussion arose on the definition of the disturbed zone. I think this area has probably been neglected and I think there is the need to determine whether there is a disturbed zone or not. It seems to be open to discussion. I was asked when criteria were going to be available on, for instance, design. We are in the process of writing the first draft of the standard format content guide for site characterization which will include some design aspects and we had hoped to have this available next week.

T.O. HUNTER, United States

Everybody seems to agree that it is the overall system performance that first has to be identified, and that the actual risk associated with geologic disposal should be determined by a systems view. Further components such as borehole plugs will perhaps serve as independent multiple barriers. I feel that we still have to answer the question of quantifying how borehole seals are supposed to behave. Meanwhile it appears that the strategy will be to define the technology to determine what is available and what can be achieved and to try to do as good a job as we can.

R.D. ELLISON, United States

I think that the draft of the NRC's latest work needs to be changed because it doesn't always allow quantification of parameters. It leads to the conclusion that, the better the repository, the harder it is to seal it. In fact in case of a perfect repository it is almost impossible to have a seal that brings it back to the original conditions.

T.O. HUNTER, United States

The next subject is the status and importance of characterization of the borehole and shaft environment and I have asked Mr. Hamstra to speak on this issue.

J. HAMSTRA, Netherlands

I think that if we have to characterize the borehole environment we should start by characterizing the host rock formation itself. We have three types of host rock, argillaceous sediments, crystalline rocks, and rock salt. The next thing is that the scope

of the characterization in my opinion can be limited, because we are
chiefly interested in maintaining isolation of the waste, or considering release from it. If you want to characterize either the
borehole environment or the host rock itself, you should start by
paying most attention to the rock permeability values and to the
local geological condition. What I should like to stress is that
the best information always comes from in situ testing and I think
it should be site specific, or at least host-rock specific testing.
If you want to know where you stand with your host rock, about its
permeability and about ground water movement, the only reliable
answer is from in situ testing. The next thing you should realize
is that the moment that you start drilling a borehole or sinking
a shaft you disturb the original geologic environment. When drilling
boreholes, you end up with a lower hydrostatic pressure in the new-
ly made penetration and you change the stress conditions of the rock
material that forms the borehole walls such that a disturbed zone
will develop immediately around the borehole ; a disturbed zone
that to my opinion will have an increased fissure volume. We don't
know much about it. I think this disturbed zone is a very important
point relative to plugging of boreholes, because it is precisely
against that disturbed zone that we will have to make a proper seal.
A seal isn't worth more than the tightness of the system underneath
it and if there is a passageway along the seal you can have an
excellent seal but the overall system doesn't work. I think more
attention must be paid to the disturbed zone. It is typically a
rock mechanics problem to define what really happens and characterize the cataclastic response of the material to the man made
change in the local stress conditions.

On the other hand, we should not overestimate the importance of the disturbed zone. I dare to say that because, there are
many sealed boreholes, and if they are tight against an appreciable
overpressure, that proves that there is hardly any passageway around
them ; the only thing we should be worried about is whether this
disturbed zone can play a dominant role in the release of radionuclides. In future plugging tests, we should see whether we can
identify the importance of the disturbed zone in relation to the
tightness of the overall system, that is plug plus the surrounding
borehole wall. If we are going to characterize the borehole environment by means of permeability figures, you could argue that you needed
to make a plug better than the host rock itself. I think there would
then be a contradiction, because for instance the permeability of
granite is of a different order to that of rock salt. You would then
need a very good seal for rock salt whereas for granite you could
do with a rather permeable seal. I don't think that is true. For
granite, I think that you need some sort of compensation for the
higher permeability of the host rock with a better seal and a better
containment of the waste. So "where are we ?". Well, we are at the
beginning, we all realize that. I think that we need more tests such
as the Bell Canyon Test with specific attention to the importance or
even the existence of the disturbed zone. I'm sure that in different
countries, as soon as they have the opportunity to get access to
boreholes, these types of tests will be carried out. How far do we
have to go ? We should remain realistic since it is the overall
system we are examining and we shouldn't lose ourselves in too much
detail.

K. MATHER, United States

What has impressed me is how much the concepts of people
from different countries and different parts of large countries have
been affected by the regional geology and the specific geologic
histories of the areas in which people are considering siting
repositories. I think you can't say too many times that the investigation, the engineering and the adjustment of the design to the
milieu is enormously site specific. It is as specific as the

successful location of a large dam. There are some kinds of sites
where you do a great deal of repair work in the foundation and then
you construct a sophisticated multi curved arch. There are other
very, very bad foundations where you can only put in grout curtain
after grout curtain and then four extra for luck, and come along
and build a very gently sloped earth dam. Every engineering work of
man that has to do with nature has to be intimately fitted in all
of its aspects to the geologic milieu and errors in the investigations of the site will come back to haunt you and, maybe, to destroy
you.

A.M.L. BOULANGER, France

        Je voudrais faire un commentaire sur ce que vient de dire
M. Hamstra sur les granites. Je ne pense pas que l'on puisse baser
toute la sûreté du dépôt sur les matériaux de remplissage. Nous
devons démontrer au préalable que la barrière géologique est efficace et celà implique de savoir qu'elle va être l'évolution de la
perméabilité du fait de l'échauffement des déchets et l'évolution
de cette perméabilité et des matériaux de remplissage au moment du
refroidissement. L'objectif des matériaux de remplissage est simple-
ment d'éviter qu'il n'y ait pas de chemin préférentiel.

E.W. PETERSON, United States

        I mentioned in the talk about flow in the
disturbed zone around the borehole plug that in that particular
hole, which is in anhydrite, we don't know the exact permeability
value, but it looks like it is very, very low. We have also made
permeability measurements at the WIPP site in the salt layer, which
has been considered as one of the potential disposal horizons. Those
measurements were made with a system which can very accurately measure flow along the disturbed zone, because it allows to differentiate flow going horizontally out into the formation from that going
back up around the packer system, either through the formation, or
through the disturbed zone. Again, in those measurements we found
that the flow going out into the formation was large compared to
anything going through the disturbed zones. In other words if we
assigned a uniform permeability to the salt zone we know that the
flow going in the vertical direction was less, and we are talking
about microdarcy values. So there is some evidence that these
disturbed zones may not be as bad as some people think. Without
any doubt they are important to look at, but I don't think that
they are catastrophic.

        We have carried out tests in many holes and everything
changes as you move down the hole and change position in it. I
think that in many holes there are zones that just cannot be plugged effectively and others that could be plugged very well.

J. HAMSTRA, Netherlands

        I would like to comment on your remark about shafts and
boreholes. We should differentiate between these two because some
types of boreholes such as disposal holes may be very near to the
source of the radionuclides. If you have a disturbed zone there
then you must worry about the start of a pathway into your host
rock. However if you look at the overall repository system then
you only have to worry about the disturbed zone around the shaft
and be sure to use proper shaft construction methods. But in my
opinion the disturbed zone will never become an important pathway
for radionuclides to return to the biosphere. So you must be very
careful to differentiate between types of disturbed zones.

D.M. ROY, United States

I think we are using different terminologies here. When we say boreholes we usually mean the exploratory holes and not the waste disposal holes. On the question of in situ testing versus laboratory testing, I think both are valid, although of course in situ testing will have to provide the final answers. In my paper I presented data on laboratory measurements of a disturbed zone. This was from a borehole which had been plugged and subsequently redrilled. The permeability of the plug material was about three orders of magnitude greater than the permeability of the rock. There is a problem as to whether the material is restrained as it would be in situ but I think the differences in the absolute magnitudes of permeability which we found in the laboratory would still appear in situ.

T.J. CARMICHAEL, Canada

There are a few points regarding the fractured or disturbed zone that have not been discussed. The fractured zone is continuous and is short in its total length compared to the length of pathways, usually considered, from the repository to the surface. Also the fractures are fresh, not weathered, therefore, they would not have the clay-like deposits within them which would retard nuclide migration. A third point is that current engineering concepts frequently show a continuous concrete lining. With such a lining, I am sure that in conditions of stagnant ground water a very high alkalinity would be preserved within the cracks and that this could have an affect on the sorption capability of the rock fissures. The concrete lining therefore might have an effect on the transport of nuclides.

T.O. HUNTER, United States

Does anyone know of a radionuclide sorption experiment on concrete ?

D.M. ROY, United States

We have done a little but I wouldn't want to rely upon it. I think more needs to be done.

K. MATHER, United States

I think that if you look into the literature on radiation shielding you find quite a lot of information about neutron and gamma-rays transmission, but much less about radionuclides transport.

W. FISCHLE, Federal Republic of Germany

In Germany we hope to start testing at the beginning of next year. The waste is incorporated in concrete and we will immerse it in a barrel filled with brine to investigate its behaviour.

T.O. HUNTER, United States

An important point appears to be that the borehole environment and the design of plugs are very site specific and can only be addressed in a site specific sense. Laboratory and in situ measurements are very different, but we should make sure to consider both in our assessments since they complement each other. Furthermore we need to address the difference between shafts and boreholes, but not to over-emphasize the disturbed zone, since it may be unwarranted in terms of radionuclide release.

R.H. GOODWIN, United States

I would like to comment on the regulatory aspect. We had a lot of discussion on what constitutes a disturbed zone and how extensive it is. I think it would be fair to say that there is some kind of a disturbed zone which implies an increase in permeability of the area around the borehole and I would suggest that there is little opportunity to eliminate that totally from the system. It has been suggested in some of the documentation that has come out, that criteria for a borehole plug should include returning the rock to its original permeability. Let us recognize that if that stands, we are defeated before we start.

T.O. HUNTER, United States

Let us close the discussion on this topic. The third topic is the role and function of the different components in the seal system and how much importance should be allocated to them, given adequate site and borehole environment characterization and some identification of the criteria. Also we will consider which operational factors have to be considered. Mr. Ellison will introduce this topic.

R.D. ELLISON, United States

The first of the components that was listed was seal geometry. We should conclude that at least in a number of cases there will be some disturbed zones. We should not over-emphasize their importance but we cannot ignore them in the analyses. For shafts and tunnels there seems to be a consensus that disturbed zones should be expected, owing to the stress configuration around the excavations. For these there will be some requirement for multiple seal geometry. We need to show the extent and importance of variable seal geometry. We can do this by analysis and by simulated field tests. ONWI has a request for proposals for performing simulated field tests in the laboratory to evaluate the disturbed zone in relation to the importance of seal geometry, but eventually we will have to move out into the field. Any parametric studies designed to see if the flow path through a system is adequate, or if the stress resistance of a system is adequate, should include some changes in seal geometry. It would appear that if complex seal geometry is a potential need, then it should be included at least in the initial phases of the analysis, and in simulated field tests and equipment evaluations.

The second topic was multiple layering. The consensus seems to be that the basic benefits are evident. They are being expressed in the United States by the regulatory agency, and it would appear that maybe the greatest benefit is that the multiple layering system is going to be a prime way to satisfy the longevity requirements. In single component seals we either don't have enough data or enough time to demonstrate that any single material would withstand all scenarios. Multiple seals provide redundancy, and that seems to be a very important issue. Most investigators are leaning in the direction of a multi-component seal system. We need to decide on the combinations of materials that are appropriate for individual sites or for generalized site conditions and then on the design that can satisfy the criteria existing at that time.

The next step is to proceed with material studies, on the basis of some priority order, so that the materials that are definitely unsuitable are discarded, and those that are suitable, either by themselves, or as part of a multi-layer system, can be incorporated into seal design. We need to perform preconceptual design studies of multi-component seals immediately, and also to carry out performance analyses to see if the results relate to the established criteria.

The next question is on nuclide sorption by seals. Some people felt that in a multi-layer seal materials with high sorption capacity should be included ; others expressed the opinion that an absorbing layer may build up an undesirable accumulation of radio-nuclides, or in fact it may work against the basic premise of mini-mizing flow. We need to determine those materials that can be suitable for sorption purposes in site specific environments. We need to per-form analyses to answer the serious concerns some people have raised and then we need to perform detailed material studies on options that appear viable.

The next issue was retrievability. We are working on a number of different repository projects, and in these projects we are using requirements that vary between retrievability for five years after a given canister is in place, to retrievability for a 50 year period after the entire repository is completed. Obviously they are very different. It is going to be extremely important to define what retrievability means. Looking at NRC's most recent po-sition, perhaps people in the United States should start to take the more conservative approach. We need to evaluate the impact of retrievability on sealing options, particularly in relation to phasing and staging the sealing of shafts and tunnels.

I have less to say on operational requirements. It would appear that the European designs require that the sealing operations take place in direct conjunction with repository operation, so there is a much closer tie. In the United States concepts a major opera-tional aspect is what must be known about sealing at the time the shaft is constructed and lined in order to avoid problems later on.

W. FISCHLE, Federal Republic of Germany

I want to ask you a question on your first point on geometry. I understood that you may want to change the geometry of the holes. My question is why ? Is it then a problem of how to prepare another sort of hole for a complex seal geometry, since boreholes normally have standard dimensions ?

R.D. ELLISON, United States

My point was that if there is a disturbed zone one way to compensate for it is to change the seal geometry.

W. FISCHLE, Federal Republic of Germany

There would always be a disturbed zone, even if you change the hole geometry.

R.D. ELLISON, United States

If the disturbed zone is time or environment dependent, then by changing the geometry immediately before placing the seal, I may have eliminated those factors. If the disturbed zone is related to drilling methodology, and if it is very costly to drill without a disturbed zone I may elect to drill by normal boring using an inexpensive method, then come back with a refined method

to rework sections that require great precision. Enlarging boreholes
is a routine operation. In some rock types it will be more difficult
than in others, but what we need is the evaluation that existing
methods are suitable for this purpose.

J. HAMSTRA, Netherlands

Concerning the geometry and especially the diameter of
the waste emplacement boreholes, I think that the controlling input
data concern the thermal loading of the canisters. The bigger the
diameter, the greater the linear heat source will become ; over a
certain level the stress loading caused by the temperature rise of
the rock will also create a disturbed zone.

I think that the licensing people who require retrievability should firstly require an in situ demonstration of the
repository system to see whether or not the design functions properly. If retrievability for 5 or 50 years is required a short-term
test analysis should be performed and I fully agree that such an
analysis is completely site specific. Flooding of the mine during
the operational period might have to be considered, and in that
scenario the design changes imposed by the need to ensure retrievability might be responsible for more serious consequences. That
should be weighed against the requirement to ensure retrievability.
In our site specific or generic analyses for repositories in salt
domes where the hypothetical possibility of flooding must be considered we have eventually rejected retrievability. If the geologic
disposal system cannot be fully trusted the waste should be kept
in storage in a surface facility.

M.J. SMITH, United States

I hope that the regulatory requirements will not establish
how each of the individual components should behave. I would prefer
criteria that establish how the whole system is to perform. If I
know the expectations for the performance of the whole system, then
I can determine the performance required of the individual components
in order to meet the overall objectives.

J. HAMSTRA, Netherlands

We shouldn't compare a repository with, for example, an
aircraft, because an aircraft is made of many parts and everybody
knows how to meet the standards for each part. Repositories will be
all different and in each case we need the freedom, almost as
artists do, to use the correct approach. There will never be enough
repositories that specifications can be defined as is the case
with boilers or airplanes.

T.O. HUNTER, United States

I'd like to move on to the fourth topic which is the
status of grouts, concretes and cements, in relation to their use
in borehole plugs. Mr. Boa from Waterways Experiment Station will
begin with some comments.

J.A. BOA, United States

The statement that has been made all morning is that everything that we do is site specific and host rock specific. I'm in
the business of designing grout mixtures and we were actively involved in the Bell Canyon Test. There has been a lot of discussion

on the definition of criteria. Component criteria and overall criteria with respect to the design of grout mixtures for site specific applications. We define the criteria for each component in order to meet the overall criteria. We design grouts that are highly pumpable, that have a minimum amount of workability time, that are level-seeking, that are expansive, that have a high strength and have a variety of other physical properties. We do some work with concretes and other materials but our main interest is in grouts. The boreholes that we are involved with are boreholes that happen to be in the same general area of the repository, but they do not intersect the repository. They are former exploratory holes or oil wells and new exploratory holes. In these cases we are not concerned with the heat that would be generated by the waste.

The mixture that was used at Carlsbad was composed, as you saw, of specific materials. The cement was a class H Portland cement. It is an oil-field type cement that has a very low fineness, approximately 2400 cm^2 per gram. Consequently it has a low water demand. We will be using in the future, and have used in the past, type K shrinkage compensated cement which is now becoming available again in the quantities that we need. We will also, I hope, be able to use in the future some self stressed cement that currently is no longer available. We are also using a proprietary expansive system from one of the large oil-well cementing companies. We also have available a variety of rock types from the Carlsbad area in the event there is other work in that area where we can definitely tailor the grout mixtures to the particular rock type at whatever location we happen to be working. We consider this a very important factor. It appears that the Bell Canyon anhydrite is probably one of the most difficult to work with. We don't anticipate the same problems with limestone or another rock type, the Bell Canyon anhydrite just happens to be a particularly tough rock type to match in order to get a good chemical bond at the interface. We got a good mechanical bond, upwards of 900 psi, but the chemical bond, the interaction of the grout with the wall rock, has not been exactly what we would have liked.

Where do we need to go with respect to our specific grout ? We want grouts with low water demands. Water/cement ratios in the 0.3 or less range. We want our grouts to be expansive. We want them to be geochemically stable with few or, hopefully, no phase changes ; but that is probably not possible. We would like the grout to be impermeable. We would like to have an interface that allows no passage of water and we would like to have a grout for each rock type. We think it is possible. The next step is that we need the simplest cementing or grouting system to facilitate placement and minimize chances of error. If we could have a system of cement and water with all the required properties that would be wonderful, but we have to put so many components together to arrive at the desired results that sometimes there is a chance for error. For example, we will add some chemical ingredients to the mixing water to improve the chemical bond of the particular grout that we are using right now, which is a fresh water grout as opposed to a salt water saturated grout. We hope this will improve the chemical bond with the anhydrite ; we may use other additives for other rock types that will be coming up in the future.

T.O. HUNTER, United States

Is the overall status of available technology adequate for emplacing grouts and sealing penetrations with the sizes of shafts and boreholes ? Finally what is your opinion as far as the availability of technology to successfully monitor the emplacement and test the quality of the seals ?

<u>J.A. BOA</u>, United States

We are also actively involved with the work at the Nevada
Test Site ; there we have grouted tunnels that have large diameters,
of the order of the excavations that may exist in repositories :
5 to 10 m in diameter. The technology exists to pump these grouts
vertically down holes, or horizontally up a slight incline, through
small diameter pipes. You can fill up, for instance, a tunnel 6 meters
in diameter, 30 or 60 m long, through a 7.5 cm line in a very short
period of time by mixing 30 to 45 m^3 of grout ; mixing, pumping and
placing that amount of material takes about an hour. So the technology exists in the hands of the large oil-well cementing companies.
The size of the holes that can be filled varies within a wide range.
It can be done, and can be done with ease, if the grouts are properly
designed and properly supervised in the field with respect to dry
batching and wet mixing. I might point out that generally these
grouts are composed of something other than just cement and water.
There are other ingredients that give the fluid properties that are
needed for the emplacement.

<u>W. FISCHLE</u>, Federal Republic of Germany

You said that you are looking for stable materials. Does
that mean that you are at the beginning of these tests ? What do
you mean by the term "stable" ?

<u>J.A. BOA</u>, United States

The initial request was that they should be stable for
100,000 years. Unfortunately there is no way to accelerate real
time. We can apply heat to give us some early results that would
indicate the strength after 28 or 56 days. With respect to simulating
years, we have not found a solution yet. We have specimens, some
salt saturated mixtures and some mixtures that are not salt saturated,
that have been produced approximately 4 years ago, and that we are
still testing in non-destructive and some destructive tests.

<u>C.W. GULICK</u>, United States

Concrete and grout can be emplaced at almost the same rate.
There are a number of techniques for emplacing cementitious plugs
that could probably meet any of the criteria.

<u>K. MATHER</u>, United States

The American Society for Testing and Materials has
standardized 3 methods of accelerated curing of concrete from which
calibration curves to predict, at least, strength can be set up.
I like the one which is the gentlest. It does not change the hydration conditions from those which you would find in reality. If the
concrete is going to be exposed to elevated temperatures this needs
to be reflected in the mix composition. With short tests we can
predict the strength after 28 days, and in a little longer, what we
will get in 90 days ; if we characterize the mixture well in terms
of elastic modulus and creep characteristics we can probably make
some rational projections up to several years in the future.

<u>G.M. IDORN</u>, Denmark

In Denmark we are developing a hydration technology that
allows to simulate the hydration in concrete. This is a new approach
that in this particular form has not been fully tested.

T.O. HUNTER, United States

In conclusion, grout, cement and concrete technology appears to be adequate for whatever role is needed in waste repositories in terms of emplacement and in terms of matching properties to the site specific nature of the shafts or boreholes. The principal questions remaining are those of defining the overall criteria which seals have to meet and of long term stability which we will discuss next.

To summarize we've talked about requirements and criteria and, even in a philosophical sense, tried to narrow down what needs to be done in these areas. We have talked about borehole environment characterization and the role and function of components. We had a discussion on cement technology. There are four things we want to cover in the second half of this session. The first question is how to address long-term stability. We want to spend some time on methods of analysis, and also devote a little time to the question of the role of field and in situ testing and how they support the development effort. And finally we want to pick up the question of further international co-operation. Let me open with the first topic which is that of long-term stability and I ask Dr. Chapman to provide us with an introductory statement.

N.A. CHAPMAN, United Kingdom

This seems to be one of the most intractable problems of them all and I think I'm being asked to pay some kind of penance by introducing it. First of all, longevity of which particular property are we trying to assure ? I think that what this boils down to is the longevity of the hydraulic properties of whatever plug or seal we're considering. So, for example, mineralogical stability, which we can predict in some ways, may be no guarantee of long term bonding properties of a plug and these are the properties which have a great bearing on hydraulic properties, because essentially what we're trying to do is prevent, or at least control, flow through or around the plug or seal. From the mineralogical point of view, I think there are three ways of assessing the longevity of the seal : the first one of these we touched on briefly already, and that was the archaeological record of various synthetic materials. This is slightly difficult I think. We can only go back maybe two thousand years at the most in looking at cements, and the conditions which ancient cements have had to endure are considerably different from any that might exist in a rather harsh repository environment, where the geochemistry is entirely different. So I don't know how much store we can set by the archaeological evidence. It is used by some people to look at the behaviour of potential canister materials, if ancient artifacts are available, and even at potential waste forms by looking at the stability of man made glasses. But as far as seals are concerned, I am rather skeptical. The second way of approaching the mineralogy of seals is basically an experimental one by tailoring the behavior of various synthetic materials to specific, known physico-chemical conditions in a repository. I think this area shows considerable promise, but it must be backed up with field evidence. There we are faced by the problem of time and kinetics ; since we cannot mess around with thermodynamics, and for example we cannot push up the temperature in an experiment to induce ageing. If we did that the system would enter a different reaction field, and the results might not have anything to do with what might happen at lower temperatures in the long term. A possible solution, which was suggested by one of the papers, is to take a pure thermodynamics approach to predicting possible reactions which might occur in a synthetic material. But the kinetics of this are the big problem, and I don't know if the thermodynamic data, that is enthalpy, entropy and ΔG functions, are available in a rigorous enough form. We know them for a few pure end members, but for mixtures, solid solutions and such like, we have very little data at all. However, if we combine the experimental and the thermodynamic approaches we have a basic predictive tool and this is probably the best method for looking at synthetic materials. The last approach and probably the most convincing is to use natural materials which have proven longevity. Various talks have suggested the use of crushed basalt, basalt-cement, halite, bitumen and last, but not least, highly compacted bentonite. All are proposed either for use alone or in combination with other natural or synthetic materials. What clearly must be guarded against when using a natural material is putting it in an environment in which it does not occur naturally. Modelling the behavior of natural materials requires caution, but they still seem to offer the best chance of ensuring the desired longevity.

So far, we have talked about the problem from the mineralogical point of view, but the mineralogical prediction is probably the easiest to do. The major problem, as far as ensuring the longevity of the hydraulic properties, is in the mechanical stability of the seals. We have to consider the differential stresses to which they might be exposed as a result of the different mechanical properties of the rock compared to the seal and we must look at the loss of bonding. The latter is going to be very difficult to assess in the long term. It was pointed out that we should look at seismicity as well and I wish to add high hydraulic gradients. These are all quite difficult to assess, but they are the main factors that might lead to loss of integrity, rather than mineralogical breakdown.

Where do we go from here ? The first step is to define what exactly we need in terms of longevity, and this brings us back to the old problem, since the requirements are site specific. I think this is the truism emerging from the meeting, that the assessment of a seal or a plug is not only site specific, but it is specific to a particular borehole, or a particular adit, in that specific environment. For example, in crystalline rocks, one of the things we want to guard against is short-circuiting of hydrogeological environments, such as flow-cells. The behavior of a plug in a horizontal tunnel at depth, which, in case of failure, could connect two different flow-cells must be different from the behavior of a plug in a permeable medium which stretches down from the surface, where we are simply trying to replicate the hydraulic properties of the surrounding rock.

Perhaps our primary aim, then, should be to ensure a longevity which is of the same order as of other parts of the local geological environment. In other words, we should ensure that the presence of a sealed cavity does not adversely affect the bulk properties of the rock which we may have used in assessing the safety of the repository. Now, some people who are involved in, for example, the assessment of the radiological consequences of geologic disposal propose some sort of time cut-off in the future, beyond which they feel it is impracticable to make a valid assessment. Some suggest this might be 10,000 years and some suggest shorter periods of time. The reason for this is that longer term realistic dose calculations are not possible due to our lack of knowledge of what people will be eating and drinking that far in the future. Is this sort of approach applicable to the assessment of plugs and seals ? Should we also have a cut-off point ? I think this is a contentious point and something we might want to discuss. However, when we try to put the borehole seal in its geological environment, what we are trying to do, rightly or wrongly, is to compare the expected life of a sealed cavity with that of the geological barrier itself.

I wish to point out that nobody has presented any hydrogeological modelling to show what exactly is the effect of an open borehole, or an open shaft, on flow within a particular environment, be it a generic environment or a specific environment. We've had some cases presented for doses which could be assumed for a repository located in a stratified deposit, but for crystalline rocks we don't really know what we need to protect against or how far into the future we must do this. This is something that should be done right at the outset before we can start thinking in terms of longevity. We need to see what proportion of the overall radiological risk can be attributed to open or poorly sealed boreholes. I think the results will vary from rock type to rock type.

I am not going to draw any conclusions, but one direction we should take is a joint approach to instrumentation of long-term field tests. I think we need to start, now, on some long-term field testing of various sealing options. We have seen some excellent results from the Bell Canyon Test. We know there is going to be a full-scale test at Stripa of back-filling, using bentonite in a horizontal gallery. What we need to do is to adapt our approach to tests which last for a period of years rather than months and to do this we need to develop some sort of instrumentation for measuring the behavior of plugs and we have to look specifically at bonding properties. Even within the time scale of these experiments the geochemical behaviour is not going to be particularly important. I don't think you can draw any conclusions on this issue of longevity, simply because we have no criteria for sealing and we have no basis on which to decide what we require in terms of longevity. Those criteria, as I said before, could be based on some contribution to the overall risk.

W. FISCHLE, Federal Republic of Germany

I have two comments. You asked the question : what happens when a shaft is open ? We have made measurements over a few years as to what happens in an open hole. We have found that there are stable brine pockets within several meters of the wall. Secondly, for long-term stability determination we had a program in which we tested concrete products. Most measurements disturbed the probe and, therefore, we could measure only once before replacing the probe, which is clearly impractical. Is there anybody here who can advise on making undisturbed measurements which are really useful in this context ?

C.W. GULICK, United States

During the Bell Canyon Test the volume of samples we took was much greater than the volume of the plug ; therefore we have a large number of samples that could be used to carry out tests that simulate long periods of time. In addition, as the work continues, we are using what material is available for non-destructive testing, and primarily for sonic velocity and dynamic modulus measurements on the same sample over several years. We are also monitoring weight changes and in some cases chemical changes in the water associated with the specimens. We are open to suggestions on both our non-destructive and destructive testing programmes.

W. FISCHLE, Federal Republic of Germany

We are doing the same, but we cannot yet say what has happened. We find changes of weight for example, but we cannot say whether this is good or bad for the material under test. All our other tests are at the beginning.

T.O. HUNTER, United States

This is one of the last points that Dr. Chapman raised, that is instrumentation, particularly in situ instrumentation, to monitor these transitions, if they occur. Going back to the beginning one thing that he brought up was the question of the period of time with which we are really concerned. I think that he mentioned a 10,000 year period. I know that in the U.S. the 10,000 year period has been a base-line case because of comparisons between repositories and natural ore bodies. I would be interested in hearing some other comments on that period of time.

J. HAMSTRA, Netherlands

I think that the time period is determined by the system performance criteria. If the risk analysis shows that release is at an acceptable rate after a few centuries or a thousand years, then a thousand years is the target time. There is no need to put a factor of 10 on it and say that we have to ensure a plug life of 10,000 years. In our opinion, it is something near to a thousand years and maybe even less, because you need not worry after such a time period since releases are so limited. If the licensing people accept our system performance analyses, as far as releases are concerned, then they should accept this time period as a longevity requirement.

T.O. HUNTER, United States

Is that time period based on spent fuel or on reprocessed waste ?

J. HAMSTRA, Netherlands

It is based on reprocessed waste. However, it doesn't really make much difference, since it is a question of pathways and release rates. There is a difference in fission products content but that is not too important.

F. GERA, NEA

I think this is a reasonable approach but of course the result would be site specific. It would depend on the particular pathways that are important at a particular site and, in general, on the results of the safety assessment. If we are looking for a generally acceptable criterion for a time cut-off, we are probably forced, before site specific assessments become possible, to compare the toxicity of the waste with the toxicity of natural radioactive elements, for example in uranium deposits. But I agree with you that in the end, when we have a site, specific pathways can be analyzed, the radiological consequences of plug failure evaluated and the future time when we can stop worrying about this particular transport system determined.

J. HAMSTRA, Netherlands

I would like to check with Dr. Chapman, but I think that Ms. Hill also took a generic approach and the time limit beyond which we don't need to worry at all was also something like 1,000 years.

N.A. CHAPMAN, United Kingdom

I think that the figure was 10,000 years, which was suggested as a cut-off point for population dose assessment.

T.G. CLENDENNING, Canada

I am very unsophisticated in the technology of concrete, but I would like to make a few comments regarding the leaching of concrete. I would like to point out that the larger the mass of the concrete, the more it would influence the surrounding environment. In other words, a thin wafer of cementitious material would have little impact on the hydrology or the geochemistry of the area, but a large mass would have a major effect. The Swedish program proposed that, in the disposal of hulls or metal components of spent fuel assemblies, a large mass of concrete should be used. In the process, they predicted a retention of alkalinity which was beneficial to prevent the escape of Ni-59. They predicted 100,000 year retention at an alkalinity in the order of ten. Now there is another consideration, and that is what is left after leaching has taken place. For example the properties of the residues of leaching and their nuclide sorption capabilities, or the permeability of the mass that would remain when most of the soluble ingredients of the system have leached away, might be worth consideration since they might determine the long-term behaviour of the plugs.

T.O. HUNTER, United States

One of the other comments that Dr. Chapman offered was that we need to concentrate on the properties we are really concerned about. I like the idea of concentrating on the hydraulic properties, because they are of primary interest, although there are others of course. Does someone have a comment on what properties should really be addressed in the long term ?

J.C. WRIGHT, United States

I guess it is a matter of semantics, but I would not use the word "properties". I think a better word would be "functions". What we want to do is to define the functions that are required of plugs or seals at a given location, whether it is to sorb, or whether is is to be impermeable. Then we need to assess what properties bear upon the material being able to fulfill that function. The longevity of these properties will determine the duration of time during which the seals can be expected to perform their function.

T.O. HUNTER, United States

Yes, I think that is appropriately stated. I am reminded of a comment that I think Bernard Cohen made, that even if the material degrades, it doesn't go anywhere, and the hole is still filled with a mass which should provide some limitation to the potential flow through the hole.

J.C. WRIGHT, United States

Maybe we need to define what the function of the plug should be as a function of time. In other words, does the plug have to perform in the same way in the first 300 years as between 300 and 3000 years, or between 3000 and 300,000 years ? Does the function of the seal have to stay the same ?

T.O. HUNTER, United States

Does anyone have a feeling about this particular aspect ? I don't know if we have reached the point where the required function can be defined as a function of time.

D.M. ROY, United States

Maybe the numbers could change depending upon the waste form and the decay constants of the radionuclides.

G.M. IDORN, Denmark

It may be interesting to observe that much Roman concrete which, from a concrete science point of view, has degraded with the course of time, in fact has become less permeable because the pores have been filled with degradation products.

D.K. SHUKLA, United States

We have to remember that if the permeability increases with time, the radioactivity decreases. So it might be possible to allow a higher flow and still maintain the function of the seal.

J. HAMSTRA, Netherlands

We should be careful with that assumption. Whilst the fission products decay in the first few hundred years, there is a very slow decrease in activity thereafter so I don't think we can allow flow to increase.

D.K. SHUKLA, United States

That is why there is a case for multiple plugs. Some layers have very low permeability for the first 300 years, but they may disintegrate afterwards, then some other components, which might have somewhat greater permeability, will ensure the plug function over the long term.

T.O. HUNTER, United States

Dr. Chapman raised two other points that I think we have time to get a few brief comments on; the first one is the use of natural materials. Would someone care to comment on an overall assessment of their potential stability and the appropriate technology for using them ? The second point addressed the methods by which synthetic materials may be analyzed.

F. GERA, NEA

From a conceptual point of view I like very much Dr. Chapman's suggestion that natural materials are the only ones that give you some feeling about their possible durability. This will come out again when we talk about field testing. If we are really concerned about the durability of the plugs over long time periods, we will have an impossible task to demonstrate that the plugs will last unless natural materials are used. In the end if for practical or technical reasons it is necessary to use cement in a part of the plug, other sections will be made out of natural materials that we know are geochemically compatible with the geological environment.

J. HAMSTRA, Netherlands

In rock salt we have to worry about convergence of the boreholes. If we could define the worst condition and subject that

amount of seal material to the lithostatic pressure that is to be
expected in the plug area, then we might come up with permeability
data that are acceptable, although if the material deforms and
remains in position we may have no cause to worry.

T.O. HUNTER, United States

Yes, deforming materials definitely have some advantages
from that standpoint.

W. FISCHLE, Federal Republic of Germany

It is good when you have a material whose permeability
becomes lower when it is under pressure, but as it comes under
pressure water enters, and becomes brine, and brine is very agressive
to certain plugs.

D.M. ROY, United States

I agree with Dr. Chapman's comments ; however I wish to
point out that most natural materials are not composed by a single
mineral. The cation exchange properties of clays are well known.
The older clays in equilibrium with brines and salts are generally
non-expanding ones ; so we have to take all these things into
account in our models of the long-term behaviour of the plugs.

M.J. SMITH, United States

Somebody asked a question about emplacement techniques
and machinery. We are sure of the ability to emplace the materials
and to plug according to the pre-conceptual designs that I showed
the other day. We made a fairly extensive study of existing machinery
and equipment, and we identified no problem with materials emplacement. There is a document on this work coming out soon.

An additional comment is about the required functions of
borehole plugs. How to estimate what the function of a plug is, is
made even more difficult by the fact that we don't know what are
the general requirements for penetration sealing, not to mention
the required performance as a function of time. I think the functions
that we have come up with are rather generic and simplistic. But
they do bring up questions as to whether or not we need certain
components in the plugs. For example, one function that might be
required of the plugging system is to restrict ground-water transport of toxic materials by both chemical and physical retardation.
We might also want the plugging system to isolate water bearing
zones from the repository and to maximize the travel time of migrating radionuclides, if the plug were to be the primary pathway to
the biosphere. Another reasonable requirement might be that the
plug has structural strength, in order to remain stable. The system
should also exhibit compatibility with the host geologic environment.
Finally, the plugging function must last for the required period of
isolation. Those are the functions that we have come up with. After
the generic functions have been defined, we can begin to address
their required performance as a function of time. For example, during
the period of time when the temperature is high, the materials must
exhibit the right properties in order to survive to those particular
physico-chemical conditions.

A.M.L. BOULANGER, France

Je voudrais faire un commentaire sur l'utilisation des
matériaux naturels. Les matériaux naturels ont une durée de vie qui

est déjà prouvée mais dans des conditions physico-chimiques bien précises et il faut prendre en compte les modifications qui sont amenées par les ouvrages qui sont faits. Prenons un exemple : excaver des granites et les stocker en conditions atmosphériques pendant la période d'exploitation du dépôt peut perturber ses propriétés et modifier sa perméabilité.

R.H. GOODWIN, United States

        I would like to inject a note of caution. I think we more or less agree that natural materials will remain mineralogically stable. Dr. Chapman alluded to the fact that if we put a natural material into a borehole to create an impermeable layer or to perform some other function, the longevity of that function is not guaranteed by the fact that the material will remain mineralogically stable. In other words we may still have clay, or whatever material we put in, not altered mineralogically, but the bonding properties may have been lost. So the longevity of the bond becomes the question. Proving that it will still be an impermeable bond over some period of time may not be straight-forward, and we should not take the attitude that because the material will remain mineralogically stable, our problem is solved.

T.O. HUNTER, United States

        I think that Dr. Chapman's introduction pointed out the essential elements of this problem. The identification of the longevity that we are trying to achieve should be foremost in our minds ; if we couple that with the identification of the function and of the properties of the plug system, not just the plug material, we might be able to develop a basis for the evaluation of the long-term stability. We did not address the question of techniques for carrying out verifications of the longevity in situ. Dr. Chapman mentioned that we should discuss that, so perhaps we can comment when we talk about field testing later.

        It would be appropriate now to discuss where we stand on methods of analysis for plug performance and I ask Mr. Peterson to introduce the subject.

E.W. PETERSON, United States

        I would like to talk about methods of analysis for plugs and any other type of barrier. For any of these methods of analysis, first you want to be able to use them to do system design, second for system evaluation such as the evaluation of plug performance, and third for long-term performance analysis. Conceptually there are many methods of analysis now available that can do almost anything one can imagine. There are very large multidimensional codes, that can be used to perform structural analysis, rock mechanics analysis and flow analysis. Almost anything can be considered, such as overburden stresses, creep phenomena, relaxation phenomena and the effect of thermal loading or pressure loading. The current state of the art allows to couple rock mechanics calculations with fluid flow calculations so the effects of changing stresses on permeabilities can be evaluated. For example, in analysing plug performance, one can put in anisotropic permeabilities, or permeabilities that are functions of pressure and stress fields. There are,however, a couple of problems that arise. One of the things we have talked about,in terms of well-bore damage, is microfractures. While these models can treat fracturing conceptually, they have been notoriously unsuccessful in prediction of fracturing, even on a large scale such as in the case of hydrofracturing, and it is much more difficult to model the microfractures that might be generated by

drilling. It is possible to do a calculation that will predict fracturing, and from the fracturing we can predict the porosity, and
from the porosity we can predict the permeability. The question is :
what is the meaning of all this ? Our models cannot predict the fracturing or the disturbance that is going to occur in a repository.
There are several reasons for this. The main problem is that realistic constitutive relations for rocks, grouts, and cements, such
as expansive properties and stress relaxation properties, are not
available. They are not available, because it is almost impossible
to measure them. We can make laboratory measurements and get numbers
that might not have anything to do with reality. The other critical
thing is that, with any of these calculations, even if we couple
them all together and do fluid mechanics and fluid flow and nuclide
migration analyses along with structural analysis, we are never
going to know the in situ stress conditions that exist either all
the way down the borehole or throughout the repository. There is
no way that you can ever determine all the in situ stress conditions.
Small flows through slightly fractured media, are dependent on the
initial in situ stresses and the stress history that has been imposed
on the formation over the time period under consideration, which
includes the thermal and pressure transients. So although the models
can handle all these effects, for a long time yet we are not going
to have the required input data. Our next step should be to start
developing methods of analysis that incorporate the uncertainty in
the data. We could have put much more detail in our plug performance
analysis, but I do not think it was warranted.

T.O. HUNTER, United States

        I would be interested in other comments on the subject
of the role of methods of analysis in defining either how plugs
should be made or how they should perform.

        Dr. Chapman raised the question of perturbation of the
hydrologic system by the drill holes. Some studies have been done
on that aspect, but we have not considered it. Perhaps in many formations, it could be an important contributor to radionuclide release
scenarios. We must also consider how we would determine flows into
a repository under operational conditions. Certainly, sinking a shaft
through water bearing materials is bound to have an impact and to
create the need for a certain type of analysis. Moving on to the next
topic, which is that of field testing of seals, I have asked
Mr. Christensen to give us an opening comment.

C.L. CHRISTENSEN, United States

        Whilst we are playing a game of which we do not yet know
the rules, I feel we can make some recommendations as to what those
rules ought to be. If we make recommendations that are based on what
we know or what we believe we can do, and the regulatory agencies
do not accept them, it is unlikely that anyone else is going to come
up with solutions which meet their requirements. However, it is not
clear that everybody here today agrees as to what those requirements
should be, so how can we be expected to be able to defend our designs ?
Within the borehole plugging or repository design group, we should
start directing our efforts towards the things that we feel are
important, such as modelling, and determining the limits to our data.
To do this we need to get more information to support our models.
First of all though, we ought to agree what the function of the
borehole seal is. Perhaps we should not worry too much about the
effects of fluid flow. Perhaps the disturbed zone, if there is one,
is not really a problem at all. But we need to find out whether it
is a problem before the regulatory agencies tell us that it is.

Laboratory tests and analytical models give us some of the data, but they are far removed from reality. We need to do more in situ testing, since that will be the basis of our case when we describe what we can do. We should go over the topics we have been discussing and agree among ourselves as to whether or not they are significant issues. I think the geochemical problem is going to be one of the major issues. I do not know how we are ever going to answer that. I think we should be candid when we go to the regulatory agencies and tell them what we have done, what we believe, what we have evidence for, and that there are some things we just do not know, nor can we define ways of finding them out.

Coming back to field testing and specifically to the Bell Canyon Test, this was not intended to shed any great light on the possible performance of a plug. It was meant to take what we thought was a suitable material, emplace it in the ground and examine its behaviour. The environment was very hostile for a cement plug, and we were not even sure it was going to stick but it did, and it gave us data. The numbers were 50 times higher than those we predicted, but that is useful information. Can we get the same numbers the next time ? We will probably do another test shortly to find out ; if we don't we can take the average and get a feeling for the variability of values. I think the first test was exceedingly simple, just emplacing a plug and measuring how much flow comes through. We could do much more exotic things, but the interpretation of results might become terribly complicated. So my recommendation would be to carry out fairly straightforward, simple experiments. There will be enough problems with analysing the information from those. I have made a list of the kinds of experiments that we are interested in doing over the next few years. I would welcome comments as to whether or not they are reasonable since they are only concepts right now. I invite you to join us. I would like to have you critically involved and I would also like the same opportunity to look over your experiments and to participate. Field experiments are extremely expensive, and they take a lot of time and have a lot of problems associated with them, therefore co-operation would be very beneficial for everybody. Data that come out of the ground seem to be widely acceptable ; they may cost a lot of money but they are extremely valuable. Thus I would like to recommend that we try to establish an international consensus of opinion on the various aspects of penetration sealing. I would also like to see an international distribution list in order to channel available information in the most efficient way. Finally I would recommend that a small expert group is formed with the task of defining some of the ground rules on the issues that we are concerned about. It would also help to define some of the experiments and methods of analysis and standardise the terminology in this field of work.

Regarding the status of present field testing technology, we planned the Bell Canyon Test saying that we would use existing oil field technology and if it proved inadequate we would try to modify it on the spot. The experiment was relatively simple and did not require much in the way of new instrumentation. What was not available was a detailed specification of what the oil-field equipment could do. We were really at the limit, trying to measure cupfuls of water at the bottom of a 350 barrel column underneath a packer. According to the oil industry, packers do not leak. But according to us a leak of a cupful in one day would ruin the experiment ; so we must be careful that we understand exactly what the experts claim. Many times they do not know because they are not used to these kinds of measurements. In general the technology in the oil industry has the capability and can be easily modified. The instrumentation is fairly straightforward, but some of it is not compatible with a 4-inch-diameter borehole.

There is an old saying about experiments : "you probably can't get what you want because as soon as you try to get it, it is gone". However, we should design the experiments in order to give the best possible data, but at the same time we should try to recognise the limitations.

T.O. HUNTER, United States

I would be interested in hearing comments about any European experimental or field programmes which are either envisaged or underway.

J. HAMSTRA, Netherlands

We have a combined effort with a German group on a borehole from the 750 to the 1050 m level in the Asse II mine. We are trying to do borehole convergence measurements ; next step will be to add a heat source and carry out additional measurements. That borehole is available now and will be used this year and next year. It will then remain available for a flooding test. The work is carried out under a contract with the Commission of the European Communities. We have no access to underground testing facilities in the Netherlands. That is all I can say about our situation at present.

T.O. HUNTER, United States

Are there plans for salt compaction and reconsolidation experiments using rock salt as a sealing material, and taking advantage of the borehole convergence ?

J. HAMSTRA, Netherlands

No.

N.A. CHAPMAN, United Kingdom

I would like to comment briefly on the U.K. situation. Borehole sealing is something that we are just beginning to consider now. We are looking at a programme, part of which is modelling and part of which is the design of a field testing programme, for borehole sealing in crystalline rocks. Some form of international cooperation on field testing would thus be of considerable interest to us.

F.P. MUZZI, Italy

In Italy we are planning to investigate clay formations with the use of a matrix of boreholes for waste disposal. We discovered that the host environment was difficult to investigate. One of the main features that we discovered was a lack of uniformity in the permeability and hydraulic gradient in a supposedly uniform clay unit. In the future the main field activity will be to find a site with an overburden pressure equivalent to approximately 400 m, and then to construct a gallery in which to perform experiments. We are now looking for a company to manage the whole waste disposal programme.

F. GERA, NEA

The Stripa Project includes activities that might have some relation to shaft sealing. Mines that have been developed on different levels offer favorable conditions for some highly instrumented and very sophisticated plugging experiments. For example, if

we had two rooms on different levels and measured the hydraulic conductivity of the intervening rock, then drilled a hole and plugged
it in such a way that we were sure that there was no flow through
the hole, and measured the hydraulic conductivity again, we might
obtain data on the importance of the disturbed zone. It would be a
much more flexible system than just plugging a deep hole from the
surface. If funds became available it might be possible to include
some plugging experiment in the Stripa Project.

C.L. CHRISTENSEN, United States

        I agree. When we first started this programme a year ago
we had precisely this in mind. One experiment was called the shallow
hole test, in which we had access to the bottom of the plug in a
mine about 1100 feet deep, and we could carry out the measurements
at the surface. We decided that that was beyond our capabilities at
the time. We had several suggestions for measuring the in situ
permeability in a mine ; for example drilling a hole through a pillar
and pressurising on one side to see what the flow was on the other
side. We want to do between-borehole testing in exactly the way you
have discussed. Our problem is that at present we do not have access
to mines where these experiments could be performed. The Stripa
facility is interesting but not for our present work, which concerns
salt deposits.

T.O. HUNTER, United States

        Let us move on to the final topic on my list which is
international cooperation. I have asked Dr. Gera to introduce this
subject.

F. GERA, NEA

        This meeting is the first attempt to bring together the
people working in this field in order to exchange information. We
need to continue and expand the exchange of information. We should
encourage the exchange of scientists between groups working on
similar problems. We should make sure that all the people attending
this workshop receive any reports or documents dealing with penetration sealing.

        In addition there is an obvious need to reach some kind
of agreement, possibly at the international level, on what are the
requirements and criteria for borehole and shaft sealing. This could
well be a valuable undertaking for the regulatory authorities of the
various countries involved in geologic disposal. They should discuss
the problem and see if they can define criteria that are generally
acceptable.

        One further suggestion is to create an overview group with
the task of ensuring coordination of effort and cooperation at an
operational level. I would like to conclude by asking if there are
any recommendations that we might come up with as a result of the
meeting, and if there are any specific actions that you think NEA
could undertake to promote cooperation in this field.

T.G. CLENDENNING, Canada

        I think all countries are experiencing the same difficulties
in getting access to sites for various activities. In Canada we even
have to seek permission to fly over areas. So I think that it would
be of particular value if there was information available on mines
where there is the possibility of access for experimentation. I realise

that the Stripa cooperative effort is excellent, but we might speed up field testing if some centralised information was available.

D.E. STEPHENSON, United States

I would like to see if we, as an international group, could agree on what the functions of sealing are and define those functions. If we can agree on the functions then we will avoid the risk that what is done in one country might turn out to be detrimental to the programmes of other countries.

T.O. HUNTER, United States

The recommendation for setting up a group of experts merits some consideration. Were we specific enough on the last issue to identify conclusions for this Workshop ?

F. GERA, NEA

I think NEA could look into the possibility of setting up a small working group to try to come up with recommendations and conclusions and, if it is possible, to reach an agreement on what are the functions and requirements for penetration seals.

T.O. HUNTER, United States

I would like to close the Workshop by thanking all the persons who volunteered to introduce the discussion topics. I appreciate your support in guiding the discussion and leading us through a fruitful exchange of information. I thank all the participants for the comments made from the floor, and further I thank our sponsors, OECD/NEA and USDOE, and finally OWNI which has been such a gracious host for the meeting.

LIST OF PARTICIPANTS

LISTE DES PARTICIPANTS

BELGIUM - BELGIQUE

COLE-BAKER, J.R., D'Appolonia Consulting Engineers, Inc., Boulevard du Souverain 100, B-1170 Bruxelles

MANFROY, P., Centre d'Etude de l'Energie Nucléaire, CEN/SCK, Boeretang 200, B-2400 Mol

CANADA

CARMICHAEL, T.J., Ontario Hydro, Dobson Research Laboratory, R 290, 800 Kipling Avenue, Toronto, Ontario M8Z 5S4

CLENDENNING, T.G., Ontario Hydro, 700 University Avenue, Toronto, Ontario M5G 1X6

GYENGE, M., Canada Centre for Mineral and Energy Technology, Mining Research Laboratories, 555 Booth Street, Ottawa, Ontario K1A 0G1

DENMARK - DANEMARK

IDORN, G.M., Special Consultant to Danish Elsam in charge of Nuclear Waste Deposition Project, Tovesuej 14B, DK-2850 Naerum

JENSEN, A., Inspectorate of Nuclear Installations, P.O. Box 217, Risø Huse 11, DK-4000 Roskilde

FRANCE

BOULANGER, A.-M.L., Geostock, Tour Aurore, Cedex 5, 92080 Paris Défense 2

FEDERAL REPUBLIC OF GERMANY - REPUBLIQUE FEDERALE D'ALLEMAGNE

EICHMEYER, H., Department of Mining and Geosciences, Technical University of Berlin, Strasse des 17. Juni 135, D-1000 Berlin 12

FISCHLE, W., Gesellschaft für Strahlen- und Umweltforschung mbH., Institut für Tieflagerung - Wissenschaftliche Abteilung, Schachtanlage Asse, D-3346 Remlingen

ITALY - ITALIE

MUZZI, F.P., Comitato Nazionale per l'Energia Nucleare, DISP, Viale Regina Margherita 125, I-00198 Rome

JAPAN - JAPON

IWAMOTO, F., Japan Gas Corporation, 2-1, Ohtemachi, 2-Chome, Chiyoda-ku, Tokyo

MASUDA, S., Power Reactor & Nuclear Fuel Development Corporation,
9-13, 1-Chome, Akasaka, Minato-ko, Tokyo

YAHIRO, T., Kajima Institute of Construction Technology,
2-19-1 Tobitakyu, Chofu-shi, Tokyo-to

THE NETHERLANDS - PAYS-BAS

HAMSTRA, J., Netherlands Energy Research Foundation, Technical
Service Department, 3 Westerduinweg, 1755 Petten (N.H.)

NORWAY - NORVEGE

HUSEBY, S., Geological Survey of Norway, Drammensveien 230, Oslo 2

SWEDEN - SUEDE

BERGSTRÖM, A.C.J., Swedish Nuclear Fuel Supply Co., Nuclear Fuel
Safety Project, SKBF/KBS, Box 5864, S-102 48 Stockholm

PUSCH, R., Division of Soil Mechanics, University of Luleå,
S-951 87 Luleå

SWITZERLAND - SUISSE

COLOMBI, C., Abt. für Sicherheit von Kernanlagen (ASK), CSD Colombi
Schmutz Dorthe AG, Kirchstr. 22, CH-3097 Liebefeld

THURY, M.F., NAGRA, National Cooperative for the Storage of
Radioactive Waste, Parkstrasse 23, CH-5401 Baden

UNITED KINGDOM - ROYAUME-UNI

CHAPMAN, N.A., Institute of Geological Sciences, Building 151,
Harwell Laboratory, Didcot, Oxon

UNITED STATES - ETATS-UNIS

BARAINCA, M.J., Columbus Program Office, Richland Operations Office,
U.S. Department of Energy, 505 King Avenue, Columbus, Ohio 43201

BASHAM Jr., S.J., Engineering Development Department, Office of
Nuclear Waste Isolation, Battelle Memorial Institute, Columbus,
Ohio 43201

BOA, J.A., U.S. Army Engineers, Waterway Experiment Station,
P.O. Box 631, Vicksburg, MS 39180

BROCE, R., Systems, Science and Software, P.O. Box 1620, La Jolla, California 92038

BURNS, F.L., Engineering Development Department, Office of Nuclear Waste Isolation, Battelle Memorial Institute, Columbus, Ohio 43201

CARTER, N.E., Battelle, Office of Nuclear Waste Isolation, 505 King Avenue, Columbus, Ohio 43201

CASEY, L., Columbus Program Office, Richland Operations Office, U.S. Department of Energy, 505 King Avenue, Columbus, Ohio 43201

CHABANNES, C.R. de St. J., D'Appolonia Consulting Engineers Inc., 2350 Alamo S.E., Suite 103, Albuquerque, New Mexico 87106

CHRISTENSEN, C.L., Experimental Programs Division 4512, Sandia National Laboratories, P.O. Box 5800, Albuquerque, New Mexico 87185

COOK, C.W., Division 1116, Sandia National Laboratories, P.O. Box 5800, Albuquerque, New Mexico 87185

DAEMEN, J., University of Arizona, Tucson, Arizona 85721

ELLISON, R.D., D'Appolonia Consulting Engineers, Inc., 10 Duff Road, Pittsburgh, Pennsylvania 15235

GOODWIN, R.H., D'Appolonia Consulting Engineers, Inc., 2350 Alamo S.E., Suite 103, Albuquerque, New Mexico 87106

GRUTZECK, M.W., Materials Research Laboratory, The Pennsylvania State University, University Park, Pennsylvania 16802

GULICK, C.W., Division 1133, Sandia National Laboratories, P.O. Box 5800, Albuquerque, New Mexico 87185

HARTUNG, L.F., Nuclear Regulatory Commission, Division of Waste Management, Mail Stop 905-SS, Washington, D.C. 20555

HODGES, F.N., Rockwell Hanford Operations, Energy Systems Group, P.O. Box 800, Richland, Wa. 99352

HOFFMANN, P.L., Battelle, Office of Nuclear Waste Isolation, 505 King Avenue, Columbus, Ohio 43201

HUNTER, T.O., Experimental Programs Division 4512, Sandia National Laboratories, P.O. Box 5800, Albuquerque, New Mexico 87185

KELLER, D.L., Battelle, Office of Nuclear Waste Isolation, 505 King Avenue, Columbus, Ohio 43201

KELSALL, P.C., D'Appolonia Consulting Engineers, Inc., 2350 Alamo S.E., Suite 103, Albuquerque, New Mexico 87106

LAMBERT, S.J., Nuclear Waste Technology Division 4511, Sandia National Laboratories, P.O. Box 5800, Albuquerque, New Mexico 87185

LANKARD, D.L., Battelle, Office of Nuclear Waste Isolation, 505 King Avenue, Columbus, Ohio 43201

LICASTRO, P.H., Materials Research Laboratory, The Pennsylvania State University, University Park, Pennsylvania 16802

MATHER, K., U.S. Army Engineer, Waterways Experiment Station, Structures Laboratory, P.O. Box 631, Vicksburg, Mississippi 39180

MEYER, D., D'Appolonia Consulting Engineers, Inc., 2350 Alamo S.E., Suite 103, Albuquerque, New Mexico 87106

NEFF, J.O., Columbus Program Office, Richland Operations Office, U.S. Department of Energy, 505 King Avenue, Columbus, Ohio 43201

PARDUE, W.M., Battelle, Office of Nuclear Waste Isolation, 505 King Avenue, Columbus, Ohio 43201

PETERSON, E.W., Systems, Science and Software, P.O. Box 1620, La Jolla, California 92038

ROY, D.M., Materials Research Laboratory, The Pennsylvania State University, University Park, Pennsylvania 16802

SHUKLA, D.K., D'Appolonia Consulting Engineers, Inc., 2350 Alamo S.E., Suite 103, Albuquerque, New Mexico 87106

SMITH, M.J., Rockwell Hanford Operations, Energy Systems Group, P.O. Box 800, Richland, Wa. 99352

STATLER, R.D., Engineering Project Division 1133, Sandia National Laboratories, P.O. Box 5800, Albuquerque, New Mexico 87185

STEPHENSON, D.E., D'Appolonia Consulting Engineers, Inc., 2350 Alamo S.E., Suite 103, Albuquerque, New Mexico 87106

WILKINSON, L.E., Foundation Sciences, Inc., 1630 S.W. Morrison, Portland, Oregon 97205

WRIGHT, J.C., D'Appolonia Consulting Engineers, Inc., 10 Duff Road, Pittsburg, Pennsylvania 15235

WUNDERLICH, R., Columbus Program Office, Richland Operations Office, U.S. Department of Energy, 505 King Avenue, Columbus, Ohio 43201

SECRETARIAT

GERA, F., OECD Nuclear Energy Agency, 38 boulevard Suchet, 75016 Paris (France)

SOME
NEW PUBLICATIONS
OF NEA

QUELQUES
NOUVELLES PUBLICATIONS
DE L'AEN

ACTIVITY
REPORTS

RAPPORTS
D'ACTIVITÉ

Activity Reports of the OECD Nuclear Energy Agency (NEA)
- 7th Activity Report (1978)
- 8th Activity Report (1979)

Rapports d'activité de l'Agence de l'OCDE pour l' Énergie Nucléaire (AEN)
- 7e Rapport d'Activité (1978)
- 8e Rapport d'Activité (1979)

Free on request — Gratuits sur demande

Annual Reports of the OECD HALDEN Reactor Project
- 18th Annual Report (1977)
- 19th Annual Report (1978)

Rapports annuels du Projet OCDE de réacteurs de HALDEN
- 18e Rapport annuel (1977)
- 19e Rapport annuel (1978)

Free on request — Gratuits sur demande

● ● ●

NEA at a Glance

Coup d'œil sur l'AEN

Free on request — Gratuit sur demande

OECD Nuclear Energy Agency: Functions and Main Activities

Agence de l'OCDE pour l'Énergie Nucléaire : Rôle et principales activités

Free on request — Gratuit sur demande

Twentieth Anniversary of the OECD Nuclear Energy Agency
- Proceedings on the NEA Symposium on International Co-operation in the Nuclear Field: Perspectives and Prospects

Vingtième Anniversaire de l'Agence de l'OCDE pour l'Énergie Nucléaire
- Compte rendu du Symposium de l'AEN sur la coopération internationale dans le domaine nucléaire: bilan et perspectives

Free on request — Gratuit sur demande

SCIENTIFIC AND TECHNICAL PUBLICATIONS

PUBLICATIONS SCIENTIFIQUES ET TECHNIQUES

NUCLEAR FUEL CYCLE

LE CYCLE DU COMBUSTIBLE NUCLÉAIRE

Reprocessing of Spent Nuclear Fuels in OECD Countries (1977)

Retraitement du combustible nucléaire dans les pays de l'OCDE (1977)

£2.50   US$5.00   F20,00

Nuclear Fuel Cycle Requirements and Supply Considerations, Through the Long-Term (1978)

Besoins liés au cycle du combustible nucléaire et considérations sur l'approvisionnement à long terme (1978)

£4.30   US$8.75   F35,00

World Uranium Potential — An International Evaluation (1978)

Potentiel mondial en uranium — Une évaluation internationale (1978)

£7.80   US$16.00   F64.00

Uranium — Resources, Production and Demand (1979)

Uranium — ressources, production et demande (1979)

£8.70   US$19.50   F78,00

● ● ●

RADIATION PROTECTION

RADIOPROTECTION

Iodine-129
(Proceedings of an NEA Specialist Meeting, Paris, 1977)

Iode-129
(Compte rendu d'une réunion de spécialistes de l'AEN, Paris, 1977)

£3.40   US$7.00   F28,00

Recommendations for Ionization Chamber Smoke Detectors in Implementation of Radiation Protection Standards (1977)

Recommandations relatives aux détecteurs de fumée à chambre d'ionisation en application des normes de radioprotection (1977)

Free on request — Gratuit sur demande

Radon Monitoring
(Proceedings of the NEA Specialist Meeting, Paris, 1978)

Surveillance du radon
(Compte rendu d'une réunion de spécialistes de l'AEN, Paris, 1978)

£8.00   US16.50   F66,00

Management, Stabilisation and Environmental Impact of Uranium Mill Tailings (Proceedings of the Albuquerque Seminar, United States, 1978)

Gestion, stabilisation et incidence sur l'environnement des résidus de traitement de l'uranium
(Compte rendu du Séminaire d'Albuquerque, États-Unis, 1978)

£9.80   US$20.00   F80,00

Exposure to Radiation from the Natural
Radioactivity in Building Materials
(Report by an NEA Group of Experts,
1979)

Exposition aux rayonnements due à la ra-
dioactivité naturelle des matériaux de
construction
(Rapport établi par un Groupe d'experts
de l'AEN, 1979)

Free on request — Gratuit sur demande

Marine Radioecology
(Proceedings of the Tokyo Seminar,
1979)

Radioécologie marine
(Compte rendu du Colloque de Tokyo,
1979)

£9.60   US$21.50   F86.00

Radiological Significance and
Management of Tritium, Carbon-14,
Krypton-85 and Iodine-129 arising
from the Nuclear Fuel Cycle
(Report by an NEA Group of Experts,
1980)

Importance radiologique et gestion des
radionucléides : tritium, carbone-14,
krypton-85 et iode-129, produits au cours
du cycle du combustible nucléaire
(Rapport établi par un Groupe d'experts
de l'AEN, 1980)

£8.40   US$19.00   F76,00

● ● ●

RADIOACTIVE WASTE MANAGEMENT

GESTION DES DÉCHETS RADIOACTIFS

Objectives, Concepts and Strategies for
the Management of Radioactive Waste
Arising from Nuclear Power Programmes
(Report by an NEA Group of Experts,
1977)

Objectifs, concepts et stratégies en matière de gestion des déchets radioactifs résultant des programmes nucléaires de
puissance
(Rapport établi par un Groupe d'experts
de l'AEN, 1977)

£8.50   US$17.50   F70,00

Treatment, Conditioning and Storage of
Solid Alpha-Bearing Waste and Cladding
Hulls
(Proceedings of the NEA/IAEA Technical
Seminar, Paris, 1977)

Traitement, conditionnement et stockage
des déchets solides alpha et des coques
de dégainage
(Compte rendu du Séminaire technique
AEN/AIEA, Paris, 1977)

£7.30   US$15.00   F60,00

Storage of Spent Fuel Elements
(Proceedings of the Madrid Seminar,
1978)

Stockage des éléments combustibles
irradiés (Compte rendu du Séminaire
de Madrid, 1978)

£7.30   US$15.00   F60,00

In Situ Heating Experiments in Geological
Formations
(Proceedings of the Ludvika Seminar,
Sweden, 1978)

Expériences de dégagement de chaleur in
situ dans les formations géologiques
(Compte rendu du Séminaire de Ludvika,
Suède, 1978)

£8.00   US$16.50   F66,00

Migration of Long-lived Radionuclides in
the Geosphere
(Proceedings of the Brussels Workshop,
1979)

Migration des radionucléides à vie longue
dans la géosphère
(Compte rendu de la réunion de travail de
Bruxelles, 1979)

£8.30   US$17.00   F68,00

Low-Flow, Low-Permeability Measurements in Largely Impermeable Rocks (Proceedings of the Paris Workshop, 1979)

Mesures des faibles écoulements et des faibles perméabilités dans des roches relativement imperméables (Compte rendu de la réunion de travail de Paris, 1979)

£7.80   US$16.00   64,00

On-Site Management of Power Reactor Wastes (Proceedings of the Zurich Symposium, 1979)

Gestion des déchets en provenance des réacteurs de puissance sur le site de la centrale (Compte rendu du Colloque de Zurich, 1979)

£11.00   US$22.50   F90,00

Recommended Operational Procedures for Sea Dumping of Radioactive Waste (1979)

Recommandations relatives aux procédures d'exécution des opérations d'immersion de déchets radioactifs en mer (1979)

Free on request — Gratuit sur demande

Guidelines for Sea Dumping Packages of Radioactive Waste (Revised version, 1979)

Guide relatif aux conteneurs de déchets radioactifs destinés au rejet en mer (Version révisée, 1979)

Free on request — Gratuit sur demande

Use of Argillaceous Materials for the Isolation of Radioactive Waste (Proceedings of the Paris Workshop, 1979)

Utilisation des matériaux argileux pour l'isolement des déchets radioactifs (Compte rendu de la Réunion de travail de Paris, 1979)

£7.60   US$17.00   F68,00

Review of the Continued Suitability of the Dumping Site for Radioactive Waste in the North-East Atlantic (1980)

Réévaluation de la validité du site d'immersion de déchets radioactifs dans la région nord-est de l'Atlantique (1980)

Free on request — Gratuit sur demande

Decommissioning Requirements in the Design of Nuclear Facilities (Proceedings of the NEA Specialist Meeting, Paris, 1980)

Déclassement des installations nucléaires : exigences à prendre en compte au stade de la conception (Compte rendu d'une réunion de spécialistes de l'AEN, Paris, 1980)

£7.80   US$17.50   F70,00

● ● ●

SAFETY

Safety of Nuclear Ships
(Proceedings of the Hamburg Symposium,
1977)

£17.00   US$35.00   F140,00

Nuclear Aerosols in Reactor Safety
(A State-of-the-Art Report by a Group of
Experts, 1979)

£8.30   US$18.75   F75,00

Plate Inspection Programme
(Report from the Plate Inspection
Steering Committee — PISC — on the
Ultrasonic Examination of Three
Test Plates), 1980

£3.30   US$7.50   F30.00

Reference Seismic Ground Motions
in Nuclear Safety Assessments
(A State-of-the-Art Report by a
Group of Experts, 1980)

£ 7.00   US$16.00   F64,00

SÛRETÉ

Sûreté des navires nucléaires
(Compte rendu du Symposium de
Hambourg, 1977)

Les aérosols nucléaires dans la sûreté
des réacteurs
(Rapport sur l'état des connaissances
établi par un Groupe d'Experts, 1979)

Programme d'inspection des tôles
(Rapport du Comité de Direction sur
l'inspection des tôles — PISC — sur l'examen
par ultrasons de trois tôles d'essai au moyen
de la procédure «PISC» basée sur le code
ASME XI), 1980

Les mouvements sismiques de référence
du sol dans l'évaluation de la sûreté
des installations nucléaires
(Rapport sur l'état des connaissances
établi par un Groupe d'experts, 1980)

• • •

SCIENTIFIC INFORMATION

Neutron Physics and Nuclear Data for Reactors and other Applied Purposes
(Proceedings of the Harwell International Conference, 1978)

£26.80   US$55.00   F220,00

Calculation of 3-Dimensional Rating Distributions in Operating Reactors
(Proceedings of the Paris Specialists' Meeting, 1979)

£9.60   US$21.50   F86.00

INFORMATION SCIENTIFIQUE

La physique neutronique et les données nucléaires pour les réacteurs et autres applications
(Compte rendu de la Conférence International de Harwell, 1978)

Calcul des distributions tri-dimensionnelles de densité de puissance dans les réacteurs en cours d'exploitation
(Compte rendu de la Réunion de spécialistes de Paris, 1979)

LEGAL PUBLICATIONS

PUBLICATIONS JURIDIQUES

Convention on Third Party Liability in the Field of Nuclear Energy — incorporating the provisions of Additional Protocol of January 1964

Free on request — Gratuit sur demande

Convention sur la responsabilité civile dans le domaine de l'énergie nucléaire — Texte incluant les dispositions du Protocole additionnel de janvier 1964

Nuclear Legislation, Analytical Study: "Nuclear Third Party Liability" (revised version, 1976)

£6.00   US$12.50   F50,00

Législations nucléaires, étude analytique: "Responsabilité civile nucléaire" (version révisée, 1976)

Nuclear Law Bulletin
(Annual Subscription — two issues and supplements)

£5.60   US$12.50   F50,00

Bulletin de Droit Nucléaire
(Abonnement annuel — deux numéros et suppléments)

Index of the first twenty five issues of the Nuclear Law Bulletin

Free on request — Gratuit sur demande

Index des vingt-cinq premiers numéros du Bulletin de Droit Nucléaire

Licensing Systems and Inspection of Nuclear Installations in NEA Member Countries (two volumes)

Free on request — Gratuit sur demande

Régime d'autorisation et d'inspection des installations nucléaires dans les pays de l'AEN (deux volumes)

NEA Statute

Free on request — Gratuit sur demande

Statuts de l'AEN

• • •

OECD SALES AGENTS
DÉPOSITAIRES DES PUBLICATIONS DE L'OCDE

ARGENTINA – ARGENTINE
Carlos Hirsch S.R.L., Florida 165, 4° Piso (Galería Guemes)
1333 BUENOS AIRES, Tel. 33.1787.2391 y 30.7122

AUSTRALIA – AUSTRALIE
Australia & New Zealand Book Company Pty Ltd.,
23 Cross Street, (P.O.B. 459)
BROOKVALE NSW 2100. Tel. 938.2244

AUSTRIA – AUTRICHE
OECD Publications and Information Center
4 Simrockstrasse 5300 BONN. Tel. (0228) 21.60.45
Local Agent/Agent local :
Gerold and Co., Graben 31, WIEN 1. Tel. 52.22.35

BELGIUM – BELGIQUE
LCLS
44 rue Otlet, B 1070 BRUXELLES. Tel. 02.521.28.13

BRAZIL – BRÉSIL
Mestre Jou S.A., Rua Guaipa 518,
Caixa Postal 24090, 05089 SAO PAULO 10. Tel. 261.1920
Rua Senador Dantas 19 s/205-6, RIO DE JANEIRO GB.
Tel. 232.07.32

CANADA
Renouf Publishing Company Limited,
2182 St. Catherine Street West,
MONTREAL, Quebec H3H 1M7 Tel. (514) 937.3519

DENMARK – DANEMARK
Munksgaard Export and Subscription Service
35, Nørre Søgade
DK 1370 KØBENHAVN K. Tel. +45.1.12.85.70

FINLAND – FINLANDE
Akateeminen Kirjakauppa
Keskuskatu 1, 00100 HELSINKI 10. Tel. 65.11.22

FRANCE
Bureau des Publications de l'OCDE,
2 rue André-Pascal, 75775 PARIS CEDEX 16. Tel. (1) 524.81.67
Principal correspondant :
13602 AIX-EN-PROVENCE : Librairie de l'Université.
Tel. 26.18.08

GERMANY – ALLEMAGNE
OECD Publications and Information Center
4 Simrockstrasse 5300 BONN Tel. (0228) 21.60.45

GREECE – GRÈCE
Librairie Kauffmann, 28 rue du Stade,
ATHÈNES 132. Tel. 322.21.60

HONG-KONG
Government Information Services,
Sales and Publications Office, Baskerville House, 2nd floor,
13 Duddell Street, Central. Tel. 5.214375

ICELAND – ISLANDE
Snaebjörn Jönsson and Co., h.f.,
Hafnarstraeti 4 and 9, P.O.B. 1131, REYKJAVIK.
Tel. 13133/14281/11936

INDIA – INDE
Oxford Book and Stationery Co. :
NEW DELHI, Scindia House. Tel. 45896
CALCUTTA, 17 Park Street. Tel. 240832

INDONESIA – INDONÉSIE
PDIN-LIPI, P.O. Box 3065/JKT., JAKARTA, Tel. 583467

IRELAND – IRLANDE
TDC Publishers – Library Suppliers
12 North Frederick Street, DUBLIN 1 Tel. 744835-749677

ITALY – ITALIE
Libreria Commissionaria Sansoni :
Via Lamarmora 45, 50121 FIRENZE. Tel. 579751
Via Bartolini 29, 20155 MILANO. Tel. 365083
Sub-depositari :
Editrice e Libreria Herder,
Piazza Montecitorio 120, 00 186 ROMA. Tel. 6794628
Libreria Hoepli, Via Hoepli 5, 20121 MILANO. Tel. 865446
Libreria Lattes, Via Garibaldi 3, 10122 TORINO. Tel. 519274
La diffusione delle edizioni OCSE è inoltre assicurata dalle migliori
librerie nelle città più importanti.

JAPAN – JAPON
OECD Publications and Information Center,
Landic Akasaka Bldg., 2-3-4 Akasaka,
Minato-ku, TOKYO 107 Tel. 586.2016

KOREA – CORÉE
Pan Korea Book Corporation,
P.O. Box n° 101 Kwangwhamun, SÉOUL. Tel. 72.7369

LEBANON – LIBAN
Documenta Scientifica/Redico,
Edison Building, Bliss Street, P.O. Box 5641, BEIRUT.
Tel. 354429 – 344425

MALAYSIA – MALAISIE
and/et SINGAPORE - SINGAPOUR
University of Malaysia Co-operative Bookshop Ltd.
P.O. Box 1127, Jalan Pantai Baru
KUALA LUMPUR. Tel. 51425, 54058, 54361

THE NETHERLANDS – PAYS-BAS
Staatsuitgeverij
Verzendboekhandel Chr. Plantijnnstraat
S-GRAVENAGE. Tel. nr. 070.789911
Voor bestellingen: Tel. 070.789208

NEW ZEALAND – NOUVELLE-ZÉLANDE
Publications Section,
Government Printing Office,
WELLINGTON: Walter Street. Tel. 847.679
Mulgrave Street, Private Bag. Tel. 737.320
World Trade Building, Cubacade, Cuba Street. Tel. 849.572
AUCKLAND: Hannaford Burton Building,
Rutland Street, Private Bag. Tel. 32.919
CHRISTCHURCH: 159 Hereford Street, Private Bag. Tel. 797.142
HAMILTON: Alexandra Street, P.O. Box 857. Tel. 80.103
DUNEDIN: T & G Building, Princes Street, P.O. Box 1104.
Tel. 778.294

NORWAY – NORVÈGE
J.G. TANUM A/S Karl Johansgate 43
P.O. Box 1177 Sentrum OSLO 1. Tel. (02) 80.12.60

PAKISTAN
Mirza Book Agency, 65 Shahrah Quaid-E-Azam, LAHORE 3.
Tel. 66839

PHILIPPINES
National Book Store, Inc.
Library Services Division, P.O. Box 1934, MANILA.
Tel. Nos. 49.43.06 to 09, 40.53.45, 49.45.12

PORTUGAL
Livraria Portugal, Rua do Carmo 70-74,
1117 LISBOA CODEX. Tel. 360582/3

SPAIN – ESPAGNE
Mundi-Prensa Libros, S.A.
Castello 37, Apartado 1223, MADRID-1. Tel. 275.46.55
Libreria Bastinos, Pelayo 52, BARCELONA 1. Tel. 222.06.00

SWEDEN – SUÈDE
AB CE Fritzes Kungl Hovbokhandel,
Box 16 356, S 103 27 STH, Regeringsgatan 12,
DS STOCKHOLM. Tel. 08/23.89.00

SWITZERLAND – SUISSE
OECD Publications and Information Center
4 Simrockstrasse 5300 BONN. Tel. (0228) 21.60.45
Local Agents/Agents locaux
Librairie Payot, 6 rue Grenus, 1211 GENÈVE 11. Tel. 022.31.89.50
Freihofer A.G., Weinbergstr. 109, CH-8006 ZÜRICH.
Tel. 01.3624282

TAIWAN – FORMOSE
National Book Company,
84-5 Sing Sung South Rd, Sec. 3, TAIPEI 107. Tel. 321.0698

THAILAND – THAILANDE
Suksit Siam Co., Ltd., 1715 Rama IV Rd,
Samyan, BANGKOK 5. Tel. 2511630

UNITED KINGDOM – ROYAUME-UNI
H.M. Stationery Office, P.O.B. 569,
LONDON SE1 9NH. Tel. 01.928.6977, Ext. 410 or
49 High Holborn, LONDON WC1V 6 HB (personal callers)
Branches at: EDINBURGH, BIRMINGHAM, BRISTOL,
MANCHESTER, CARDIFF, BELFAST.

UNITED STATES OF AMERICA – ÉTATS-UNIS
OECD Publications and Information Center, Suite 1207,
1750 Pennsylvania Ave., N.W. WASHINGTON D.C.20006.
Tel. (202) 724.1857

VENEZUELA
Libreria del Este, Avda. F. Miranda 52, Edificio Galipan,
CARACAS 106. Tel. 32.23.01/33.26.04/33.24.73

YUGOSLAVIA – YOUGOSLAVIE
Jugoslovenska Knjiga, Terazije 27, P.O.B. 36, BEOGRAD.
Tel. 621.992

Les commandes provenant de pays où l'OCDE n'a pas encore désigné de dépositaire peuvent être adressées à :
OCDE, Bureau des Publications, 2, rue André-Pascal, 75775 PARIS CEDEX 16.

Orders and inquiries from countries where sales agents have not yet been appointed may be sent to:
OECD, Publications Office, 2 rue André-Pascal, 75775 PARIS CEDEX 16.

63309-10-1980

PUBLICATIONS DE L'OCDE, 2 rue André-Pascal, 75775 PARIS CEDEX 16 - N° 41 732 1980
IMPRIMÉ EN FRANCE
(3000 UD-66 80 09 3) ISBN 92-64-02114-0